普通高等教育“十三五”规划教材（软件工程专业）

# Android 应用开发基础教程

主　编　巫湘林

副主编　陈　彬　胡世洋　黄华升

中国水利水电出版社
www.waterpub.com.cn

·北京·

## 内 容 提 要

本书主要介绍运用 Eclipse 对 Android 应用程序进行开发的相关知识与技能，以及如何使用 PHP 动态网页开发技术、MUI 框架和 MySQL 数据库等。通过本书读者可全面了解 Android 应用程序开发的基本流程与方法，熟练掌握 Android 应用程序开发技能。

本书共 12 章：Android 基础入门、Android 应用结构分析、基本控件和事件处理、布局、高级控件、菜单与相关控件、Activity、Fragment、Android 后台处理、Android 数据存储、网络编程、应用项目开发实例，内容符合 Android 应用程序开发认知体系，先了解基本理论知识，再通过实际案例掌握开发技能。

本书内容全面、注重实践，理论深浅适宜、条理清晰，精编案例图文并茂、易于理解，适合 Android 应用程序开发的初学者使用，可作为各高校及 IT 培训学校的教材，也可供 Android 应用程序开发人员参考。

**本书配有电子教案和源代码，读者可以到中国水利水电出版社网站和万水书苑上免费下载，网址为 http://www.waterpub.com.cn/softdown/和 http://www.wsbookshow.com。**


**图书在版编目（CIP）数据**

Android应用开发基础教程 / 巫湘林主编. -- 北京 : 中国水利水电出版社, 2017.11
普通高等教育“十三五”规划教材. 软件工程专业
ISBN 978-7-5170-5815-1

Ⅰ. ①A… Ⅱ. ①巫… Ⅲ. ①移动终端－应用程序－程序设计－高等学校－教材 Ⅳ. ①TN929.53

中国版本图书馆CIP数据核字(2017)第218814号

策划编辑：石永峰　责任编辑：周益丹　加工编辑：周莹　张溯源　封面设计：李佳

| | |
|---|---|
| 书　名 | 普通高等教育“十三五”规划教材（软件工程专业）<br>Android 应用开发基础教程<br>Android YINGYONG KAIFA JICHU JIAOCHENG |
| 作　者 | 主　编　巫湘林<br>副主编　陈　彬　胡世洋　黄华升 |
| 出版发行 | 中国水利水电出版社<br>（北京市海淀区玉渊潭南路 1 号 D 座　100038）<br>网址：www.waterpub.com.cn<br>E-mail：mchannel@263.net（万水）<br>sales@waterpub.com.cn<br>电话：（010）68367658（营销中心）、82562819（万水） |
| 经　售 | 全国各地新华书店和相关出版物销售网点 |
| 排　版 | 北京万水电子信息有限公司 |
| 印　刷 | 三河市鑫金马印装有限公司 |
| 规　格 | 184mm×260mm　16 开本　19.5 印张　480 千字 |
| 版　次 | 2017 年 11 月第 1 版　2017 年 11 月第 1 次印刷 |
| 印　数 | 0001—3000 册 |
| 定　价 | 40.00 元 |

# 前　言

在移动互联网时代，人们的工作、生活等各方面都与移动终端（如微信、支付宝等）紧密关联。这个巨大的新兴市场吸引着成千上万的开发者不断加入其中。Android 现已成为占市场份额第一的操作系统，三星、华为、小米、魅族等手机生产厂商通过 Android 定制手机获得了巨大成功。随着 Android 手机在国内销量的不断提升，基于 Android 系统的应用程序开发成为了我国移动互联应用程序开发最重要的组成部分。

本书的编写宗旨是培养具有创新和创业能力的应用型人才，特点是面向应用、内容全面、注重实践、易于掌握，每一章都配有实际案例，既可作为教师的教学案例，又可供学生实践练习。本书的作者有经验丰富的一线教师，也有企业级导师，在编写过程中既吸收了 Android 开发设计类书籍的优点，又总结了一些培训机构的教学方法。

本书主要介绍如何运用 Eclipse 进行 Android 应用程序开发的相关知识与技能，同时还包含 MUI 框架、jQuery 组件、WAMP5 工具和 MySQL 数据库的相关知识。第 1 章介绍 Android 的基本发展情况、开发环境的搭建、DDMS 的使用、Android 程序开发的基本流程以及程序的调试过程，为后续 Android 应用程序开发的学习做准备。第 2 章介绍 Android 应用程序目录结构与 Android 应用程序中各文件的基本属性与使用方法。第 3 章介绍 Android 各类基本控件的使用与 Android 事件处理机制。第 4 章介绍各类布局的基本属性与使用方法，以及嵌套布局的使用。第 5 章介绍各类高级控件的使用场景与使用方法。第 6 章介绍菜单的基本使用方法、ActionBar 的基本属性与 Dialog、Toast 等对话框的使用场景。第 7 章介绍 Activity 的四种状态与生命周期、Intent 和 Bundle 的相关属性与使用方法。第 8 章介绍 Fragment 的生命周期、管理与通信。第 9 章介绍 Service、Notification、BroadcastReceiver 的基本使用方法。第 10 章介绍 SharedPreferences、ContentProvider、SQLite 的基本语法和相关操作。第 11 章介绍 HTTP 协议、Handler 与 Asynctask 的使用场景、网络状态判断、HttpURLConnection 和 JSON 的使用。第 12 章介绍一个应用项目开发实例——贺州旅游新闻系统的开发。

本书面向初学者，既可作为本科、高职高专院校和计算机培训机构相关课程的教材，又可作为 Android 系统开发设计人员的参考书。

本书由巫湘林任主编，陈彬、胡世洋、黄华升任副主编。其中巫湘林主持全书的编写及审稿工作，并编写第 1 章至第 5 章、第 8 章至第 10 章，胡世洋编写第 6 章，黄华升编写第 7 章，陈彬编写第 11 章和第 12 章。

由于作者经验和水平有限，书中难免有疏漏和不足之处，恳请广大读者和专家批评指正。

编　者

2017 年 9 月

# 目　　录

# 第 1 章　Android 基础入门

（1）了解 Android 的基本发展情况。
（2）掌握 Android 开发环境的配置和 DDMS 的使用。
（3）掌握 Android 应用程序开发的基本流程。
（4）掌握 Android 应用程序的调试过程。

随着移动互联网时代的到来，Android 操作系统受到广大用户的青睐，Android 应用程序的开发也因此越来越受欢迎。本章将从 Android 操作系统的基本发展情况开始，逐步讲解 Android 应用程序开发的各个过程。

## 1.1　Android 简介

Android 操作系统现已成为全世界最流行的操作系统，三星、华为、小米、魅族等手机生产厂商通过推广 Android 定制手机获得了巨大的成功。Android 手机销量的提升，也促进了对 Android 开发人才需求的增长。Android 手机除了已成为人手必备的通信工具外，也将会成为未来智能家居、智能监控等的通信入口。从发展趋势上看，未来对 Android 开发人才的需求量会急剧上升。

### 1.1.1　初识 Android

Android 是一种基于Linux的自由及开放源代码的操作系统，主要用于移动设备，如智能手机和平板电脑，由Google公司和开放手机联盟（Open Handset Alliance，OHA）领导及开发，目前尚未有统一的中文名称，很多人称之为“安卓”或“安致”。Android 操作系统最初由一家名为 Android 的公司研发，主要支持手机。2005 年 8 月，这家仅成立了 22 个月的公司被Google公司全资收购，原公司的 CEO 安迪鲁宾成为 Google 公司工程部副总裁，继续负责 Android 项目。

2007 年 11 月 5 日，Google公司正式向外界展示了这款名为 Android 的操作系统，并且在当天宣布建立一个全球性的联盟组织——开放手机联盟，该组织由 34 家手机制造商、软件开发商、电信运营商以及芯片制造商组成，共同研发改良 Android 系统。这一联盟将支持 Google 发布的手机操作系统和应用软件。Google 公司以 Apache 免费开源许可证的授权方式发布了 Android 的源代码。2008 年 9 月，Google 公司正式发布 Android 1.0 系统，这也是 Android 系统最早的版本。

### 1.1.2 Android发展历史

2007年11月，Google公司宣布开发基于Linux平台的开源手机操作系统，项目代号为Android。

- Android 1.0：于2008年9月发布。
- Android 1.5：于2009年4月30日发布，命名为Cupcake（纸杯蛋糕），这是Android较为稳定的一个版本，支持立体声蓝牙耳机和屏幕虚拟键盘，采用WebKit技术的浏览器，支持复制、粘贴、页面中搜索等功能。
- Android 1.6：于2009年9月15日发布，命名为Donut（甜甜圈），主要更新包括重新设计的Android Market，支持CDMA网络、OpenCore 2媒体引擎、更高的屏幕分辨率等。
- Android 2.0/2.0.1/2.1：于2009年10月26日发布，命名为Éclair（松饼），其改良了用户界面，使用新的浏览器用户接口，支持内置相机闪光灯、蓝牙2.1、动态桌面的设计和HTML5等。
- Android 2.2：于2010年5月20日发布，命名为Froyo（冻酸奶），其整体性能大幅提升，支持3G网络共享功能和Flash，而且提供了更多的Web应用API接口的开发等。
- Android 2.3：于2010年12月7日发布，命名为Gingerbread（姜饼），增加了新的垃圾回收和优化处理事件，支持VP8和WebM视频格式，提供AAC、AMR音频编码和新的音频效果器，支持前置摄像头、SIP、VOIP和NFC（近场通信）。
- Android 3.0：于2011年2月3日发布，命名为Honeycomb（蜂巢），支持多任务处理，拥有硬件加速功能、3D功能、视频通话功能等，最大的特点是支持平板电脑。
- Android 4.0：于2011年10月19日在香港发布，命名为Ice Cream Sandwich（冰激凌三明治），使用了全新的界面和全新的Chrome Lite浏览器，有离线阅读、标签页、隐身浏览模式和更强大的图片编辑功能，新增了流量管理工具，可具体查看每个应用产生的流量。
- Android 4.1/4.2/4.3：于2012年6月28日发布，命名为Jelly Bean（果冻豆），其特效动画的帧速提高至60fps，增加了三倍缓冲，改进了通知栏，支持智能语音搜索和Google Now应用，拥有全景拍照、改进的锁屏、可扩展通知和允许用户直接打开应用等功能。
- Android 4.4：于2013年9月4日发布，命名为KitKat（奇巧），整体改进了桌面图标、锁屏、开机动画、来电显示和配色方案等功能，支持无线打印、屏幕录像、计步器应用、低功耗音频和定位模式等。
- Android 5.0：于2014年11月3日发布，命名为Lollipop（棒棒糖），其使用一种新的Material Design设计风格，使系统整体界面的显示更加漂亮，支持整合碎片化、64位处理器，使用ART虚拟机。
- Android 6.0：于2015年5月28日发布，命名为Marshmallow（棉花糖），其继续使用扁平化的Material Design设计风格，在软件体验与运行性能上进行了大幅度优化，并提升了设备的续航能力。

- Android 7.0：于 2016 年 3 月 10 日发布，命名为 Nougat（牛轧糖），其支持分屏多任务、全新下拉快捷开关页、通知消息快捷回复、夜间模式、流量保护模式、菜单键快速应用切换等功能。

随着 Android 版本的不断更新，对计算机硬件的配置要求也越来越高，因为 Android 5.0 平台可以使用 Eclipse 开发完成且对硬件配置要求相对较低，符合大部分学习者的要求，所以本书使用 Android 5.0 平台进行应用开发。

### 1.1.3　Android 应用场景

Android 从最初的手机操作系统，逐渐成为平板电脑、智能手表、智能电视、智能眼镜、智能汽车等设备的操作系统。随着这些智能硬件产品的推出，在该系统上的相应软件应用这块市场（通信、教育、监控、打车、购物、餐饮、娱乐等）也吸引了更多创新创业人才来开发。手机作为当今主要的通信工具，在移动互联网方面的发展趋势如表 1-1 所示（表中数据由中国互联网络信息中心提供）。

表 1-1　2015.12—2016.6 中国网民各类手机互联网应用的使用率

| 应用 | 2016.6 | | 2015.12 | | 半年增长率 |
|---|---|---|---|---|---|
| | 用户规模（万） | 网民使用率 | 用户规模（万） | 网民使用率 | |
| 手机即时通信 | 60346 | 91.9% | 55719 | 89.9% | 8.3% |
| 手机网络新闻 | 51800 | 78.9% | 48165 | 77.7% | 7.5% |
| 手机搜索 | 52409 | 79.8% | 47784 | 77.1% | 9.7% |
| 手机网络音乐 | 44346 | 67.6% | 41640 | 67.2% | 6.5% |
| 手机网络视频 | 44022 | 67.1% | 40508 | 65.4% | 8.7% |
| 手机网上支付 | 42445 | 64.7% | 35771 | 57.7% | 18.7% |
| 手机网上购物 | 40070 | 61.0% | 33967 | 54.8% | 18.0% |
| 手机网络游戏 | 30239 | 46.1% | 27928 | 45.1% | 8.3% |
| 手机网上银行 | 30459 | 46.4% | 27675 | 44.6% | 10.1% |
| 手机网络文学 | 28118 | 42.8% | 25908 | 41.8% | 8.5% |
| 手机旅行预定 | 23226 | 35.4% | 20990 | 33.9% | 10.7% |
| 手机邮件 | 17343 | 26.4% | 16671 | 26.9% | 4.0% |
| 手机网上外卖 | 14627 | 22.3% | 10413 | 16.8% | 40.5% |
| 手机论坛/BBS | 8462 | 12.9% | 8604 | 13.9% | -1.7% |
| 手机网上炒股 | 4815 | 7.3% | 4293 | 6.9% | 12.1% |
| 手机在线教育 | 6987 | 10.6% | 5303 | 8.6% | 31.8% |

从表 1-1 可以看出，手机即时通信是移动互联网终端最主要的应用，微信、米聊、陌陌、钉钉等手机软件的出现给大家带来了方便，相信未来会有更多的手机即时通信软件出现。使用率排名靠前的还有手机网络新闻，其在城市智能手机用户中已基本普及，在农村智能手机用户

中的使用率也在逐步增加，手机用户在工作之余阅读新闻就像是品尝一道文化大餐，较受欢迎的有微博、今日头条、一点资讯、网易新闻和腾讯新闻等手机客户端。另外，手机网络音乐和网络视频的使用率也占较大的比重，给人们带来最快最直接的视听享受。

未来手机的投影技术与相应的配套软件将可能成为手机发展的一个新方向，用户体验至上的时代正在到来。AR 与 VR 技术融入智能手机生态圈将会给即时通信、实时新闻和网络购物等带来全新的体验。智能硬件（小到智能纽扣、智能手环，大到智能家居等）与手机应用软件的结合丰富了手机的应用领域。

由中国手机网民的增多与手机应用领域的不断丰富可以看出，Android 手机应用在未来会有更大的发展空间。

### 1.1.4 Android 体系结构

Android 是一个开放的软件系统，它采用分层的结构思想，由上到下分为 4 个层次，分别是：应用程序层（Application）、应用程序框架层（Application Framework）、核心类库（Libraries）和 Linux 内核层（Linux Kernel），如图 1-1 所示。

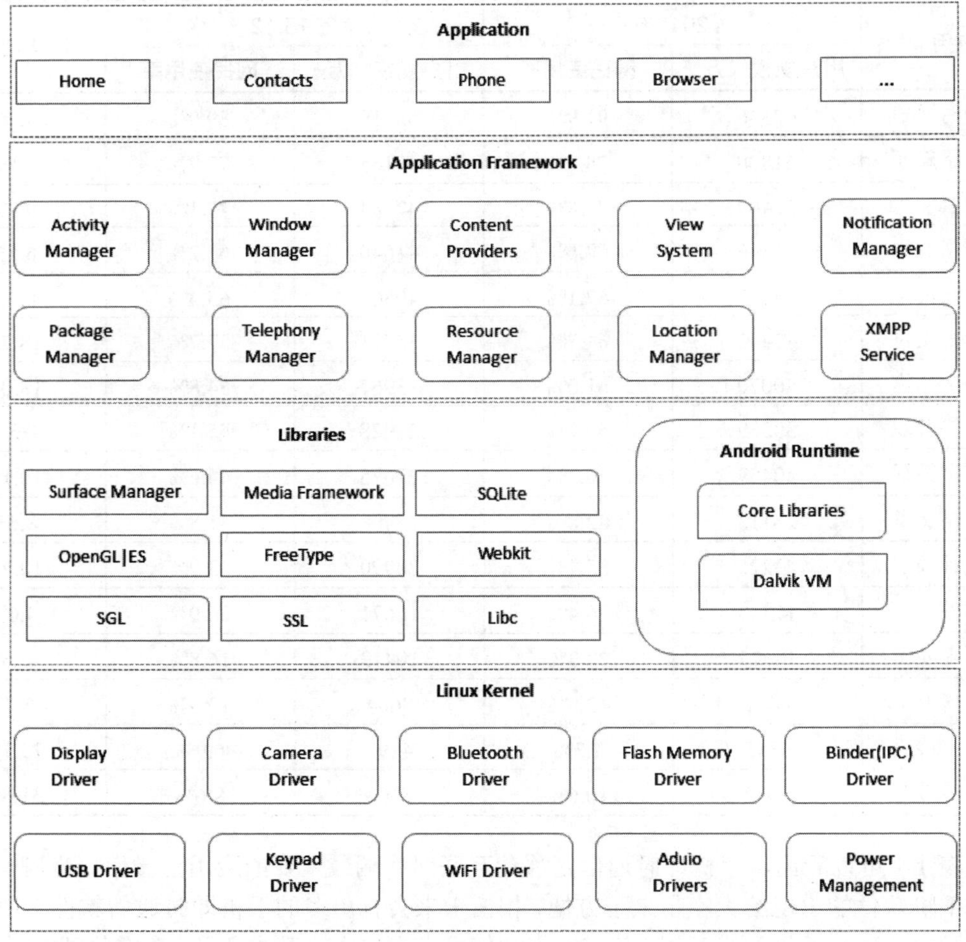


图 1-1 Android 体系结构

从图 1-1 可以看出 Android 体系结构的各个层次，接下来分别对这 4 个层次进行介绍。

1. 应用程序层

Android 不单是手机操作系统，还预先在系统里安装了一组常用应用程序，包括联系人程序、短信程序、日历程序、浏览器程序等，这些应用程序都是使用 Java 语言编写的。Android 系统的开放性使得开发者可以自己编写程序并部署到 Android 系统中，这些程序与系统自带的程序是彼此平等、友好共处的，这样做的优势是 Android 系统更加灵活。

2. 应用程序框架层

应用程序框架层是 Android 系统提供给应用程序层所使用的 API 框架，在进行应用程序开发的过程中要使用到这些 API 框架，但是必须遵守其开发原则，此做法的目的是减少重用组件时代码编写的工作量。

从图 1-1 可以看出，应用程序框架层主要提供了一系列服务来管理应用程序，主要包括如下 9 部分：

（1）Activity Manager（活动管理器）：主要负责管理应用程序的生命周期，提供常用的导航与返回功能。

（2）Window Manager（窗口管理器）：主要负责管理所有的窗口程序。

（3）Content Manager（内容管理器）：主要用于一个应用程序访问另外一个应用程序的数据，当然第三方应用程序也可以通过它来实现自己内部数据的共享。

（4）View System（视图系统）：主要提供一套功能强大可扩展的视图组件。它包括文本框（TextView）、按钮（Button）、输入框（EditText）、列表（Lists）等在程序设计界面中经常用到的组件，还提供了给开发者使用的自定义视图接口。

（5）Notification Manager（通知管理器）：主要负责应用程序在状态栏中的显示信息。

（6）Package Manager（包管理器）：主要负责 Android 系统内的程序管理。

（7）Telephony Manager（电话管理器）：主要负责 Android 系统中与手机通话相关的管理，如电话的接听与挂断、手机信息获取等。

（8）Resource Manager（资源管理器）：主要负责 Android 系统中非编码资源的访问，如本地化字符串、图形与布局文件等。

（9）Location Manager（位置管理器）：主要负责用户位置信息的管理，如无线信息热点、精准位置定位的 GPS 信息等。

3. 核心类库

核心类库由系统库与运行环境两部分组成，系统库由 C/C++库为 Android 系统提供主要的特性支持，如 SQLite 库提供了一个对所有应用程序可用的功能强大的关系型数据库引擎，Webkit 库提供了对浏览器内核的支持等。

Android 运行时库（Android Runtime Library）提供的一些核心库允许开发者使用 Java 语言开发 Android 应用，此外 Android 运行时库还有一个 Dalvik 虚拟机，使得每一个 Android 应用都能运行在独立的进程中，Dalvik 虚拟机不同于 Java 虚拟机，它针对手机制定了一些特定的手机内存和 CPU 处理等优化操作。

4. Linux 内核层

Android 系统基于 Linux 2.6 内核，并针对手机进行了特定的裁剪与优化，如电源管理、进程管理、网络协议等，此外还增加了 WiFi 驱动、蓝牙驱动、Binder IPC 驱动等。

# 1.2 Android 开发环境

我们已经知道了 Android 发展历史与体系结构，以及 Android 应用开发的广阔前景，接下来就进行 Android 开发环境的搭建。

Android 开发可以在 Linux 操作系统和 Windows 操作系统上完成，这两个操作系统上的开发环境搭建过程大致相同，当前比较流行的开发工具分别是 Android Studio 与 Eclipse。Android Studio 对计算机的配置要求较高且消耗内存较大，而 Eclipse 则对计算机的配置要求较低，所以本书选用 Eclipse 进行开发。

## 1.2.1 Java 下载安装

1. JDK 下载

在进行 Android 开发之前先要安装 JDK，在 Oracle 官方网站下载最新版本的 JDK1.8，其下载地址为http://www.oracle.com/technetwork/java/javase/downloads/jdk8-downloads-2133151.html，下载过程中注意选择与自己计算机操作系统相对应的 JDK 版本，如图 1-2 所示。

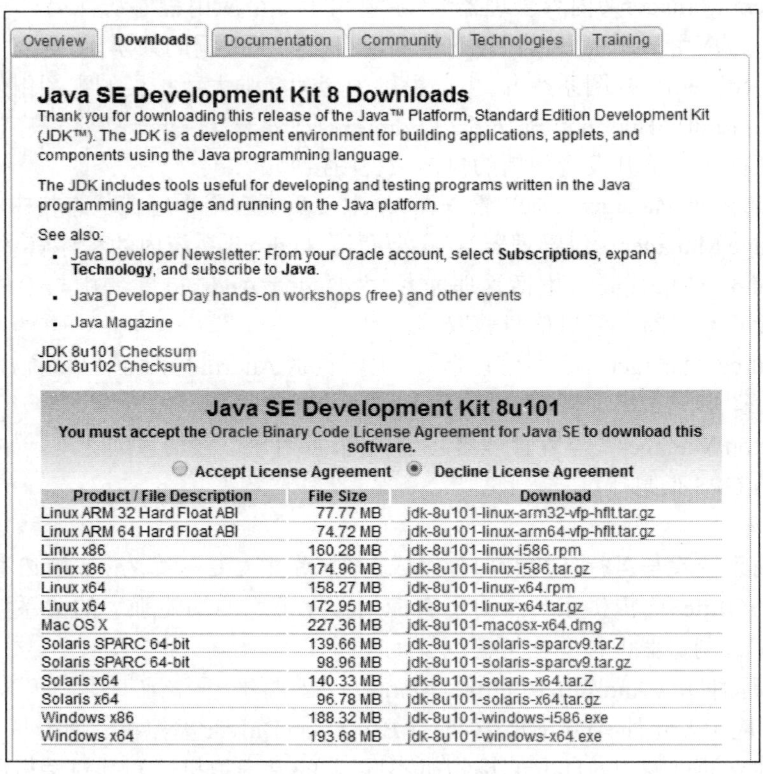


| Product / File Description | File Size | Download |
|---|---|---|
| Linux ARM 32 Hard Float ABI | 77.77 MB | jdk-8u101-linux-arm32-vfp-hflt.tar.gz |
| Linux ARM 64 Hard Float ABI | 74.72 MB | jdk-8u101-linux-arm64-vfp-hflt.tar.gz |
| Linux x86 | 160.28 MB | jdk-8u101-linux-i586.rpm |
| Linux x86 | 174.96 MB | jdk-8u101-linux-i586.tar.gz |
| Linux x64 | 158.27 MB | jdk-8u101-linux-x64.rpm |
| Linux x64 | 172.95 MB | jdk-8u101-linux-x64.tar.gz |
| Mac OS X | 227.36 MB | jdk-8u101-macosx-x64.dmg |
| Solaris SPARC 64-bit | 139.66 MB | jdk-8u101-solaris-sparcv9.tar.Z |
| Solaris SPARC 64-bit | 98.96 MB | jdk-8u101-solaris-sparcv9.tar.gz |
| Solaris x64 | 140.33 MB | jdk-8u101-solaris-x64.tar.Z |
| Solaris x64 | 96.78 MB | jdk-8u101-solaris-x64.tar.gz |
| Windows x86 | 188.32 MB | jdk-8u101-windows-i586.exe |
| Windows x64 | 193.68 MB | jdk-8u101-windows-x64.exe |

图 1-2 JDK 下载页面

2. JDK 安装

下载完 JDK 安装包后，在 Windows 操作系统上双击进行安装与环境设置，在此不再详细说明。

3. Java 环境检测

在 Windows 命令行模式下输入 java –version，然后按回车键查看 JDK 的版本信息，如果安装成功，则会出现如图 1-3 所示的信息。

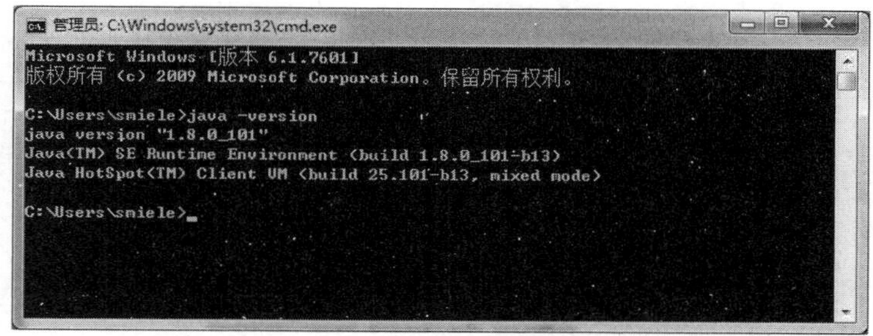


图 1-3　检查 JDK 安装情况

### 1.2.2　ADT Bundle 下载

最初进行 Android 开发时，先要下载 Eclipse 开发工具，Eclipse 官方下载地址为 https://www.eclipse.org/downloads/packages/eclipse-ide-java-ee-developers/oxygen1a，如图 1-4 所示，接着登录 http://www.android-studio.org 下载 Android SDK 工具包，最后在 Eclipse 中安装 ADT 插件。这样复杂的环境搭建让很多初学者感到非常困难。读者可以登录作者的百度云盘（地址为 http://pan.baidu.com/s/1sl4EW13）下载已经做好的 Android 集成开发环境。

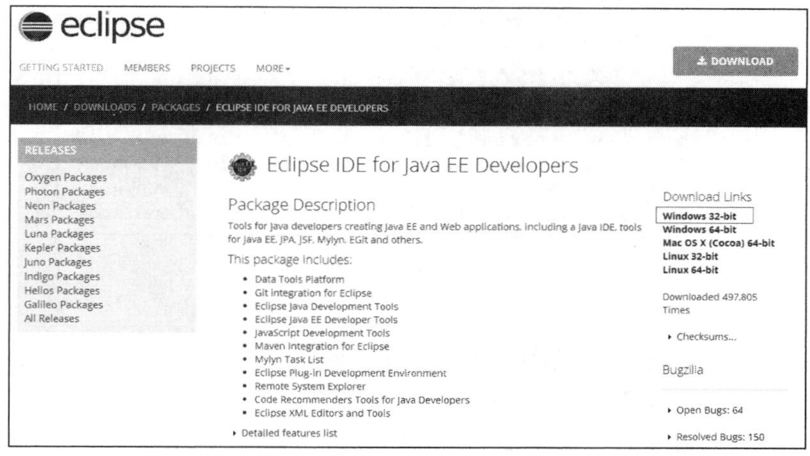


图 1-4　下载 Eclipse 开发工具

将下载完的压缩包 adt-bundle-windows-x86-20140702.zip 进行解压，可以看到 SDK Manager.exe 文件、SDK 文件夹和 Eclipse 文件夹。接下来将分别对这 3 种开发工具进行详细介绍。

1. SDK Manager.exe

SDK Manager.exe 负责管理计算机上目前安装的各种版本的 Android SDK。双击这个文件可以查看当前可用的 Android SDK 版本，如图 1-5 所示。由于 Android 的版本众多，我们只下载当前需要开发的 Android SDK 版本即可。因为 Google 公司官方更新 Android SDK 版本的网址被屏蔽，我们要用到国内的代理来更新。在 Android SDK Manager 窗口的 Tools 菜单下选择 Options，在弹出的 Android SDK Manager-Settings 对话框中进行代理设置，如图 1-6 所示。

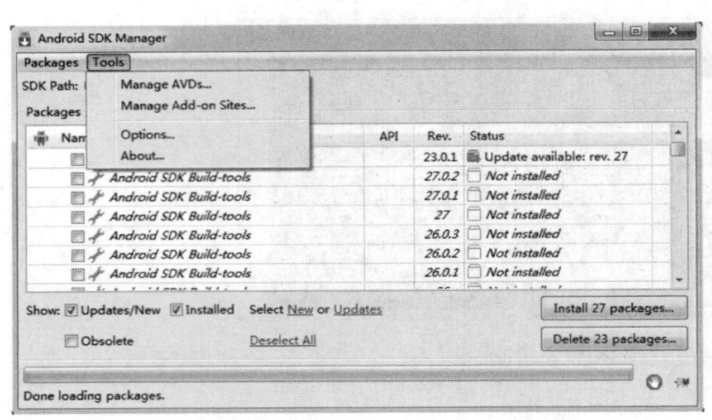


图 1-5 Android SDK Manager 窗口

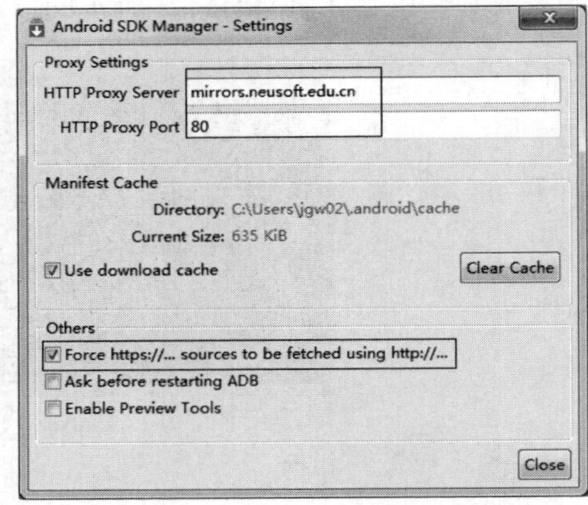


图 1-6 Android SDK Manager 代理设置

因为集成的 SDK 工具包没有 Android 5.0 版本，所以我们要更新如图 1-7 所示的内容。因为资源量比较大，所以下载耗时会比较长。

| Android 5.0.1 (API 21) | | | |
|---|---|---|---|
| SDK Platform | 21 | 2 | Not installed |
| Android TV ARM EABI v7a System Image | 21 | 3 | Not installed |
| Android TV Intel x86 Atom System Image | 21 | 3 | Not installed |
| Android Wear ARM EABI v7a System Image | 21 | 3 | Not installed |
| Android Wear Intel x86 Atom System Image | 21 | 3 | Not installed |
| ARM EABI v7a System Image | 21 | 3 | Not installed |
| Intel x86 Atom_64 System Image | 21 | 4 | Not installed |
| Intel x86 Atom System Image | 21 | 4 | Not installed |
| Google APIs ARM EABI v7a System Image | 21 | 15 | Not installed |
| Google APIs Intel x86 Atom_64 System Imag | 21 | 15 | Not installed |
| Google APIs Intel x86 Atom System Image | 21 | 15 | Not installed |
| Google APIs | 21 | 1 | Not installed |
| Sources for Android SDK | 21 | 1 | Not installed |

图 1-7 Android 5.0 更新部分

2. SDK

SDK 为开发者进行软件开发提供了丰富的库文件和其他开发工具。整个 SDK 文件夹下包括多个子文件夹，如图 1-8 所示。

| 名称 | 修改日期 | 类型 |
|---|---|---|
| build-tools | 2014/7/2 12:22 | 文件夹 |
| extras | 2014/7/2 12:22 | 文件夹 |
| platforms | 2014/7/2 12:22 | 文件夹 |
| platform-tools | 2014/7/2 12:21 | 文件夹 |
| tools | 2014/7/1 16:14 | 文件夹 |

图 1-8 SDK 文件夹目录

SDK 文件夹下的子文件夹各有不同的用途，下面进行简要介绍。

- build-tools：各版本 SDK 的编译工具。
- extras：扩展开发包，如高版本的 API 在低版本中开发时使用。

- platforms：各版本的 SDK。根据 API Level 划分的 SDK 版本，当前如果更新了 Android 5.0 的 SDK，那么在该目录下就会有一个 android-21 文件夹。
- platform-tools：各版本 SDK 的通用工具，比如 adb、fastboot 和 sqlite3 等文件。
- tools：各版本 SDK 的自带工具，如 DDMS 是用于启动 Android 的调试工具，draw9patch 是绘制 Android 平台的可缩放 png 图片的工具，mksdcard 是模拟器 SD 镜像的创建工具等。

3. Eclipse

Eclipse 是进行 Android 应用程序开发的一种工具，在 Eclipse 上需要安装 ADT 插件来为用户提供便捷的图标按钮操作。ADT 插件安装成功后，在 Java-Eclipse 窗口中有如图 1-9 所示的 3 个图标。方框中最左边的图标表示 Android SDK 管理器，它和 SDK Manager.exe 功能相同；中间的图标是配置与启动 Android 模拟器的，Android 模拟器的大部分功能与真机效果是相同的；最右边的图标是用来检测 Android 程序代码的。点击中间的图标，出现如图 1-10 所示的窗口。

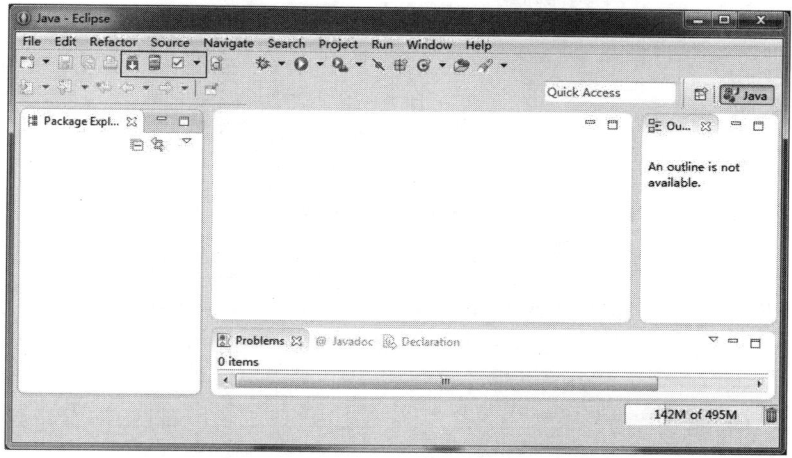


图 1-9　Java-Eclipse 窗口

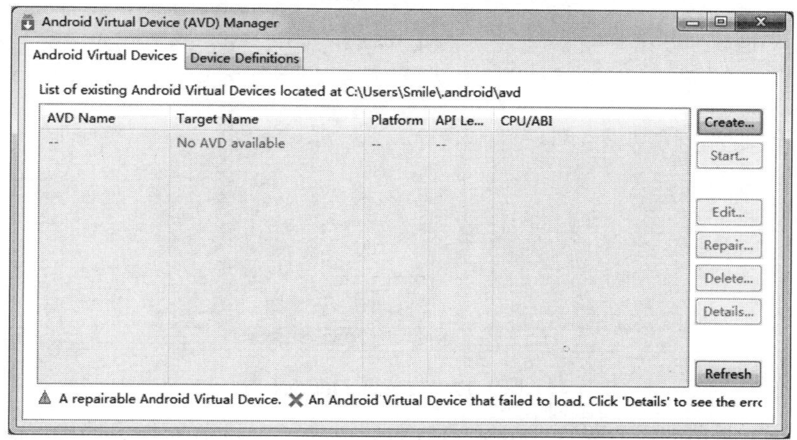


图 1-10　Android Virtual Device（AVD）Manager

在图 1-10 中，点击右侧的 Create 按钮，弹出如图 1-11 所示的对话框，然后创建一个名为 android 5.0 的模拟器，选择 3.2 英寸屏幕的手机，目标 SDK 版本为 Android 5.0，接着再指定手机内存与 SD 卡的存储空间。创建成功后，在 Android Virtual Device（AVD）Manager 窗口中会显示刚才创建的模拟器信息，然后点击右侧的 Start 按钮，弹出 Launch Options 对话框，如图 1-12 所示。

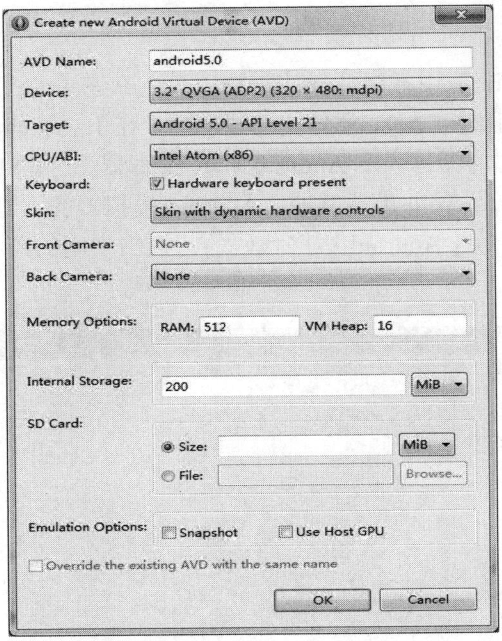

图 1-11　创建模拟器

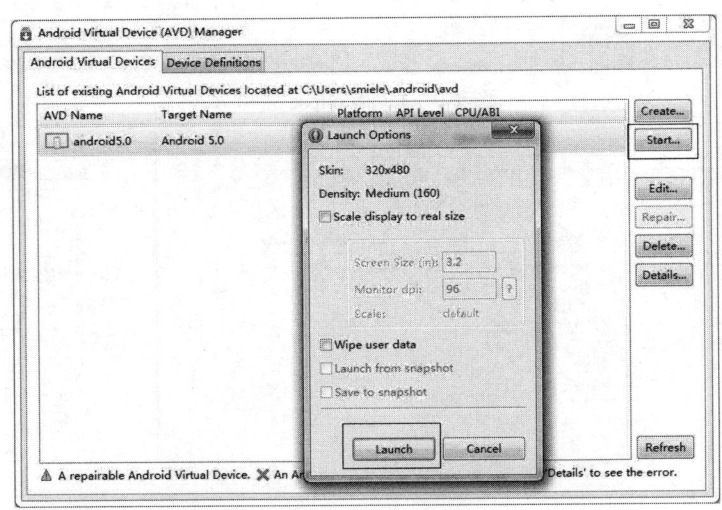


图 1-12　Launch Options

点击图 1-12 中的 Launch 按钮，出现的模拟器界面和真机界面基本相同。启动成功的模拟器界面如图 1-13 所示。

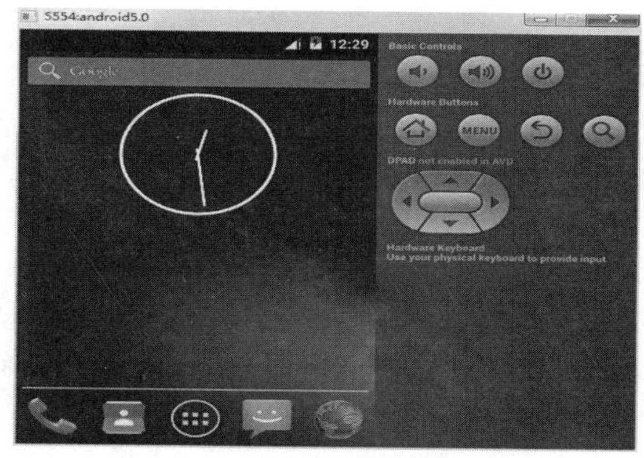


图 1-13　模拟器界面

通过上述步骤，我们成功地在 Windows 平台上搭建好了 Android 应用开发环境。在整个过程中需要下载 JDK 和 Android 集成的 SDK 工具包，进行 SDK 更新以及模拟器的创建。

### 1.2.3　Android 调试工具

Android 的调试工具是指位于 SDK 的 platform-tools 文件夹下的 adb.exe 文件。它的任务是完成开发者与模拟器或真机的通信。为了方便快捷地使用这个工具，需要将 adb.exe 所在的位置添加到 PATH 路径中。

adb.exe 的主要功能有：运行设备的 shell（命令行），管理模拟器或设备的端口映射，计算机和设备之间上传/下载文件，将本地 apk 软件安装至模拟器或 Android 设备。

adb 常用命令如下：

- adb start-server：开启 adb 服务。
- adb kill-server：关闭 adb 服务。
- adb devices：查看当前连接的设备。
- adb install<应用程序名>：安装 apk 程序。
- adb uninstall<应用程序名>：卸载 apk 程序。
- adb push<本地路径><远程路径>：上传文件到设备
- adb pull<远程路径><本地路径>：下载文件到设备。

接下来以 adb devices 命令为例，显示当前连接计算机的手机设备。打开 Windows 命令行窗口，输入 adb devices 并按回车键，显示结果如图 1-14 所示。因为只开启了一个模拟器，所以只显示一个设备。

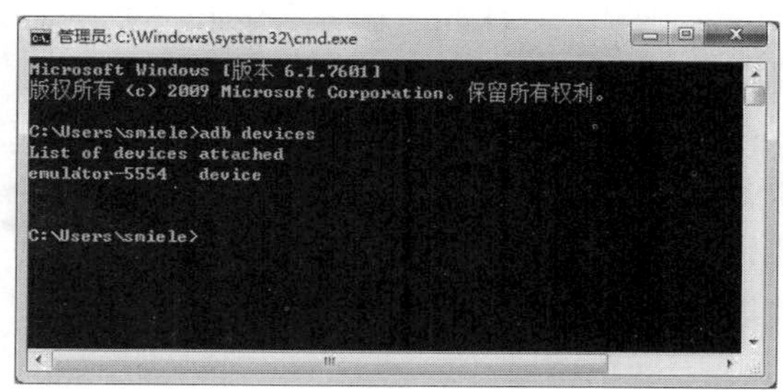


图 1-14　adb devices 命令

### 1.2.4 DDMS 的使用

DDMS 的全称是 Dalvik Debug Monitor Service，其为 IDE、模拟器与真机设备构建了一座桥梁，开发者可以通过 DDMS 看到目标机器上运行的进程/线程状态，如查看进程的 Heap 信息、LogCat 信息和分配内存的情况，还可以模拟拨入电话、模拟接收短信等。

DDMS 启动后会与 ADB 之间建立一个 Device Monitoring Service 用于监控设备，当设备断开连接时，这个 Service 就会通知 DDMS。当一个设备连接上时，DDSM 和 ADB 之间又会建立 VM Monitoring Service 用于监控设备上的虚拟机，通过 ADB Deamon 与设备上虚拟机的 debugger 建立连接，这样 DDMS 就开始与虚拟机对话了。将 Eclipse 从 Java 视图转换到 DDMS 视图，如图 1-15 所示。

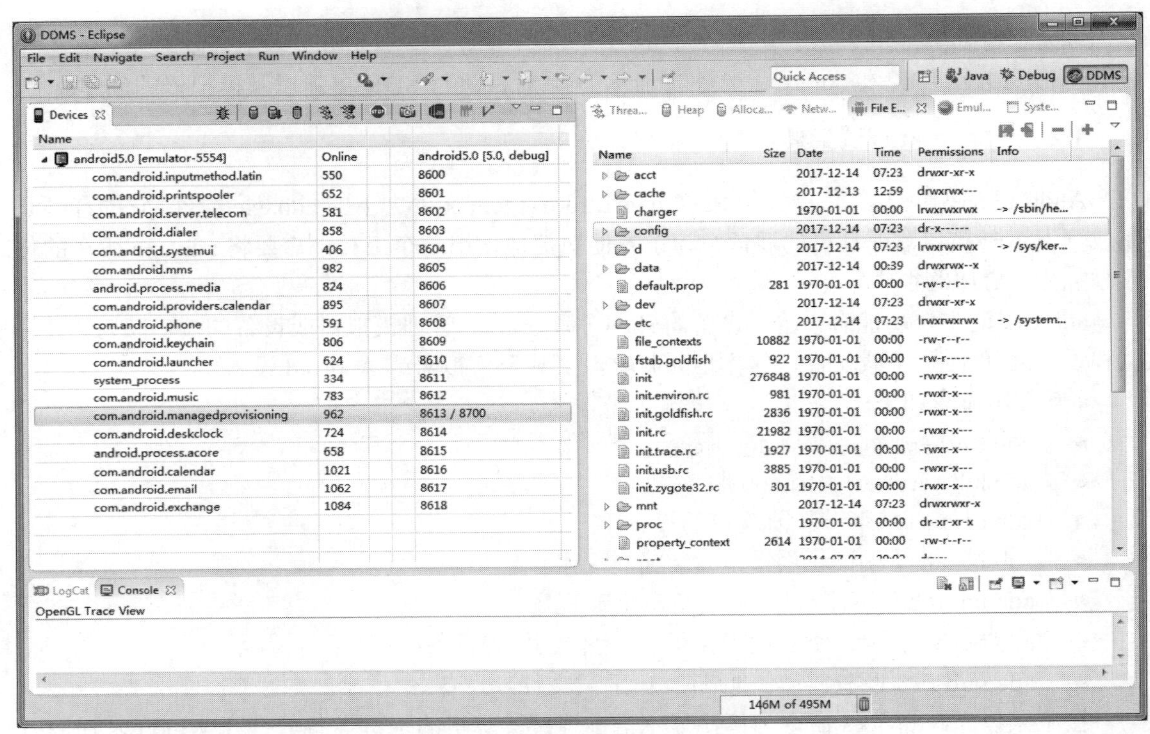

图 1-15 DDMS 视图

在图 1-15 所示的 DDMS 视图中，可以看到模拟器 emulator-5554 的运行状态和运行进程，显示进程时会显示进程 ID（左侧区域中 Online 那一列显示的即是终端上运行的进程 ID）和与进程相关联的端口号，端口号从 8600 依次往后增加。在右侧区域可以看到 Threads、Heap、Allocation Tracker、Network Statistics、File Explorer、Emulator Control、System Information 选项卡。接下来对这些选项卡进行介绍。

- Threads：表示当前进程中的所有线程状态。
- Heap：表示当前进程堆的使用情况。
- Allocation Tracker：分配跟踪器。
- Network Statistics：网络分析功能。

- File Explorer：浏览终端的文件系统，进行文件相关操作，可以将外部文件导入到终端中，也可将终端文件导出或删除。
- Emulator Control：可以实现往模拟器中打电话、发送短信、发送地理位置坐标等功能。
- System Information：查看终端的 CPU 负载以及内存使用情况。

### 1.2.5　使用 adb 命令安装与卸载 Android 应用程序

1. 安装 Android 应用程序

以安装和卸载 QQ 拼音输入法为例。在计算机的“开始”菜单→“附件”→“管理员：命令提示符”窗口中把路径切换到 Android SDK 安装目录下的 platform-tools 文件夹中。为了安装方便，直接将 QQ 拼音输入法（QQshurufa_1928.apk）文件放到 platform-tools 文件夹中，然后使用 adb install 命令将以上 apk 文件安装到 Android 模拟器中，安装完成后将显示 Success 信息，如图 1-16 所示，在模拟器中显示的结果如图 1-17 所示。

图 1-16　使用 adb 命令安装 Android 应用程序到模拟器

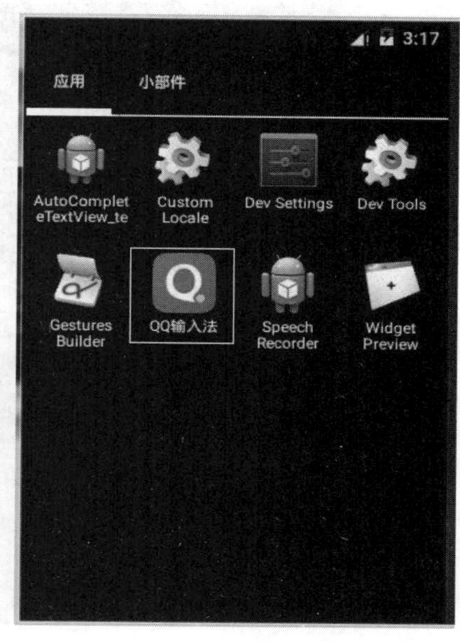


图 1-17　安装效果

2. 卸载 Android 应用程序

在计算机的“开始”菜单→“附件”→“管理员：命令提示符”窗口中把路径切换到 Android SDK 安装目录下的 platform-tools 文件夹中，使用 adb uninstall 命令卸载指定的 Android 应用程序，卸载 QQ 拼音输入法，如图 1-18 所示。

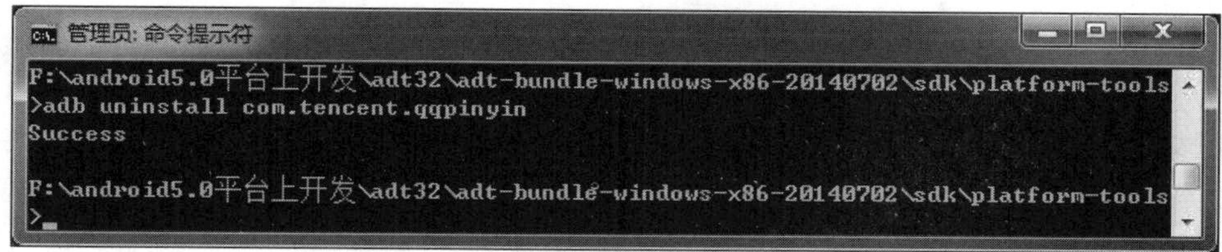


图 1-18　卸载 QQ 拼音输入法

**注意**：使用卸载命令 adb uninstall 卸载 Android 应用程序时，后面跟的是应用程序包名，而不是 apk 安装文件名。如果要查看已经安装成功的应用程序的完整包名，可以在命令提示符下使用以下几个命令：

- adb shell
- cd data
- cd app
- exit

例如查找 QQ 拼音输入法的完整包名，如图 1-19 所示。

```
管理员: 命令提示符
Success

F:\android5.0平台上开发\adt32\adt-bundle-windows-x86-20140702\sdk\platform-tools
>adb shell
root@generic:/ # cd data
cd data
root@generic:/data # cd app
cd app
root@generic:/data/app # ls
ls
ApiDemos
CubeLiveWallpapers
GestureBuilder
SmokeTest
SmokeTestApp
SoftKeyboard
WidgetPreview
com.example.autocompletetextview_test-1
com.hzu.autocompletetextview_test-1
com.tencent.qqpinyin-1
root@generic:/data/app # exit
exit
```

图 1-19 查看 apk 文件的完整包名

## 1.3 开始第一个 Android 应用

Android 应用程序开发是建立在应用程序框架之上的，Android 编程是面向应用程序框架 API 的编程，这种编程方式与 Java SE 应用程序编程区别不大，只是 Android 自己新增了一些 API，需要开发者自行理解这些 API 后再进行编程。

### 1.3.1 创建 HelloWorld 项目

在 Eclipse 中创建第一个 Android 工程，以 HelloWorld 项目为例。

（1）在 Eclipse 的菜单栏中选择 File→New，在子菜单中选择 Android Application Project 命令，如果没有这个命令，则选择 Other，如图 1-20 所示。

（2）在弹出的窗口中选择 Android，显示多个 Android 项目类型，选择 Android Application Project，如图 1-21 所示。

（3）点击 Android Application Project 后，弹出如图 1-22 所示的界面。

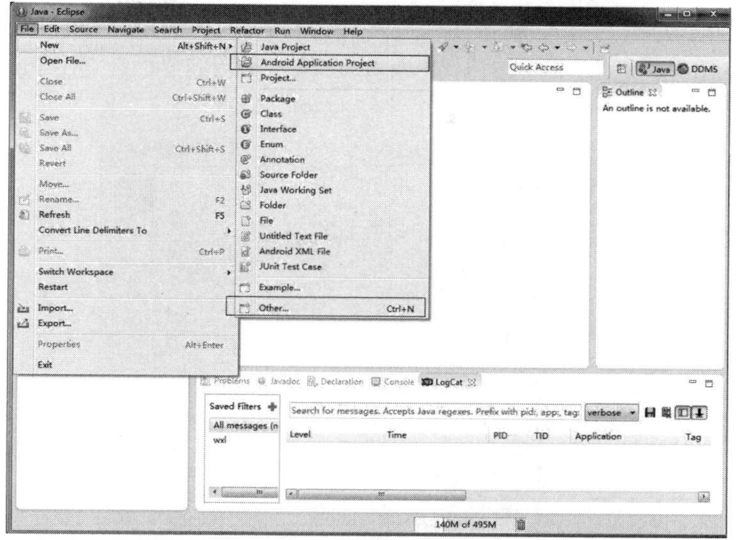


图 1-20　创建项目

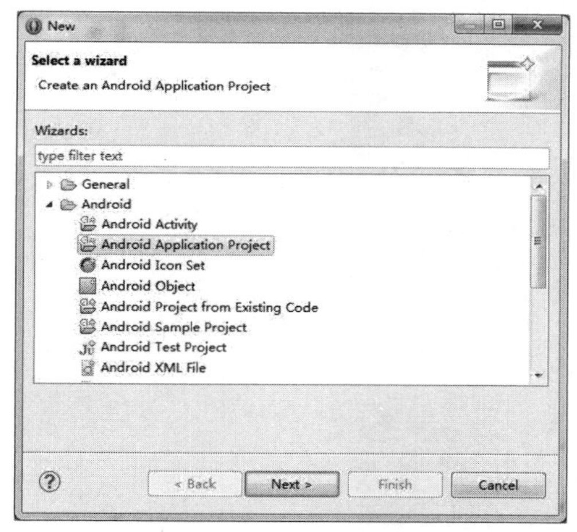


图 1-21　项目类型

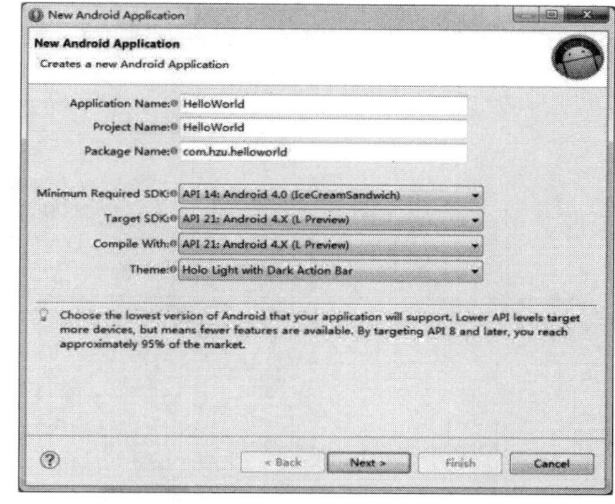


图 1-22　新建项目 1

图 1-22 由两部分组成，上半部分是与 Android 程序相关的设置，下半部分是与 SDK 版本相关的设置。接下来分别对这两部分进行介绍。

- Application Name：表示应用程序名称，是应用程序安装到手机后显示的名称。
- Project Name：表示项目名称，项目创建好后在 Eclipse 中显示。
- Package Name：表示项目中的包名，在当前创建的 HelloWorld 项目里显示的包名为 com.hzu.helloworld。
- Minimum Required SDK：表示应用程序能兼容的最低版本，当前选择 Android 4.0。
- Target SDK：表示应用程序最匹配的目标版本，不支持此版本的向上兼容，当前选择 Android 4.X。
- Compile With：表示应用程序使用哪个版本进行编译，当前选择 Android 4.X。
- Theme：表示应用程序所使用的主题，当前选择 Holo Light with Dark Action Bar。

点击图 1-23 所示界面中的 Next 按钮，进入项目的各个初始化配置过程，如图 1-24 所示，这里全部选择默认配置，点击 Next 按钮，出现如图 1-25 所示的界面，点击 Finish 按钮完成项目的创建。

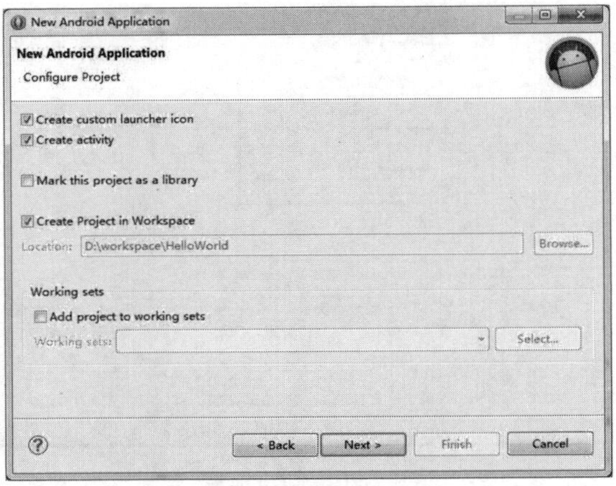

图 1-23　新建项目 2

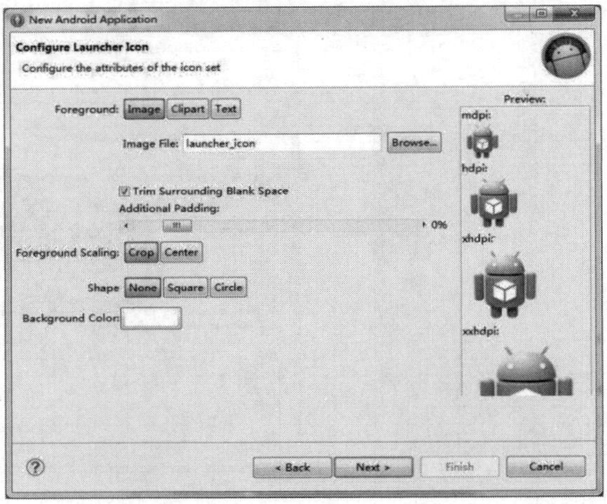

图 1-24　新建项目 3

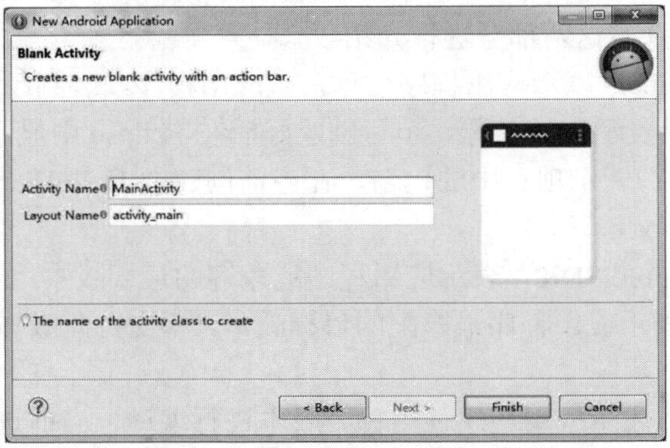

图 1-25　新建项目 4

项目创建好之后，出现如图 1-26 所示的 Eclipse 主界面。

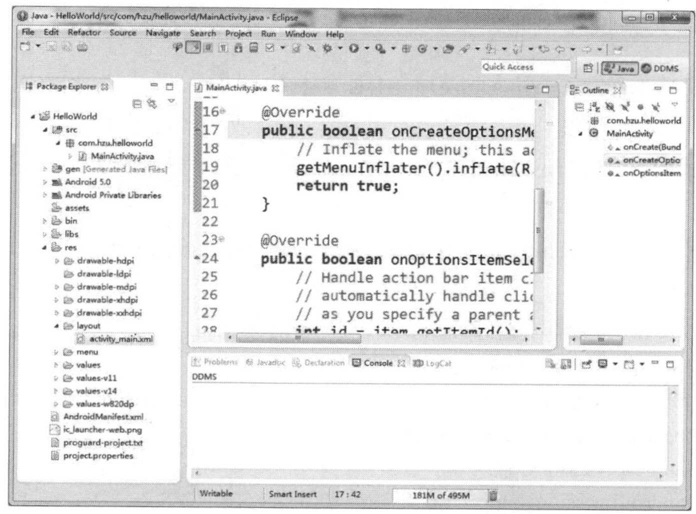


图 1-26　Eclipse 主界面

### 1.3.2　运行程序

启动模拟器，然后在 Eclipse 主界面中的 Package Explorer 区域选择要运行的项目名 HelloWorld，右击并在弹出的菜单中选择 Run As→Android Application，即可在模拟器中看到项目运行的界面，运行结果如图 1-27 所示。如果需要横屏显示结果，可以在计算机键盘上使用 Ctrl+F12 组合键，运行结果如图 1-28 所示。

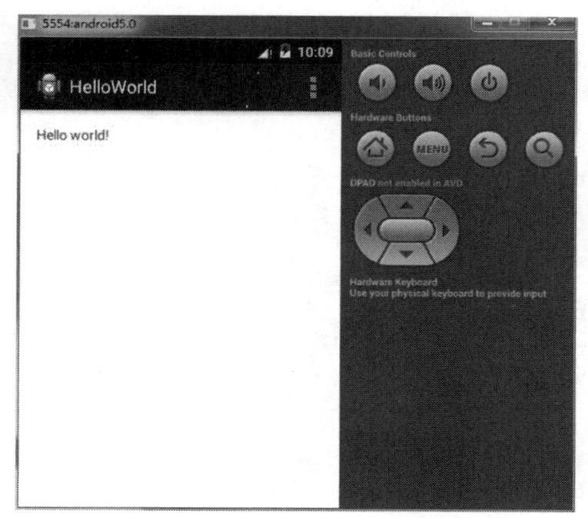


图 1-27　HelloWorld 运行结果

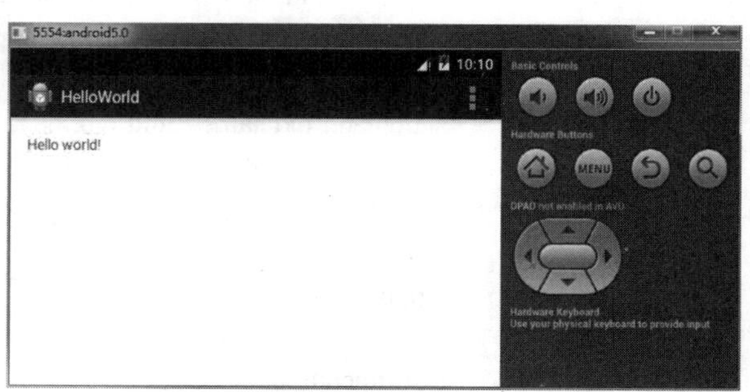


图 1-28　HelloWorld 横屏运行结果

## 1.4　程序调试

每个 Android 应用程序在正式上线前都要进行大量测试，以避免 bug 的出现，保证程序的正常平稳运行。测试 Android 应用程序一般使用 JUnit 单元测试，此外程序出错时，还可以使用 LogCat（日志控件台）进行程序调试。

### 1.4.1 JUnit 单元测试

JUnit 是一个测试框架，它在 Android SDK 1.5 中就加入了自动化测试功能，在单独完成某一个功能后就可以进行测试，不需要安装到真实手机或模拟器中，这样可以大大提高应用程序的开发速度与质量。下面对应用程序的单元测试步骤进行详细介绍。

**【开发步骤】**

（1）创建一个名为 AndroidJunit_test 的项目，包名为 com.hzu.androidjunit_test，Activity 组件名为 MainActivity。

（2）在 AndroidJunit_test 项目下的 AndroidManifest.xml 文件中编写代码，内容如下：

```
1    <?xml version="1.0" encoding="utf-8"?>
2    <manifest xmlns:android="http://schemas.android.com/apk/res/android"
3        package="com.hzu.androidjunit_test"
4        android:versionCode="1"
5        android:versionName="1.0">
6
7        <uses-sdk
8            android:minSdkVersion="14"
9            android:targetSdkVersion="21" />
10
11       <application
12           android:allowBackup="true"
13           android:icon="@drawable/ic_launcher"
14           android:label="@string/app_name"
15           android:theme="@style/AppTheme">
16       <uses-library android:name="android.test.runner" />
17       <activity
18               android:name=".MainActivity"
19               android:label="@string/app_name">
20       <intent-filter>
21       <action android:name="android.intent.action.MAIN" />
22       <category android:name="android.intent.category.LAUNCHER" />
23       </intent-filter>
24       </activity>
25       </application>
26
27       <instrumentation
28           android:name="android.test.InstrumentationTestRunner"
29           android:targetPackage="com.hzu.androidjunit_test" />
30
31   </manifest>
```

1）第 16 行在<application>标签（元素）下添加了函数库<uses-library>。

2）第 27～29 行在<manifest>标签下添加了指令集<instrumentation>。

**注意：**在<instrumentation>标签下 android:targetPackage 的名字一定要与被测试的 Android

应用程序的包名一致，否则将会出现找不到测试用例的错误。

（3）在项目 src 下 com.hzu.androidjunit_test 包中添加一个待测试类 Compute。

（4）在项目 src 下创建一个测试类 ComputeTest，此类需要继承自 AndroidTestCase 类，代码内容如下：

```
1    package com.hzu.androidjunit_test;
2
3    import android.test.AndroidTestCase;
4    import junit.framework.Assert;
5
6    public class ComputeTest extends AndroidTestCase {
7        private Compute compute;
8
9        @Override
10       protected void setUp() throws Exception {
11           compute = new Compute();
12           super.setUp();
13       }
14
15       public void testAdd() throws Exception {
16           int result = compute.add(2, 10);
17           Assert.assertEquals(12, result);
18       }
19
20       public void testDivide() throws Exception {
21           int result = compute.divide(10, 5);
22           Assert.assertEquals(2, result);
23       }
24
25       public void testsubtract() throws Exception {
26           int result = compute.sub(10, 5);
27           Assert.assertEquals(5, result);
28       }
29
30       public void testmultiply() throws Exception {
31           int result = compute.multiply(10, 2);
32           Assert.assertEquals(0, result);
33       }
34
35       @Override
36       protected void tearDown() throws Exception {
37           compute = null;
38           super.tearDown();
39       }
40   }
```

1）第 4 行引入 junit.framework.Assert 包，在类中测试方法内使用断言（Assert）来测试要测试的方法。

2）第 10～13 行初始化待测试类 Compute 对象。

3）第 15～33 行添加四个测试方法，每个测试方法都抛出异常（throws Exception），然后通过 Assert 对结果进行断言。

4）第 36～39 行销毁待测试类 Compute 对象。

（5）运行测试，有两种方式。一种方式是选中 ComputeTest 这个类名，然后右击选择 Run As→Android Junit Test，对该类下的所有方法进行测试；另一种方式是选中 ComputeTest 类下的某个方法（如 testAdd 方法），然后右击选择 Run As→Android Junit Test，对该类下的某个方法（如 testAdd 方法）进行测试。

当运行结果与预期相同时，JUnit 窗口会显示为绿色。上述前一种方式（测试类中所有方法）的测试效果如图 1-29 所示，后一种方式（测试 testAdd 方法）的测试效果如图 1-30 所示。

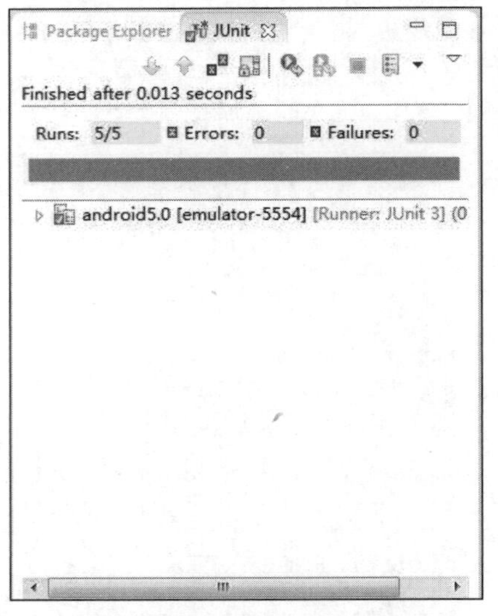


图 1-29　对所有方法进行测试

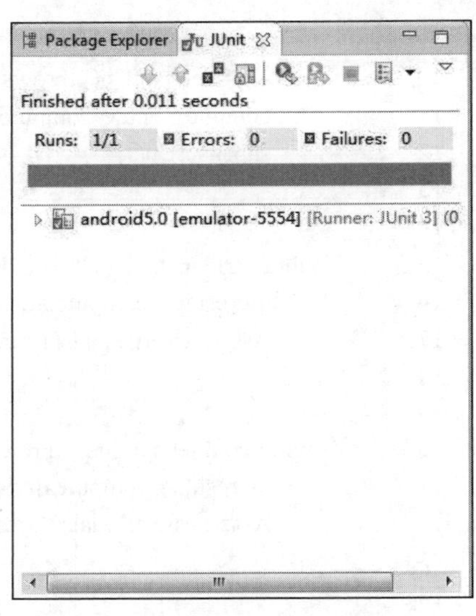


图 1-30　对单个方法进行测试

注意：如果测试出错，JUnit 窗口会显示为红色，错误的方法上会出现“x”字样，点击这个错误的方法可以查看具体的错误代码行。JUnit 测试主要针对逻辑层代码，保证代码在数据层与控制层之间正常流转。

### 1.4.2　LogCat 的使用

LogCat 是 Android 应用开发过程中用来显示打印日志的工具，特别是在后期调试程序 bug 时用得比较多，类似在 Java 中用 System.out.print()进行输出，LogCat 可以根据自己的需要定制输出结果。

在 Android 应用程序中进行信息输出时主要采用 android.util.Log 类的静态方法来实现，LogCat 划分了 5 个打印日志的级别，具体如下：

- Log.v()：用于打印那些最琐碎、最没意义的日志信息。对应级别为 verbose，是 Android 日志里面级别最低的一种，在 LogCat 控制台以黑色信息输出。
- Log.d()：用于打印一些调试信息，这些信息对开发者调试程序和分析问题应该是非常有帮助的。对应级别为 debug，在 LogCat 控制台以蓝色信息输出。

- Log.i()：用于打印一些比较重要的数据，这些数据应该是开发者非常想看到的，可以帮开发者分析用户行为。对应级别为 info，在 LogCat 控制台以绿色信息输出。
- Log.w()：用于打印一些警告信息，提示开发者此应用程序在这个地方可能存在潜在的风险。对应级别为 warn，在 LogCat 控制台以橙色信息输出。
- Log.e()：用于打印应用程序中的错误信息，提示开发者此应用程序出现了错误。对应级别为 error，在 LogCat 控制台以红色信息输出。

**【例 1.1】**设计 Log.v()、Log.d()、Log.i()、Log.w()、Log.e()五种信息输出到 LogCat 控制台。

**【开发步骤】**

创建一个名为 Logcat_activity 的项目，包名为 com.hzu.logcat_activity，Activity 组件名为 MainActivity。在 src/com/hzu/logcat_activity/MainActivity.java 中编写如下代码：

```
1    package com.hzu.logcat_activity;
2    import android.app.Activity;
3    import android.os.Bundle;
4    import android.util.Log;
5    public class MainActivity extends Activity {
6        private static final String TAG = "MainActivity";
7        protected void onCreate(Bundle savedInstanceState) {
8            super.onCreate(savedInstanceState);
9            setContentView(R.layout.activity_main);
10           Log.v(TAG, "verbose");
11           Log.d(TAG, "debug");
12           Log.i(TAG, "info");
13           Log.w(TAG, "warn");
14           Log.e(TAG, "error");
15       }
16   }
```

（1）第 4 行引入 android.util.Log 包。

（2）第 10～14 行分别使用 Log 类的 5 个静态方法。

**【运行结果】**在 Eclipse 中启动 Android 模拟器，接着运行 Logcat_activity 项目。运行结果在 LogCat 控件台的输出信息如图 1-31 所示。

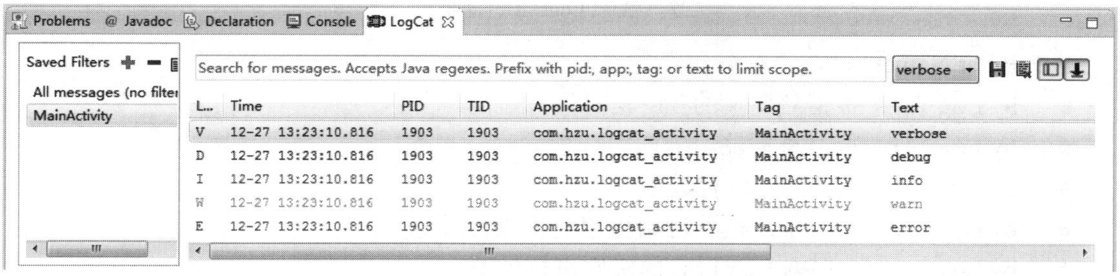

| L... | Time | PID | TID | Application | Tag | Text |
|---|---|---|---|---|---|---|
| V | 12-27 13:23:10.816 | 1903 | 1903 | com.hzu.logcat_activity | MainActivity | verbose |
| D | 12-27 13:23:10.816 | 1903 | 1903 | com.hzu.logcat_activity | MainActivity | debug |
| I | 12-27 13:23:10.816 | 1903 | 1903 | com.hzu.logcat_activity | MainActivity | info |
| W | 12-27 13:23:10.816 | 1903 | 1903 | com.hzu.logcat_activity | MainActivity | warn |
| E | 12-27 13:23:10.816 | 1903 | 1903 | com.hzu.logcat_activity | MainActivity | error |

图 1-31　LogCat 控制台

注意：如果没有找到 LogCat 控制台，可以在 Eclipse 主界面中选择 Window→ShowView→Other，在弹出的 Show View 窗口中选择 Android→LogCat，如图 1-32 所示。

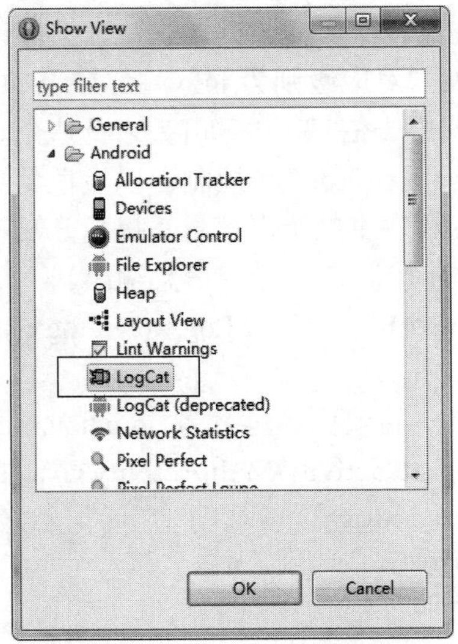


图 1-32　打开 LogCat 窗口

由于 LogCat 输出的信息很多，有时找到自己想要的 Log 信息比较困难，Eclipse 中提供了 LogCat 输出信息的过滤器，可以屏蔽不需要的信息。点击图 1-31 中 LogCat 控制台左侧的“+”，弹出如图 1-33 所示的对话框。

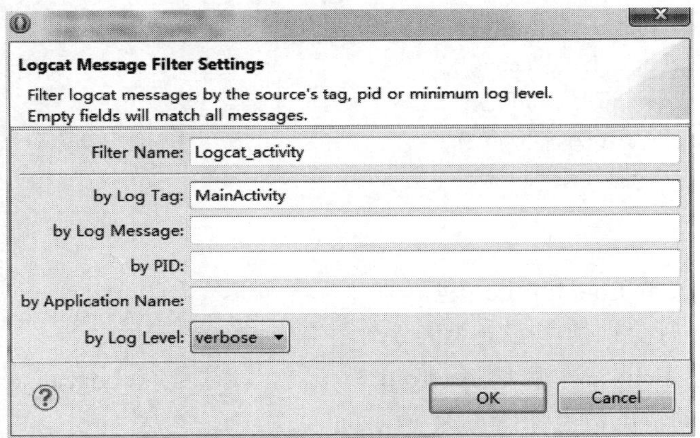


图 1-33　LogCat 过滤器

- Filter Name：过滤器的名称。
- by Log Tag：根据自定义的 Tag 信息进行过滤。
- by Log Message：根据输出的内容进行过滤。
- by PID：根据进程 ID 进行过滤。
- by Application Name：根据应用程序名称进行过滤。
- by Log Level：根据 Log 日志的级别进行过滤。

**注意**：Android 应用开发过程中除了可以使用 LogCat 进行日志过滤，也可以使用 System.out.print()进行输出，但是使用 System.out.print()时会输出很多信息，很难定位到自己需要的日志信息，所以推荐使用 LogCat。

## 1.5　本章小结

本章是学习 Android 应用程序开发的起点，首先以简介的形式介绍了 Android 操作系统的发展历史、应用场景和体系结构，接着对 Android 应用程序开发环境的搭建和相关工具的使用进行了说明，最后在 1.3 节和 1.4 节介绍了 Android 应用程序开发的基本流程以及程序开发过程中调试功能的使用。相信通过本章的学习，读者已经可以独立编写一个简单的手机应用程序了。

# 第 2 章　Android 应用结构分析

学习目标

（1）了解 Android 应用程序目录结构。

（2）掌握 Android 应用程序中各个文件的基本属性与使用方法。

（3）掌握 AndroidManifest.xml 文件结构。

（4）了解 Android 程序权限。

通过第一章的学习可以看到，在创建一个 HelloWorld 项目后，只要在 Eclipse 中按照基本规则进行一些参数填写，就可以在模拟器中运行。掌握 Android 应用程序开发是很简单的，但要设计出一个符合社会需求的产品还是有很多内容需要学习的。接下来将从 Android 应用程序目录结构开始进行介绍。

## 2.1　Android 应用程序目录结构

创建成功的 HelloWorld 项目，在 Eclipse 中形成如图 2-1 所示的目录结构，可以看出 Android 应用程序主要是使用 Java 语言与 XML 语言进行编写的，接下来对目录结构进行介绍。注意，以上两种语言都是对大小写很敏感的。

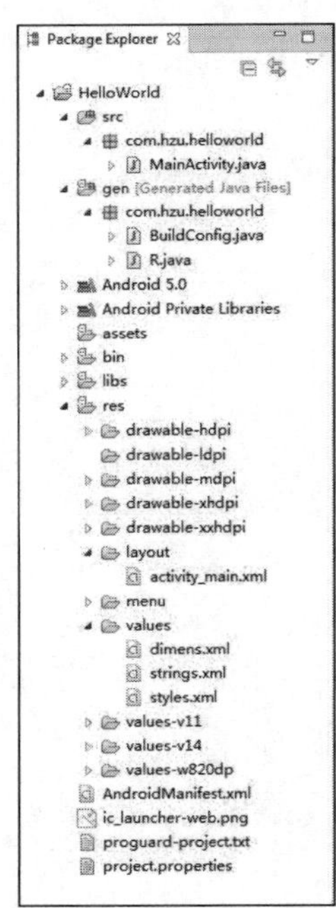


图 2-1　HelloWorld 项目目录

1. src 目录

src 目录主要用来存放 Android 应用程序中的 Java 代码，根据开发者的要求将其存放在相应的包下。在当前创建的 HelloWorld 项目下定义了一个名为 com.hzu.helloworld 的包，在其包下存放的 MainActivity.java 文件的完整路径为 com.hzu.helloworld/MainActivity.java。根据程序的 MVC 设计模式，复杂项目一般都包含多个包，每个包下有多个 Java 文件。

2. gen 目录

gen 目录下的 R.java 文件是在创建项目时由 Eclipse 的 ADT 插件自动生成的，这个文件是只读模式的，不能更改。R.java 文件中定义了一个 R 类，该类中包含很多静态类，且静态类的名字都对应 res 目录中的一个名字，即 R 类定义了该项目下所有资源的索引。通过 R.java 文件可以很快地查找到所需要的资源，另外编译器也会检查该文件列表中的资源是否被使用，没有被使用的资源不会编译到应用程序中，这样可以减少应用程序在手机中占用的空间。当 R.java 文件丢失时，可以在 Eclipse 菜单栏中选择 Project→clean 命令，来对该项目进行重新构建。

3. Android 5.0 目录

Android 5.0 目录下包含 android.jar 文件，这是一个 Java 归档文件，其中包含构建应用程序所需的所有 Android SDK 库和 APIs。通过 android.jar 可以将应用程序绑定到 Android SDK 和 Android Emulator 中，允许开发者使用所有的 Android 库和包，且保证应用程序能在适当的环境中调试。

4. Android Private Libraries 目录

Android Private Libraries 目录是 libs 目录下 Jar 包的映射。

5. assets 目录

assets 目录下存放了项目中用到的相关资源文件，如音频文件、文本文件等，并且这些资源文件都是未进行编译的原生文件。

6. bin 目录

bin 目录下存放生成的可执行文件。如果项目没有执行，则该目录为空；若执行已完成，则在该目录下生成执行文件。该目录下会生成 dex 文件、apk 文件等，dex 文件是 Android Dalvik 的执行程序，apk 文件是 Android 设备上的安装文件。

7. libs 目录

所有引入的第三方 Jar 包都会被放到 libs 目录中。

8. res 目录

res 目录存放项目经常使用的资源文件，如图片、布局等文件。其中有 5 个以 drawable 开头的子文件夹，分别用来存放高分辨率、低分辨率、中分辨率、超高分辨率、超超高分辨率的图标文件。不同的分辨率图标适应不同的屏幕分辨率和运行环境。

layout 文件夹主要用来存放 XML 语言编写的布局文件，当然也可以使用 Java 语言来动态生成布局文件。

menu 文件夹主要用来存放 XML 语言编写的菜单布局以及 Action Bar 的声明。

values 文件夹主要用来存放各种类型的数据。最常用的几种文件如下：

- dimens.xml：用来定义尺寸大小。
- strings.xml：用来定义字符串与数值。
- styles.xml：用来定义应用程序的主题样式。

以 values 开头的子文件夹，例如 values-v11，其代表在 API 11 的设备上运行项目时，使用该目录下的 styles.xml 代替 res/values/styles.xml，类似的文件夹依次按此方法处理。

9. AndroidManifest.xml 文件

AndroidManifest.xml 文件提供了应用程序的基本信息，是一个功能清单文件，在系统运行这个程序之前必须知道这些信息。注意，每个 Android 项目都必须有这个文件。

10. proguard-project.txt 文件

proguard-project.txt 文件是混淆代码的脚本配置文件，定义了 proguard 对项目代码的优化与混淆，以达到很难反编译应用程序的效果，起到保护开发者版权的作用。

11. project.properties 文件

project.properties 文件是项目的配置信息，一般不进行修改。如果导入的 Android 源代码项目在 Eclipse 中没有找到相应的 Android SDK 版本，可以通过修改 project.properties 的 target 进行修复。例如 target=android-22，该语句表示使用的 Android SDK 版本为 22，即 Android 5.1。如果 Eclipse 中没有版本 22 但有版本 21，则修改为 target=android-21，并清理（clean）该项目即可。

## 2.2 Android 应用程序分析

Android 应用程序由三个部分组成：应用程序源文件（用 Java 语言编写）、应用程序描述源文件（用 XML 语言编写）和各种资源文件（图标、Jar 包等），其中开发者的主要工作是完成逻辑层 Java 代码的编写以及各种界面的设计。接下来用 1.3 节的 HelloWorld 项目来分析 Android 应用程序。

### 2.2.1 资源描述源文件

1. string.xml 文件

在进行应用程序开发时，使用纯文本字符串时需要用到 res/values 目录中的一个 XML 文件（通常名为 res/values/string.xml），根标签为<resources>，希望编码为字符串的各类资源都有一个<string>标签。<string>标签包含 name 特性，这是此字符串唯一的名称，还有一个文本元素包含字符串的文本。以下是具体代码示例：

```
<?xml version="1.0" encoding="utf-8"?>
<resources>
<string name="hello">HelloWorld</string>
</resources>
```

以上代码的标签为<resources>，其下有子标签<string>。在<string>中有一个 name 的名称为 hello，hello 在当前这个文件中只能使用一次，对应 Java 中相当于有一个 string 类型的变量名为 hello。文本元素包含的字符串内容为 HelloWorld，对应 Java 中相当于给变量名 hello 赋初始值为 HelloWorld。

（1）在布局文件中使用 string 的资源。

在布局文件中引用 string 的资源格式为@sting/…，其中省略号表示唯一的名称，例如@sting/hello。

```
public static final class string {   //在 R.java 文件中 hello 表示的情况
    public static final int hello=0x7f050001;
}
```

以下是在布局文件中引用 hello 的案例。

```
<TextView
    android:layout_width="wrap_content"
    android:layout_height="wrap_content"
    android:text="@string/hello"      //string.xml 中 hello 的引用
/>
```

（2）在 Java 代码中使用 string 的资源。

通过使用 getString()方法和字符串资源的 ID 来获取 string.xml 的资源。

```
protected void onCreate(Bundle savedInstanceState) {
    super.onCreate(savedInstanceState);
        setContentView(R.layout.activity_main);
        String str=getString(R.string.hello);
}
```

除了通过以上 getString()方法来取得资源给 String 对象外，还可以直接设置到一些控件上，代码如下：

```
protected void onCreate(Bundle savedInstanceState) {
        super.onCreate(savedInstanceState);
        setContentView(R.layout.activity_main);
        TextView textView=(TextView) findViewById(R.id.tvtest);
        textView.setText(R.string.hello);
}
```

（3）使用资源 String 的优势。

1）字符串独立于布局和代码，便于维护。

2）全球化的 Internet 需要全球化的软件，意味着同一种版本的产品更容易适用于不同地区的市场，Android 采用资源管理的方式，可以非常方便地实现程序国际化。

（4）实现资源国际化。

**【例 2.1】**根据 Android 操作系统语言的不同，实现应用程序的中文版内容与英文版内容自由切换。

**【开发步骤】**

（1）创建一个名为 Resource_Activity 的项目，包名为 com.hzu.resource_activity，Activity 组件名为 MainActivity。

（2）在 res/values/string.xml 下的代码内容如下：

```
1  <?xml version="1.0" encoding="utf-8"?>
2  <resources>
3      <string name="app_name">Resource_Activity</string>
4      <string name="hello_world">Hello world!</string>
5      <string name="action_settings">Settings</string>
6      <string name="test">testContent</string>
7  </resources>
```

第 3～6 行声明各种英文通用资源。

（3）在 res 下创建一个目录，名为 values-zh-rCN，在此目录下新建 string.xml，其代码内容如下：

```
1  <resources>
2      <string name="app_name">资源案例</string>
3      <string name="hello_world">你好，世界！</string>
4      <string name="action_settings">设置</string>
5      <string name="test">测试内容</string>
6  </resources>
```

第 2～5 行声明各种中文通用资源。

（4）在 layout/activity_main.xml 下编写如下代码：

```
<RelativeLayout xmlns:android="http://schemas.android.com/apk/res/android"
    xmlns:tools="http://schemas.android.com/tools"
    android:layout_width="match_parent"
    android:layout_height="match_parent"
    android:paddingBottom="@dimen/activity_vertical_margin"
    android:paddingLeft="@dimen/activity_horizontal_margin"
    android:paddingRight="@dimen/activity_horizontal_margin"
```

```
8         android:paddingTop="@dimen/activity_vertical_margin">
9         <TextView
10            android:id="@+id/tv_hello"
11            android:layout_width="wrap_content"
12            android:layout_height="wrap_content"
13            android:text="@string/hello_world" />
14        <TextView
15            android:layout_width="wrap_content"
16            android:layout_height="wrap_content"
17            android:layout_below="@id/tv_hello"
18            android:text="@string/test" />
19    </RelativeLayout>
```

1）第 1～19 行声明了一个相对布局，第 5～8 行规定了其内部控件与底部、左边、右边以及顶部的边距。

2）第 9～13 行声明了一个 TextView(文本框)，其 id 为 tv_hello，第 11 行与第 12 行声明了宽度与高度都为 wrap_content，第 13 行设置了 TextView 显示的内容为 string.xml 中 hello_world 定义的内容。

3）第 14～18 行声明了另外一个 TextView，此 TextView 为了与 id 为 tv_hello 的文本框垂直排列，使用了 android:layout_below 属性。

**【运行结果】**在 Eclipse 中启动 Android 模拟器，接着运行 Resource_Activity 项目，模拟器的默认语言是英文，运行结果如图 2-2 所示，桌面程序显示如图 2-3 所示。

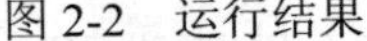

图 2-2　运行结果

图 2-3　桌面程序显示

将 Android 模拟器的语言从英文切换到中文，在桌面上找到 Settings 程序并打开，选择 Language&input，如图 2-4 所示，在弹出的列表中选择 Language，然后选择“中文（简体）”，如图 2-5 所示。

再次返回到桌面，看到 Resource_Activity 应用程序的桌面名称变为“资源案例”，如图 2-6 所示。打开程序看到里面的内容全部显示成了中文，如图 2-7 所示。

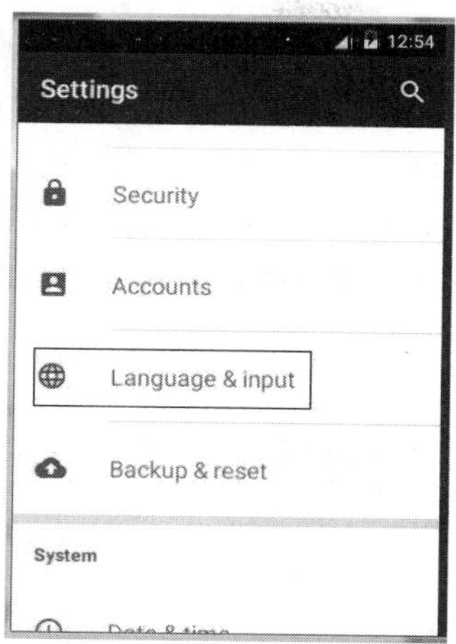


图 2-4 切换语言

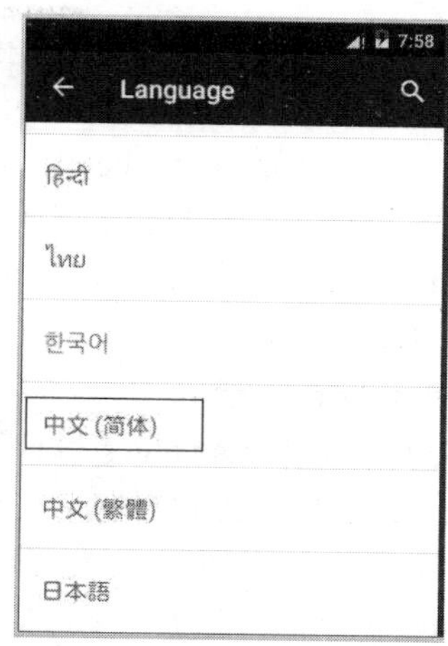


图 2-5 选择“中文（简体）”

图 2-6 中文程序名

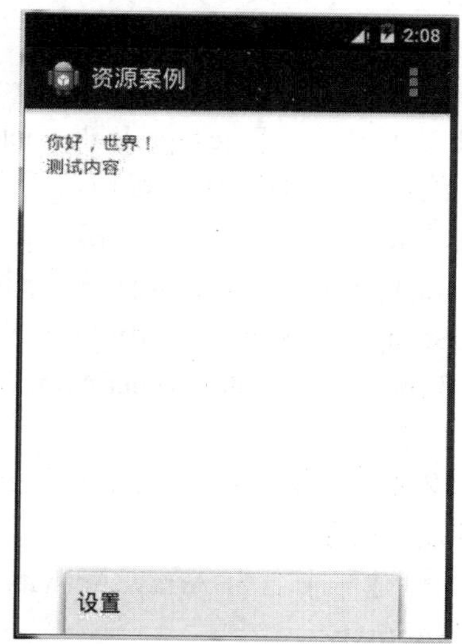


图 2-7 中文结果

对于Android的国际化，要在 res 目录下新建对应的字符串目录。例如若模拟器的语言是中文，则在项目的 res 目录下新建一个 values-zh-rCN 目录，然后将翻译好的 strings.xml 或 arrays.xml 文件放到该目录下即可。下面给出部分国家的文件目录名称（应用程序中默认文件夹名为 values，用英文表示，加上后缀后变为相应国家）。

- 中文（中国）：values-zh-rCN。
- 希腊文：values-el-rGR。
- 法文（法国）：values-fr-rFR。
- 英语（英国）：values-en-rGB。
- 英文（澳大利亚）：values-en-rAU。
- 英文（加拿大）：values-en-rCA。

注意“@+id”与“@id”的区别。

我们经常在布局文件中看到一些控件使用 id 的属性，有时使用“@+id/”，有时使用“@id/”。“@+id/”表示会在 R 文件中生成一个新的 id，变量名就是“/”后面的内容，例如，@+id/ tv_hello 会在 R.java 文件中生成 int tv_hello = value 语句，其中 value 是一个十六进制的数。如果 tv_hello 在 R.java 中已经存在同名的变量就不会生成新的变量。@id/tv_hello 表示引用已经定义好的 id，如图 2-2 中 testContent 的 TextView 控件使用属性 android:layout_below="@id/tv_hello"，表示 testContent 的控件位于“hello world!”控件的下方。

2. 数组（Array）资源

Android使用位于 res/values 目录下的 arrays.xml 文件中定义的数组资源，此文件的根元素为<resources>，它的下面可以包含三种子元素。

- <array.../>子元素，定义普通类型的数组。
- <string-array.../>子元素，定义字符串类型的数组。
- <integer-array.../>子元素，定义整数类型的数组。

在资源文件中定义好数组资源后，就可以在 Java 代码与 XML 文件中使用它了，其格式如下：

Java 代码中：[package_name.]R. array. array_name。

XML 文件中：@[package:]color/color_name。

为了能在 Java 代码中很好地访问数组中的数据，可以通过 getResources()获得 Resources 对象，然后根据需要运用相关方法进行数据访问：

- int[] getIntArray(int id)：根据资源文件中整型数组资源的名称来获得实际数据。
- String[] getStringArray(int id)：根据资源文件中字符串数组资源的名称来获得实际数据。
- TypedArray obtainTypedArray(int id)：根据资源文件中普通数组资源的名称来获得实际数据。

**【例 2.2】**编写一个读取 Array 数组中天气信息的应用程序。

**【开发步骤】**

（1）创建一个名为 Array_Activity 的项目，包名为 com.hzu.array_activity，Activity 组件名为 MainActivity。

（2）在 res/values 目录下新建一个名为 arrays.xml 的文件，其代码内容如下：

```
<?xml version="1.0" encoding="utf-8"?>
<resources>
<string-array name="weather">
<item>snow</item>
<item>rain</item>
<item>overcast</item>
<item>fog</item>
<item>sunny</item>
</string-array>
</resources>
```

（3）编写逻辑代码。打开 src/com.hzu.array_activity 下包中的 MainActivity.java 文件，根据要求编写如下代码：

```
package com.hzu.array_activity;
import android.app.Activity;
```

```
3	import android.content.res.Resources;
4	import android.os.Bundle;
5	import android.util.Log;
6	public class MainActivity extends Activity {
7		private static final String TAG = "MainActivity";
8		protected void onCreate(Bundle savedInstanceState) {
9			super.onCreate(savedInstanceState);
10				setContentView(R.layout.activity_main);
11				Resources resources = getResources();
12				String[] weather = resources.getStringArray(R.array.weather);
13				for (String st : weather) {
14					Log.i(TAG, st);
15				}
16		}
17	}
```

1）第 5 行引入 android.util.Log 资源。

2）第 7 行声明了一个字符串常量。

3）第 11 行获得 Resources 对象 resources 。

4）第 12 行通过 resources 的方法 getStringArray()取得 arrays.xml 文件中的天气数组信息。

5）第 13～15 行使用 Log 在 LogCat 控件台显示数据信息。

**【运行结果】**在 Eclipse 中启动 Android 模拟器，接着运行 Array_Activity 项目，在 Eclipse 的 LogCat 控件台显示的信息如图 2-8 所示。

| Level | Time | PID | TID | Application | Tag | Text |
|---|---|---|---|---|---|---|
| I | 12-23 03:39:01.638 | 2122 | 2122 | com.hzu.array_activity | MainActivity | snow |
| I | 12-23 03:39:01.638 | 2122 | 2122 | com.hzu.array_activity | MainActivity | rain |
| I | 12-23 03:39:01.638 | 2122 | 2122 | com.hzu.array_activity | MainActivity | overcast |
| I | 12-23 03:39:01.638 | 2122 | 2122 | com.hzu.array_activity | MainActivity | fog |
| I | 12-23 03:39:01.638 | 2122 | 2122 | com.hzu.array_activity | MainActivity | sunny |

图 2-8 Array 数组中显示的信息

### 3. 颜色（Color）资源

Android 中的颜色是十六进制的 RGB 值，也可以指定 Alpha 通道，其中 Alpha 值可以省略。如果省略了 Alpha 值，那么该颜色默认是完全不透明的。我们可以选择使用一个字符的十六进制值或是两个字符的十六进制值，提供的形式有以下四种。

- RGB：分别指定红、绿、蓝三原色的值（只支持 0～f 这 16 级颜色）来代表颜色。
- ARGB：分别指定红、绿、蓝三原色的值（只支持 0～f 这 16 级颜色）以及透明度（只支持 0～f 这 16 级透明度）来代表颜色。
- RRGGBB：分别指定红、绿、蓝三原色的值（支持 00～ff 这 256 级颜色）来代表颜色。
- AARRGGBB：分别指定红、绿、蓝三原色的值（支持 00～ff 这 256 级颜色）以及透明度（支持 00～ff 这 256 级透明度）来代表颜色。

上面四种形式中的 A、R、G、B 都代表一个十六进制的数，其中 A 代表透明度，R 代表红色的数值，G 代表绿色的数值，B 代表蓝色的数值。

【例 2.3】在 Android 应用程序中，使用 RGB 分别表示默认颜色、红色、绿色三种字体颜色。

【开发步骤】

（1）创建一个名为 Color_Activity 的项目，包名为 com.hzu.color_activity，Activity 组件名为 MainActivity。

（2）在 res/values 目录下新建一个 color_demo.xml 文件，代码如下：

```
<?xml version="1.0" encoding="utf-8"?>
<resources>
<color name="red">#ff0000</color>
<color name="blue">#00ff00</color>
</resources>
```

（3）编写逻辑代码。打开 layout/activity_main.xml 文件，根据要求编写如下代码：

```
1   <LinearLayout xmlns:android="http://schemas.android.com/apk/res/android"
2         xmlns:tools="http://schemas.android.com/tools"
3         android:layout_width="match_parent"
4         android:layout_height="match_parent"
5         android:orientation="vertical">
6   <TextView
7             android:layout_width="wrap_content"
8             android:layout_height="wrap_content"
9             android:text="默认颜色测试" />
10  <TextView
11            android:layout_width="wrap_content"
12            android:layout_height="wrap_content"
13            android:text="红色测试"
14            android:textColor="@color/red" />
15  <TextView
16            android:layout_width="wrap_content"
17            android:layout_height="wrap_content"
18            android:text="绿色测试"
19            android:textColor="@color/blue" />
20  </LinearLayout>
```

1）第 1～20 行声明了一个线性布局，方向为垂直方向。

2）第 6～19 行在其线性布局中添加了三个 TextView 控件，后面两个 TextView 控件通过使用 android:textColor="@color/来引用 color_demo.xml 的颜色。

【运行结果】在 Eclipse 中启动 Android 模拟器，然后运行 Color_Activity 项目，其效果如图 2-9 所示。

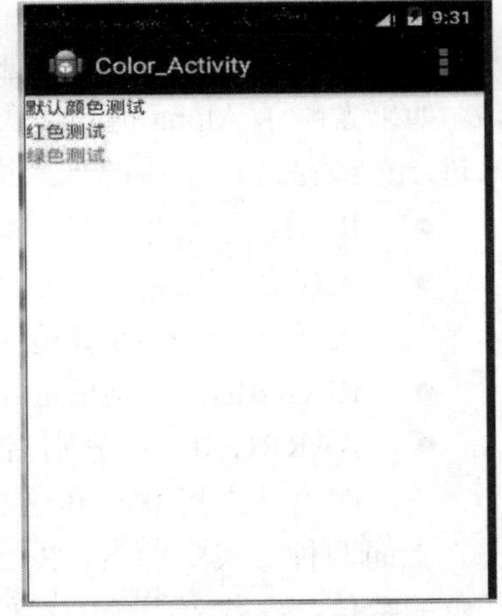


图 2-9 颜色显示效果

4. Drawable 资源

Drawable 资源是 Android 应用程序开发过程中经常使用的资源。Android 通过 Drawable 来处理图像，处理的对象可以是一张图片（*.png、*.jpg、*.gif 等），也可以是一个逐帧动画、一片红色区域，还可以是一个 XML 文件。

Drawable 资源通常保存在 res/drawable 目录下，为了适应不同尺寸手机的屏幕分辨率，具体可保存在/res/drawable-ldpi、/res/drawable-mdpi、/res/drawable-hdpi、/res/drawable-xhdpi、/res/drawable-xxhdpi 目录下。

**【例 2.4】**编写关于一个 Button 按下与抬起以及一个 CheckBox 被选中与未被选中显示效果的应用程序。

**【开发步骤】**

（1）创建一个名为 Drawable_Activity 的项目，包名为 com.example.drawable_activity，Activity 组件名为 MainActivity。

（2）准备图片资源。将图片资源复制到本项目的 res/drawable-hdpi 目录中。

（3）在 res/drawable-hdpi 目录中创建 buttonselector.xml 文件，代码内容如下：

```
1    <?xml version="1.0" encoding="utf-8"?>
2    <selector xmlns:android="http://schemas.android.com/apk/res/android">
3    <item android:drawable="@drawable/button_bg_normal"
4              android:state_pressed="false"></item><!设置未按下时的图片>
5    <item android:drawable="@drawable/button_bg_pressed"
6              android:state_pressed="true"></item><!设置按下时的图片>
7    </selector>
```

1）第 2～7 行声明图片按钮的选择标签，使用<selector >标签。

2）第 3～4 行声明按钮未被按下的图片，按钮未按下时使用 android:state_pressed="false"，图片资源用 android:drawable="@drawable/button_bg_normal"来显示未被选中的效果。

3）第 5～6 行图片资源用 android:drawable="@ drawable/button_bg_pressed "来显示被选中的效果。

（4）在 res/ drawable-hdpi 目录中创建 checkselector.xml 文件，代码内容如下：

```
1    <?xml version="1.0" encoding="utf-8"?>
2    <selector xmlns:android="http://schemas.android.com/apk/res/android">
3    <item android:drawable="@drawable/checkbox_unpressed"
4              android:state_checked="false"></item><!未选中时显示的图片>
5    <item android:drawable="@drawable/checkbox_pressed"
6              android:state_checked="true"></item><!选中时显示的图片>
7    </selector>
```

1）第 2～7 行声明图片按钮的选择标签，使用<selector >标签。

2）第 3～6 行分别表示 CheckBox 未被选中与未选中时的效果。

（5）在 layout/activity_main.xml 下，编写如下代码：

```
<RelativeLayout xmlns:android="http://schemas.android.com/apk/res/android"
    xmlns:tools="http://schemas.android.com/tools"
    android:layout_width="match_parent"
    android:layout_height="match_parent"
    android:paddingBottom="@dimen/activity_vertical_margin"
    android:paddingLeft="@dimen/activity_horizontal_margin"
    android:paddingRight="@dimen/activity_horizontal_margin"
    android:paddingTop="@dimen/activity_vertical_margin"
    tools:context="com.example.drawable_activity.MainActivity">
```

```
10          <Button
11              android:id="@+id/button1"
12              android:layout_width="wrap_content"
13              android:layout_height="wrap_content"
14              android:background="#00000000"
15              android:drawableTop="@drawable/buttonselector"
16              android:text="button"
17              android:textSize="40px" />
18          <CheckBox
19              android:id="@+id/checkBox1"
20              android:layout_width="wrap_content"
21              android:layout_height="wrap_content"
22              android:layout_below="@+id/button1"
23              android:layout_marginTop="20dp"
24              android:button="@drawable/checkselector"
25              android:text="CheckBox"
26              android:textSize="40px" />
27      </RelativeLayout>
```

1）第 1～27 行声明了一个相对布局。

2）第 10～17 行声明了一个 Button，android:background="#00000000"是为了去掉 Button 的背景色。android:drawableTop="@drawable/buttonselector"是用来加载用户按下与未按下时的图片资源，其中 android:drawableTop 表示图片资源位于“button”文字的上方。android:text 与 android:textSize 分别表示在界面上显示的文字内容与字体大小。

3）第 18～26 行声明了一个 CheckBox，其中 android:layout_below="@+id/button1"表示这个 checkBox1 位于 button1 的下方，android:layout_marginTop="20dp"表示 checkBox1 顶部边距为 20dp，android:button 引入了用户选中与未选中 CheckBox 时的图片资源。

**【运行结果】**在 Eclipse 中启动 Android 模拟器，接着运行 Drawable_Activity 项目，效果如图 2-10 所示。

图 2-10　显示效果

5. dimens.xml 文件

dimens.xml 文件位于 res/values 目录下。它的作用是在样式和布局文件中定义边界、高度与尺寸大小。其中 Android 常用的度量单位有：

- px（像素）：屏幕上的点，绝对长度，与硬件相关。
- mm（毫米）：长度单位。
- in（英寸）：长度单位。
- pt（磅）：1/72 英寸。
- dp（与密度无关的像素）：一种基于屏幕密度的抽象单位。在每英寸 160 点的屏幕上，1dp=1px。注意，随着设备的屏幕密度的改变，dp 与 px 之间的换算也会有所改变。
- sp（可伸缩像素）：使用与 dp 相同的设计理念，主要用于字体大小的显示。

6. style.xml 文件

在 Android 系统中，预先定义了很多的样式与主题，这些样式与主题使布局显示呈现在用户面前更加美观，示例代码如下：

```
1   <resources>
2   <style name="AppBaseTheme" parent="android:Theme.Light"></style>
3   <style name="AppTheme" parent="AppBaseTheme"></style>
4   <style name="Mystyle">
5   <item name="android:textColor">#885533</item>
6   <item name="android:textSize">30dp</item>
7   </style>
8   </resources>
```

1）第 1 行定义了<resources>标签。

2）第 2～3 行分别定义了两个<style>标签，并声明了它们的 name 与 parent 属性。这两行是创建好项目后系统自带的两个 style 样式。

3）第 4～7 行自定义了一个名为 Mystyle 的样式，设置应用这种样式的字体颜色与字体大小分别为#885533 和 30dp，注意<item>是<style>的子标签。

4）第 8 行使用</resources>标签作为结束标志。

style 定义的步骤：

1）判断需要统一 style 的控件有哪些属性是一致的。

2）在 value 的 style.xml 中定义自己的 style。

- style 的 name 决定了@style 后的名字。
- 每一项 item 都是一个布局中的属性，分别对应属性名和属性值。
- style 继承 parent 中定义过的属性值。

3）在 XML 布局中，控件使用自定义的 style，例如：@style/stylename。

### 2.2.2 布局文件

在创建 HelloWorld 项目成功后，activity_main.xml 在 HelloWorld 项目的 res/layout 目录下，其代码内容如下：

```
<RelativeLayout xmlns:android="http://schemas.android.com/apk/res/android"
    xmlns:tools="http://schemas.android.com/tools"
```

```
3         android:layout_width="match_parent"
4         android:layout_height="match_parent"
5         android:paddingBottom="@dimen/activity_vertical_margin"
6         android:paddingLeft="@dimen/activity_horizontal_margin"
7         android:paddingRight="@dimen/activity_horizontal_margin"
8         android:paddingTop="@dimen/activity_vertical_margin"
9         tools:context="com.hzu.helloworld.MainActivity">
10        <TextView
11            android:layout_width="wrap_content"
12            android:layout_height="wrap_content"
13            android:text="@string/hello_world" />
14    </RelativeLayout>
```

1）第 1 行定义了启动界面使用相对布局样式，并规定了当前文件的命名空间。

2）第 2 行定义 tools 可以告诉 Eclipse 哪些属性在运行的时候是被忽略的，但只在设计布局的时候有效。

3）第 3～4 行分别定义了相对布局的宽度与高度。

4）第 5～8 行分别定义了相对布局的底部、左边、右边、顶部的边距。

5）第 9 行记录当前相对布局关联到 com.hzu.helloworld.MainActivity。

6）第 10～13 行向相对布局中添加了一个文件控件 TextView，并设定了该文件控件的宽与高，显示的内容为 string.xml 中 hello_world 变量值的内容。

7）第 14 行为相对布局标签的结束标志。

## 2.3 AndroidManifest.xml 文件

AndroidManifest.xml 文件提供了应用程序的基本信息，是一个功能清单文件，相当于应用程序的全局描述。成功创建一个新项目后，Eclipse 下的 ADT 会自动创建一个 AndroidManifest.xml 文件，HelloWorld 项目下的对应文件内容如下：

```
<?xml version="1.0" encoding="utf-8"?>
<manifest xmlns:android="http://schemas.android.com/apk/res/android"
    package="com.hzu.helloworld"
    android:versionCode="1"
    android:versionName="1.0">
<uses-sdk
        android:minSdkVersion="14"
        android:targetSdkVersion="21" />
<application
        android:allowBackup="true"
        android:icon="@drawable/ic_launcher"
        android:label="@string/app_name"
        android:theme="@style/AppTheme">
<activity
```

```
15                android:name=".MainActivity"
16                android:label="@string/app_name">
17    <intent-filter>
18    <action android:name="android.intent.action.MAIN" />
19    <category android:name="android.intent.category.LAUNCHER" />
20    </intent-filter>
21    </activity>
22    </application>
23    </manifest>
```

1）第 1 行声明了 XML 的版本和编码方式。

2）第 2 行标记命名空间 xmlns:android=http://schemas.android.com/apk/res/android，这个命名空间的作用是使各种属性都可以在 AndroidManifest.xml 使用。

3）第 3 行 package 表示应用程序的包名。

4）第 4 行 android:versionCode 表示应用程序的相对版本，即版本更新过多少次。

5）第 5 行 android:versionName 表示应用程序的版本信息，需要显示给用户。

6）第 7 行 android:minSdkVersion 表示使用 Android SDK 的最低版本。

7）第 8 行 android:targetSdkVersion 表示应用程序的目标版本。

8）第 9 行定义应用程序的组件等，在应用程序下使用的各种组件都放在 application 下。

9）第 10 行允许备份应用程序的数据。

10）第 11 行 android:icon 表示指定应用程序的图标。

11）第 12 行表示指定应用程序的名称。

12）第 13 行表示指定应用程序的主题。

13）第 14～21 行定义了一个 Activity 组件，其中 android:name 表示指定当前 Activity 使用哪个类；android:label 指定应用程序名将显示在 Activity 画面上方标题栏的位置；在 Activity 中可以设置多个 intent-filter，用于 Activity 的 Intent 过滤条件；android.intent.action.MAIN 用于指定该 Activity 是应用程序的入口；android.intent.category.LAUNCHER 指定加载该应用时运行此 Activity。

14）第 22～23 行分别表示为 application 与 manifest 的结束标签。

## 2.4 应用程序权限声明

在应用程序中，遇到需要访问网络等情况时，AndroidManifest.xml 中规定要使用权限声明。这种做法的优势是可以指定其他应用程序是否有权限访问该程序。

在 AndroidManifest.xml 中声明允许访问网络的权限格式如下：

```
<uses-permission
    android:name="android.permission.INTERNET">
</uses-permission>
```

通过上面的代码可知权限声明的用法非常简单。在 Android 中有很多种权限，并且这些权限声明都放在 AndroidManifest.xml 中，下面介绍一些比较常用的权限，如表 2-1 所示。

表 2-1 Android 常用的权限

| 权限常量 | 权限使用说明 |
|---|---|
| ACCESS_NETWORK_STATE | 获取网络信息状态，如当前的网络连接是否有效 |
| ACCESS_FINE_LOCATION | 通过 GPS 芯片接收卫星的定位信息 |
| BLUETOOTH | 允许程序连接配对过的蓝牙设备 |
| BLUETOOTH_ADMIN | 允许程序发现和配对新的蓝牙设备 |
| BROADCAST_SMS | 当收到短信时触发一个广播 |
| BATTERY_STATS | 获取电池电量统计信息 |
| CALL_PHONE | 允许程序从非系统拨号器里输入电话号码 |
| CALL_PRIVILEGED | 允许程序拨打电话，替换系统的拨号器界面 |
| CAMERA | 允许访问摄像头进行拍照 |
| CHANGE_NETWORK_STATE | 改变网络状态，如是否能联网 |
| CHANGE_CONFIGURATION | 允许当前应用改变配置，如定位 |
| CHANGE_WIFI_STATE | 改变 WiFi 状态 |
| CLEAR_APP_CACHE | 清除应用缓存 |
| CLEAR_APP_USER_DATA | 清除应用的用户数据 |
| DELETE_PACKAGES | 允许程序删除应用 |
| FLASHLIGHT | 允许访问闪光灯 |
| READ_PHONE_STATE | 访问电话状态 |
| READ_CONTACTS | 允许应用访问联系人通讯录信息 |
| READ_SMS | 读取短信内容 |
| READ_CALENDAR | 允许程序读取用户的日程信息 |
| SEND_SMS | 发送短信 |
| SET_ALARM | 设置闹铃提醒 |
| SET_WALLPAPER | 设置桌面壁纸 |
| VIBRATE | 允许振动 |
| WRITE_SMS | 允许编写短信 |

## 2.5 本章小结

本章主要介绍了 Android 应用程序的目录结构，并结合 HelloWorld 项目较详细地介绍了 Android 应用程序的组成。AndroidManifest.xml 是整个应用程序非常重要的功能清单文件，此基础上我们介绍了应用程序权限声明。通过本章的学习，相信大家对 Android 应用程序的组成已经有了基本的了解。

# 第 3 章　基本控件和事件处理

（1）掌握 Android 各类基本控件的使用。

（2）理解 Android 事件处理机制。

通过第 2 章的学习，我们已经知道 Android 应用程序的基本目录结构，其中 res/layout 要放一些用户显示的基本界面，这些界面是由多个基本控件组合来完成的。本章将学习基本控件的使用以及用户点击控件产生的行为方法。

## 3.1　基本控件概述

Android 是一个基于图形用户界面（Graphical User Interface，GUI）的应用开发系统，为了设计出令用户满意的界面，使用户通过点击相关图形界面就能很好地操作应用程序，开发者设计的应用程序必须以友好性为前提。Android 提供了大量功能强大的 UI 组件，这些组件配合使用事件响应机制就能完成用户点击后的相关动作，这些组件就称为控件。

## 3.2　Android 基本控件

Android 提供了大量的控件，合理使用这些控件可以大大减少开发者的工作量。这些控件继承于 View 和 ViewGroup 这两个类，它们存放在 android.widget 包下。View 表示屏幕上的一块矩形区域，并负责绘制这个区域和事件处理。ViewGroup 是 View 类的子类，它是一个能容纳其他控件的容器，其内可以有 View 对象与 ViewGroup 对象。在学习这些类时，必须学习 Android 开发文档，这个文档存放在下载好的 sdk 目录下的 docs\reference\android\view\package-summary.html 中，当用户点击这个页面时，显示的界面如图 3-1 所示。

在使用这些控件时，可以使用 XML 语言来定义用户界面，也可以使用 Java 语言动态添加控件来完成用户界面的设计，接下来介绍 View 类常用的 XML 属性与对应的方法及说明，如表 3-1 所示。

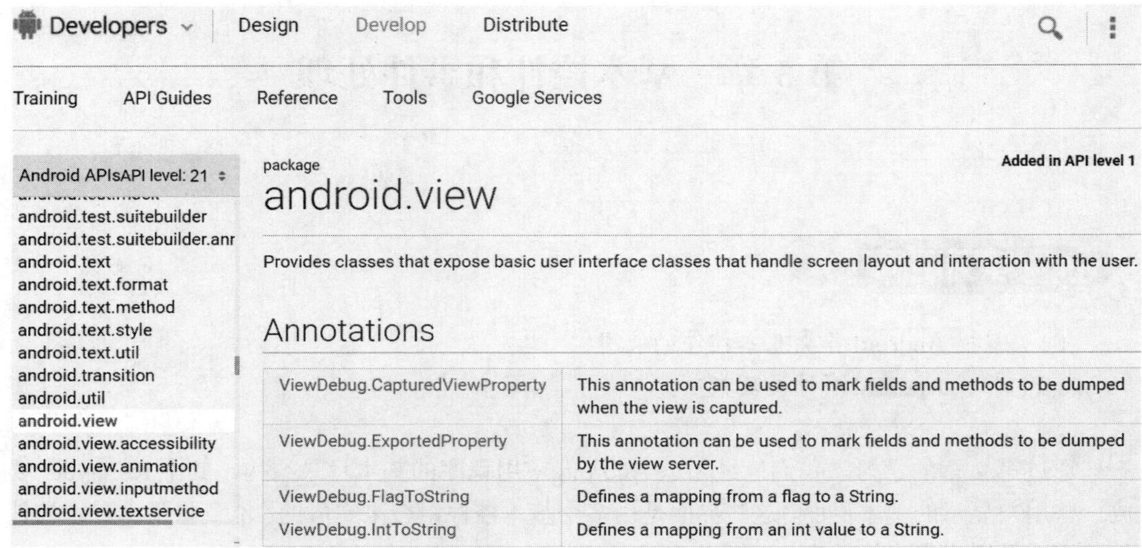


图 3-1　Android 的 API 文档

表 3-1　View 类常用的 XML 属性与对应的方法及说明

| 属性 | 方法 | 说明 |
|---|---|---|
| android:background | setBackgroundResource(int) | 设置控件背景颜色 |
| android:focusable | setFocusable(boolean) | 控制 View 是否可以获取焦点 |
| android:id | setId(int) | 为 View 设置一个唯一标识，在 Java 代码中通过 FindViewById 来获取它 |
| android:minWidth | setMinimumWidth(int) | 设置 View 的最小宽度 |
| android:minHeight | setMinimumHeight(int) | 设置 View 的最小高度 |
| android:padding | setPadding(int,int,int,int) | 设置 View 的四边填充区域 |
| android:longClickable | setLongClickable(boolean) | 设置 View 是否可以响应长点击事件 |
| android:clickable | setClickable(boolean) | 设置 View 是否可以响应点击事件 |
| android:visibility | setVisibility(int) | 设置 View 是否可见 |
| android:scaleX | setScaleX(float) | 设置 View 在水平方向的缩放比 |
| android:scaleY | setScaleY(float) | 设置 View 在垂直方向的缩放比 |

除了要熟悉以上基本 View 类的属性外，还要知道 View 类与其子类之间的关系，如图 3-2 所示。

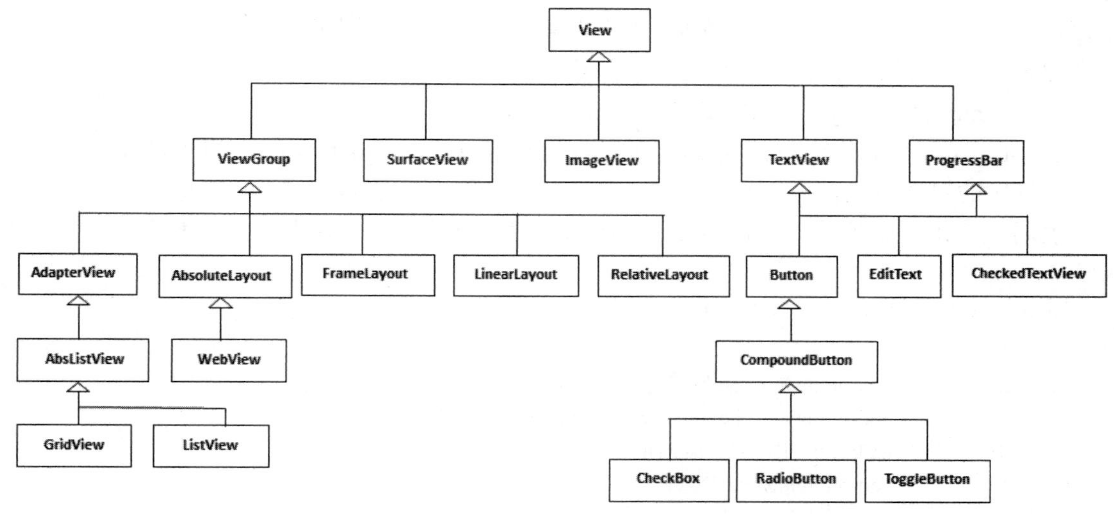


图 3-2　View 类与其子类的关系

### 3.2.1　TextView

TextView（文本框）用于显示文本给用户看，但用户不可以编辑里面的内容。它位于 Android.widget.TextView 包中。如果在 Java 代码中使用这个控件，需要在头部添加"import android.widget.TextView;"语句。

TextView 的属性包括文字内容、字体大小、字体颜色、字体样式等。TextView 的属性与对应的方法及说明如表 3-2 所示。

表 3-2　TextView 类常用的属性与对应的方法及说明

| 属性 | 方法 | 说明 |
|---|---|---|
| android:text | setText(charSequence) | 设置 TextView 显示文本的内容 |
| android:textSize | setTextSize(float) | 设置 TextView 的文本字符大小 |
| android:textColor | setTextColor(ColorStateList) | 设置 TextView 的文本颜色 |
| android:textStyle | setTextStyle(TextStyle) | 设置 TextView 的文本风格（如粗体、斜体） |
| android:textface | setTextface(Textface) | 设置 TextView 的文本字体 |
| android:gravity | setGravity(int) | 设置 TextView 在文本框内文本的对齐方式 |
| android:lines | setLines(int) | 设置 TextView 最多占几行 |
| android:height | setHeight(int) | 设置 TextView 的高度，以像素为单位 |
| android:width | setWidth(int) | 设置 TextView 的宽度，以像素为单位 |
| android:padding | SetPadding(int) | 设置 TextView 中显示文本与其父容器边界的间距 |

在使用以上属性时，还要注意一些区别。

（1）android:padding 与 android:layout_margin 的区别。

padding 是以父 View 为参考点，规定它里面的内容与这个父 View 边界的距离。layout_margin

是以自己为参考点，规定自己和其他（上下左右）View 之间的距离，如果在当前这一级只有一个 View，那它设置的效果就和 padding 一样。

（2）android:gravity 与 android:layout_gravity 的区别。

gravity 用于设置这个 View 中所有子元素的对齐方式，layout_gravity 用于设置这个 View 在父容器中的对齐方式。

**说明**：TextView 的属性 android:textSize 表示设置标签中字体的大小，其单位一般以 sp 来表示。

通过 XML 配置初始属性的代码如下：

```
<TextView
        android:layout_width="wrap_content"
        android:layout_height="wrap_content"
        android:id="@+id/username"
        android:textSize="30sp" />
```

通过 Java 代码修改属性如下：

```
protected void onCreate(Bundle savedInstanceState) {
        super.onCreate(savedInstanceState);
        setContentView(R.layout.activity_main);
        TextView tvUserName=(TextView) findViewById(R.id.username);
        tvUserName.setTextSize(30);
}
```

其中，在 XML 语言中声明的 id 为 username，在 Java 代码中使用 findViewById(R.id.username) 方法来表示这个对象。可通过 TextView 的属性 android:textColor 设置标签中字体的颜色，它使用十六进制格式的颜色代码（RGB）。可通过 XML 设置为 android:textColor="#FFFF00FF"，或使用 Java 代码修改属性为 tvUserName.setTextColor(0xffff00ff)。

**【例 3.1】** 设计如图 3-3 所示的布局文件。

**【说明】** 整个窗体背景是白色的。

**【代码】** 布局文件中 TextView 的代码部分如下：

```
1    <TextView
2            android:layout_width="wrap_content"
3            android:layout_height="wrap_content"
4            android:id="@+id/username"
5            android:textSize="25sp"
6            android:textColor="#FF0000"
7            android:background="#008000"
8            android:padding="15dip"
9            android:text="这里是一个 TextView，可以直接看到显示结果！"
10   />
```

图 3-3　TextView 的效果

（1）第 1 行声明了一个 TextView 控件。

（2）第 2～3 行分别定义 TextView 的宽度与高度。

（3）第 4 行定义 TextView 的 id 为 username。

（4）第 5 行定义文本字符大小。

（5）第 6 行定义文本的颜色，"#FF0000"表示红色。

（6）第 7 行定义文本框的背景颜色，"#008000"表示绿色。

（7）第 8 行设置文本与其父容器（即文本框）边界的间距为 15 个像素。

（8）第 9 行设置文本框内显示的文字内容。

（9）第 10 行为 TextView 控件的结束标志。

### 3.2.2　Button

Button 继承了 TextView，其主要的作用是在界面上生成一个按钮，当需要用户点击界面某个区域来产生特定的行为时，就可以使用 Button。用户点击这个按钮后将会触发一个 onClick 事件，还要为按钮添加 setOnClickListener()方法才能真正实现完整的事件监听。在本章的 3.3 节中将会详细介绍事件监听。

Button 控件位于 Android.widget.Button 类中，如果 Java 代码中使用这个控件，需要在头部添加“import android.widget.Button;”语句，Button 在 XML 语言中使用的代码如下所示：

```
1    <Button
2            android:id="@+id/button1"
3            android:layout_width="wrap_content"
4            android:layout_height="wrap_content"
5            android:text="Button1" />
```

（1）第 1 行声明了一个 Button 控件。

（2）第 2 行定义 Button 的 id 为 button1。

（3）第 3～4 行分别定义 Button 的宽度与高度。

（4）第 5 行定义 Button 按钮在界面上显示为 Button1。

### 3.2.3　EditText

EditText（输入框）可以向用户显示文本内容，也允许用户对文本内容进行编辑，它是 TextView 的子类，拥有 TextView 所有的属性。它最主要的用途是设计用户登录界面，如用户名与密码等信息的输入。EditText 控件位于 Android.widget. EditText 类中，如果 Java 代码中使用这个控件，需要在头部添加“import android.widget. EditText;”语句，它特有的属性如下：

- android:digits：指定字段只接受某些字符。
- android:hint：输入为空时的提示信息。
- android:inputType：限定输入的字符类型。
- android:singleLine：控制字段是单行输入框还是多行输入框（换句话说，按回车键是将焦点移到下一个部件还是换行）。

**【例 3.2】**设计如图 3-4 所示的布局文件。

**【说明】**EditText 在默认情况下宽度与高度都使用 wrap_content，并且使 EditText 输入框与输入框中的内容都居中显示。

图 3-4　EditText 的效果

**【代码】**布局文件中 EditText 的代码部分如下：

```
<EditText
        android:layout_width="wrap_content"
        android:layout_height="wrap_content"
```

```
4            android:text="请输入内容："
5            android:gravity="center"
6            android:layout_gravity="center_horizontal"
7     />
```

（1）第 1 行声明了一个 EditText 控件。

（2）第 2～3 行分别定义 EditText 的宽度与高度。

（3）第 4 行定义 EditText 输入框默认显示的内容。

（4）第 5 行定义 EditText 中的文本内容居中。

（5）第 6 行定义 EditText 控件与屏幕顶部中间对齐。

（6）第 7 行为 EditText 的结束标志。

### 3.2.4 ImageView

ImageView 是用于在界面展示图片的一个控件。它还可以对图片执行一些其他设置，如图片的缩放，调整图片的 Alpha 值等。

ImageView 的属性与对应的方法及说明如表 3-3 所示。

**表 3-3 ImageView 的属性与对应的方法及说明**

| 属性 | 方法 | 说明 |
|---|---|---|
| android:adjustViewBounds | setAdjustViewBounds(boolean) | 是否保持图片的宽高比 |
| android:maxHeight | setMaxHeight(int) | 设置 ImageView 的最大高度 |
| android:maxWidth | setMaxWidth(int) | 设置 ImageView 的最大宽度 |
| android:src | setImageResource(int) | 设置 ImageView 的 drawable（如图片，也可以是颜色，但是需要指定 View 的大小） |
| android:scaleType | setScaleType(ImageView.ScaleType) | 调整或移动图片来适应 ImageView 的尺寸，当 scaleType 取值为 fitXY 时拉伸图片（不按比例）以填充 View 的宽高；当 scaleType 取值为 Center 时按原图大小居中显示；当 scaleType 取值为 centerCrop 时按比例扩大图片居中显示；当 scaleType 取值为 fitCenter 时把图片按比例扩大/缩小到 ImageView 的宽度并居中显示 |

**【例 3.3】**设计如图 3-5 所示的布局文件。

**【说明】**除了在界面顶部中间显示一张图片外，在其下方还要显示一行文字。

图 3-5 ImageView 的效果

**【代码】**布局文件中相关的代码部分如下：

```
<ImageView
        android:id="@+id/image1"
        android:layout_width="wrap_content"
        android:layout_height="wrap_content"
        android:layout_gravity="center_horizontal"
```

```
6              android:src="@drawable/ic_launcher" />
7      <TextView
8              android:layout_width="wrap_content"
9              android:layout_height="wrap_content"
10             android:layout_gravity="center_horizontal"
11             android:text="这是一张 Android 图片" />
```

（1）第 1～6 行声明了一个 ImageView 控件。

（2）第 5 行设置图片水平居中显示。

（3）第 6 行定义图片的来源是 drawable 目录下的 ic_launcher.png 文件。

（4）第 7～11 行声明了一个 TextView 控件。

### 3.2.5　CheckBox

CompoundButton 是一个带有选中/未选中状态的按钮，但是作为抽象类，其不能直接使用。CheckBox（复选框）继承自 CompoundButton，表示选中与否的状态，可用于多选的情况，它有一些经常使用的方法，如表 3-4 所示。

**表 3-4　CheckBox 的常用方法及说明**

| 方法 | 说明 |
|---|---|
| isChecked() | 判断是否被选中，如选中则为 true，否则为 false |
| setChecked (boolean checked) | 通过传参来改变控件的状态 |
| performClick() | 使用代码主动去调用控件的点击事件 |
| toggle() | 取反控件选中的状态，即原来是选中则返回为未选中，原来是未选中则返回为选中 |
| setOnCheckedChangeListener(CompoundButton.OnCheckedChangeListener listener) | 为控件添加 OnCheckedChangeListener 监听器 |

**【例 3.4】**设计如图 3-6 所示的布局文件。

**【说明】**这组 CheckBox 控件是将各种颜色作为复选框的选项，并且以所见所得的形式表现，即什么颜色就以什么颜色显示。

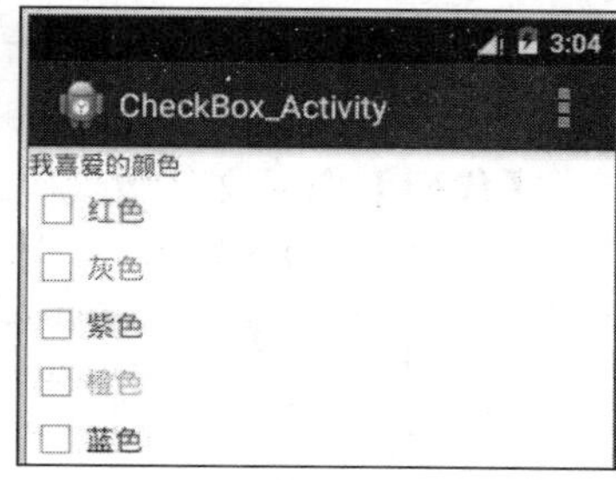

图 3-6　CheckBox 的效果

**【代码】**布局文件中代码部分如下：

```
<TextView
        android:layout_width="wrap_content"
        android:layout_height="wrap_content"
        android:text="我喜爱的颜色" />
<CheckBox
        android:layout_width="wrap_content"
        android:layout_height="wrap_content"
        android:text="红色"
        android:textColor="#FF0000" />
<CheckBox
```

```
11              android:layout_width="wrap_content"
12              android:layout_height="wrap_content"
13              android:text="灰色"
14              android:textColor="#808080" />
15      <CheckBox
16              android:layout_width="wrap_content"
17              android:layout_height="wrap_content"
18              android:text="紫色"
19              android:textColor="#800080" />
20      <CheckBox
21              android:layout_width="wrap_content"
22              android:layout_height="wrap_content"
23              android:text="橙色"
24              android:textColor="#FFA500" />
25      <CheckBox
26              android:layout_width="wrap_content"
27              android:layout_height="wrap_content"
28              android:text="蓝色"
29              android:textColor="#0000FF" />
```

（1）第 1～4 行声明了一个 TextView 控件，用于显示文本框中的内容“我喜爱的颜色”。

（2）第 5～29 行定义了 5 个 CheckBox。

### 3.2.6 RadioButton 和 RadioGroup

RadioButton 与 CheckBox 的不同在于一组 CheckBox 可以一次选中多个，而 RadioButton 只能选中一个，所以 RadioButton 通常要与 RadioGroup 一起使用，用于定义一组单选按钮。

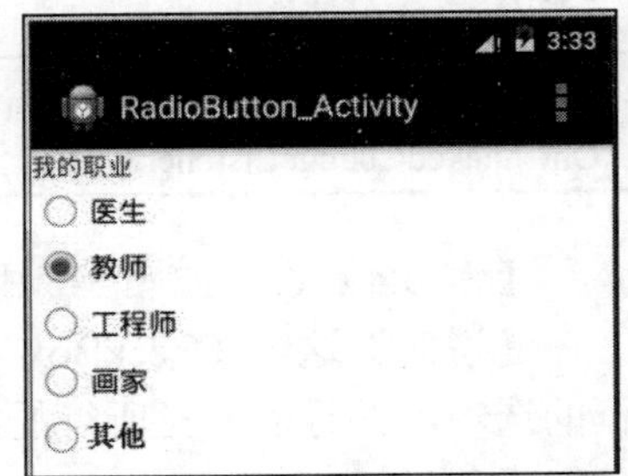


图 3-7 RadioButton 的效果

**【例 3.5】**设计如图 3-7 所示的布局文件。

**【说明】**这组 RadioButton 控件形成了一系列可供选择的职业。若有多个单选按钮，可通过使用 RadioGroup 控件把这些 RadioButton 组合成一组，这样可保证任何时候都只有一个单选按钮被选中。

**【代码】**布局文件中代码部分如下：

```
<TextView
        android:layout_width="wrap_content"
        android:layout_height="wrap_content"
        android:text="我的职业" />
<RadioGroup
        android:layout_width="fill_parent"
        android:layout_height="wrap_content"
        android:checkedButton="@+id/teacher"
        android:orientation="vertical">
<RadioButton
        android:id="@+id/doctor"
        android:layout_width="wrap_content"
```

```
13                  android:layout_height="wrap_content"
14                  android:text="医生" />
15    <RadioButton
16                  android:id="@+id/teacher"
17                  android:layout_width="wrap_content"
18                  android:layout_height="wrap_content"
19                  android:text="教师" />
20    <RadioButton
21                  android:id="@+id/engineer"
22                  android:layout_width="wrap_content"
23                  android:layout_height="wrap_content"
24                  android:text="工程师" />
25    <RadioButton
26                  android:id="@+id/painter"
27                  android:layout_width="wrap_content"
28                  android:layout_height="wrap_content"
29                  android:text="画家" />
30    <RadioButton
31                  android:id="@+id/other"
32                  android:layout_width="wrap_content"
33                  android:layout_height="wrap_content"
34                  android:text="其他" />
35    </RadioGroup>
```

（1）第 1～4 行声明了一个 TextView 控件，用于显示文本框中的内容“我的职业”。

（2）第 5～35 行，先声明了一个 RadioGroup 控件，然后默认垂直方向排列了 5 个 RadioButton，其中第 8 行表示默认选中“教师”单选按钮。

### 3.2.7　DatePicker 和 TimePicker

在 Android 应用程序开发过程中，经常需要用到与日期和时间相关的控件，接下来对日期选择控件 DatePicker 和时间选择控件 TimePicker 的基本使用方法进行介绍。

1．日期选择控件

DatePicker 是一个日期选择控件，它继承自 FrameLayout 类，主要功能是使用户方便地选择日期。如果用户对日期进行了改动，需要为 DatePicker 添加 OnDateChangedListener 监听器来捕获用户改动日期的数据，DatePicker 的常用方法及说明如表 3-5 所示。

**表 3-5　DatePicker 的常用方法及说明**

| 方法 | 说明 |
|---|---|
| getYear() | 获取当前日期的年 |
| getMonth() | 获取当前日期的月 |
| getDayOfMonth() | 获取当前日期的日 |
| setMaxDate(long maxDate) | 设置最大日期 |
| setMinDate(long minDate) | 设置最小日期 |

续表

| 方法 | 说明 |
|---|---|
| updateDate(int year,int month,int dayOfMonth) | 更新当前日期 |
| init(int year,int monthOfYear,int dayOfMonth, DatePicker.OnDateChangedListener onDateChangedListener); | 初始化日期，以 onDateChangedListener 为监听器对象来监听日期的变化 |

2. 时间选择控件

TimePicker 是一个时间选择控件，它也继承自 FrameLayout 类，可以为用户提供一天中的时间，它有 24 小时制和 AM/PM 制两种模式，并且允许用户修改。如果用户对时间进行了改动，需要为 TimePicker 添加 OnTimeChangedListener 监听器来捕获用户改动时间的数据，TimeDicker 的常用方法及说明如表 3-6 所示。

表 3-6　TimePicker 的常用方法及说明

| 方法 | 说明 |
|---|---|
| getCurrentHour() | 获取当前时间的小时 |
| getCurrentMinute() | 获取当前时间的分钟 |
| is24HourView() | 获取是否为 24 小时模式 |
| setCurrentHour(int currentHour) | 设置当前时间的小时 |
| setCurrentMinute(int currentMinute) | 设置当前时间的分钟 |
| setIs24HourView(boolean is24HourView) | 设置 24 小时模式 |
| setEnabled(boolean enabled) | 设置时间控件是否可用 |
| setOnTimeChangedListener(TimePicker.OnTimeChangedListener onTimeChangedListener) | 为时间控件添加 OnTimeChangedListener 监听器 |

【例 3.6】设计如图 3-8 所示的布局文件。

【说明】这是使用 DatePicker 和 TimePicker 控件来显示日期与时间，用户可以更改这两个控件的信息。

【代码】布局文件中代码部分如下：

```
<TextView
        android:id="@+id/textview"
        android:layout_width="fill_parent"
        android:layout_height="wrap_content"
        android:text="日期与时间控件使用" />
<DatePicker
       android:id="@+id/datepicker"
       android:layout_width="wrap_content"
       android:layout_height="wrap_content" />
<TimePicker
```

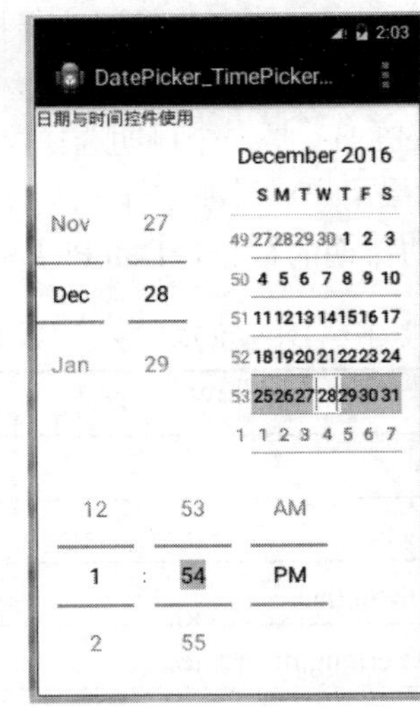


图 3-8　DatePicker 与 TimePicker 的效果

```
android:id="@+id/timepicker"
android:layout_width="wrap_content"
android:layout_height="wrap_content" />
```

## 3.3　Android 事件处理

Android 应用程序通过控件可以很好地展示漂亮的界面，但是界面上的各种控件都是固定的，不能与用户很好地互动。通过引入事件处理机制，可以让程序的控制更加精准与灵活。例如点击某个按钮来触发一个新的界面出现，或者点击某个图片从而弹出新的功能菜单。因此，事件处理机制的使用可以提高开发者的开发效率。

Android 应用程序处理得最多的就是用户请求，即经常要处理用户的各种点击行为，这种为用户各种点击行为提供响应的机制就是事件处理机制。Android 提供了如下两种类型的事件处理机制：

- 基于回调的事件处理机制。
- 基于监听的事件处理机制。

基于回调的事件处理机制，主要的实现方法是重写 Android 中特定组件的回调方法和 Activity 的回调机制方法。

基于监听的事件处理机制，主要的实现方法是为 Android 界面上的控件绑定特定的事件监听器，并为开发者处理各种逻辑操作。

一般情况下，基于回调的事件处理机制用于处理一些具有普遍性的事件，而当有些操作无法使用基于回调的事件处理机制时，就要使用基于监听的事件处理机制。

### 3.3.1　基于回调的事件处理机制

回调不是由方法实现方直接调用的，而是在某个事件发生时，由另外一方通过一个接口来调用，用于对该事件进行响应处理。回调是一种双向调用模式，被调用方在被调用时也会调用对方，这就叫回调。

在用户使用 Android 中的控件触发某个事件时，控件本身会用自己特定的方法来处理该响应。而这些方法不能动态地添加，需要使用 Android 控件类，并通过重写这个类的事件处理方法来实现。

Android 控件类提供了很多事件处理回调方法来实现回调机制，以 View 类为例，它就包含以下三种方法。

1．onKeyDown()方法

几乎所有的 View 类都包含 onKeyDown()方法，此方法用来监听手机键被按下的事件，是接口 KeyEvent.Callback 中的抽象方法，View 类实现了这个接口并重写了 onKeyDown()方法。

onKeyDown()方法的声明格式：boolean onKeyDown(int keyCode,KeyEvent event)。

参数说明：

（1）参数 keyCode。此参数为 int 类型，表示手机键盘被按下的值，手机键盘的每个键都对应一个键盘码，用户按下手机键后可通过识别应用程序中对应的键盘码进行接下来的逻辑处理。

（2）参数 KeyEvent。此参数为按键事件的对象，它包括触发事件的具体信息，例如事件的状态、类型等。当用户按下手机键盘上某个键时，系统会自动将此事件封装为 KeyEvent 对象给应用程序使用。

（3）返回值。此方法的返回值为一个 boolean 类型。当返回值为 true 时，表示已经成功地处理完了这个事件，不需要其他回调方法再进行处理；当返回值为 false 时，表示此事件还没有处理完全，还需要其他回调方法进行处理。

2. onKeyUp()方法

onKeyUp()方法用来监听手机键被抬起的事件。此方法也是接口 KeyEvent.Callback 中的抽象方法，View 类实现了这个接口并重写了 onKeyUp()方法。

onKeyUp()方法的声明格式：boolean onKeyUp (int keyCode,KeyEvent event)。

onKeyUp()方法的参数和返回值类型与 onKeyDown()方法相同，在此不再作说明。

onKeyUp()方法的使用与 onKeyDown()方法相同，只是此方法是在手机键被抬起时调用，当开发者要对抬起事件做处理时，只要重写此方法即可。

3. onTouchEvent()方法

onTouchEvent()方法用来处理用户触摸屏幕事件，View 类实现了此方法，并且所有的 View 子类都重写了这个方法。

onTouchEvent ()方法的声明格式：public boolean onTouchEvent (MotionEvent event)。

参数说明：

（1）参数 event。此参数为手机屏幕触摸事件封装类的对象，其中封装了该事件的所有信息，例如触摸的位置、触摸的类型以及触摸的时间等。该对象会在用户触摸手机屏幕时被创建。

（2）返回值。该方法的返回值原理与键盘响应事件的原理相同，即当已经完整地处理了该事件且不希望其他回调方法再次处理时返回 true，否则返回 false。

此方法不像上述两种方法那样只处理一个事件，它可以处理以下三种情况下的事件，只是这三种情况下的动作值不相同。

（1）触击屏幕。当用户的动作是触击屏幕时，程序会识别并调用该方法来处理此事件，此时 MotionEvent.getAction()的值为 MotionEvent.ACTION_DOWN。当开发者要处理应用程序的用户触击屏幕事件时，只需要重写此方法，然后在此方法中添加自己的逻辑处理代码即可。

（2）离开屏幕。当用户的动作是离开屏幕时，此事件同样需要 onTouchEvent()方法来处理，这时 MotionEvent.getAction()的值为 MotionEvent.ACTION_UP，表示离开屏幕的事件。

（3）在屏幕上滑动。当用户的动作是在屏幕上滑动时，此事件也是调用 onTouchEvent()方法来处理，这时 MotionEvent.getAction()的值为 MotionEvent.ACTION_MOVE。

onTouchEvent()方法还负责处理触控笔在屏幕上滑动的事件，同样是调用 MotionEvent.getAction()方法来判断动作值是否为 MotionEvent.ACTION_MOVE，然后再进行处理。

**【例 3.7】**在屏幕区域内捕获用户的触击、抬起与滑动事件，并记录相关事件的坐标。

**【说明】**通过学习 Android API 文档，查阅相关方法。

（1）使用 MotionEvent.getAction()来获取用户的触屏行为。

（2）使用 event.getX()、event.getY()方法来获取触屏坐标。

**【开发步骤】**

（1）创建一个名为 Touch_Activity 的项目，包名为 com.hzu.touch_activity，Activity 组件名为 MainActivity。

（2）在 res/layout 目录下的布局文件 activity_main 中编写如下代码：

```
1    <LinearLayout xmlns:android="http://schemas.android.com/apk/res/android"
2         xmlns:tools="http://schemas.android.com/tools"
3         android:layout_width="fill_parent"
4         android:layout_height="fill_parent"
5         android:orientation="vertical"
6         android:background="#FF88FF"
7    >
8    <TextView
9              android:id="@+id/touch_test"
10             android:layout_width="fill_parent"
11             android:layout_height="350dip"
12             android:text="触屏测试区"
13             android:textColor="#99FFFF"
14   />
15   <TextView
16             android:id="@+id/event_text"
17             android:layout_width="fill_parent"
18             android:layout_height="wrap_content"
19             android:text="触屏事件："
20             android:textColor="#00FF00"
21   />
22   </LinearLayout>
```

1）第 1～7 行声明了一个垂直排列的线性布局控件。

2）第 8～14 行声明了一个 TextView 控件，为了在触屏测试区下方显示触屏信息，设置此控件高度为 350dip。

3）第 15～21 行再次声明了一个 TextView 控件，声明此控件资源的 id 为 event_text。

（3）编写逻辑代码。在 src/com.hzu.touch_activity 目录下的 MainActivity.java 文件中编写如下代码：

```
package com.hzu.touch_activity;
import android.app.Activity;
import android.os.Bundle;
import android.view.MotionEvent;
import android.widget.TextView;

public class MainActivity extends Activity {

    private TextView event_show;

    @Override
    protected void onCreate(Bundle savedInstanceState) {
          super.onCreate(savedInstanceState);
          setContentView(R.layout.activity_main);
```

```
15              event_show = (TextView) findViewById(R.id.event_text);
16          }
17
18          @Override
19          public boolean onTouchEvent(MotionEvent event) {
20              switch (event.getAction()) {
21                  case MotionEvent.ACTION_DOWN:
22                  showInformation("ACTION_DOWN", event);
23                  break;
24                  case MotionEvent.ACTION_UP:
25                  showInformation("ACTION_UP", event);
26                  break;
27                  case MotionEvent.ACTION_MOVE:
28                  showInformation("ACTION_MOVE", event);
29                  break;
30                  }
31                  return super.onTouchEvent(event);
32          }
33
34          public void showInformation(String action, MotionEvent event) {
35              int x = (int) event.getX();
36              int y = (int) event.getY();
37              String message = “”;
38              message += "响应的事件为" + action + "\n";
39              message += "坐标（x,y）为" + String.valueOf(x) + "," + String.valueOf(y);
40              event_show.setText(message);
41          }
42      }
```

1）第 4 行引入了 android.view.MotionEvent 类，因在代码中使用了相关的触屏事件。

2）第 5 行引入了 android.widget.TextView 类，因在代码中使用了一个 TextView 对象。

3）第 9 行为 TextView 声明了一个 event_show 类对象。

4）第 11～16 行重写 OnCreate()方法。其中在第 15 行为 event_show 赋值，即将资源中 id 号为 event_text 的 TextView 对象赋给变量 event_show。

5）第 18～32 行重写 onTouchEvent()方法。在此方法中 event 参数为一个 MotionEvent 对象。其中第 20 行使用 getAction()来获取事件的状态。接下来通过 switch-case 语句来判断用户的触屏行为，将 ACTION_DOWN、ACTION_UP、ACTION_MOVE 三种用户行为中的一种传入到 showInformation()方法中。

6）第 34～41 行定义了一个 showInformation()方法，通过 onTouchEvent()方法中用户的行为动态调用相应的触屏行为状态信息，并将这些信息显示在 event_show 对象上。

**【运行结果】**在 Eclipse 中启动 Android 模拟器，接着运行 Touch_Activity 项目，运行结果如图 3-9 所示。

图 3-9　屏幕区域内捕获用户操作

### 3.3.2　基于监听的事件处理机制

在 Android 中基于监听的事件处理机制与 Java 中 Swing、AWT 的事件处理机制非常相似，只是对应的事件监听器和事件处理方法名不相同，在基于监听的事件处理机制中有以下三个概念要清楚：

（1）EventSource（事件源）：产生这个事件的组件，即事件发生的来源，如按钮、窗口等。

（2）Event（事件）：UI 上面事件源发生的特定事件，并且该事件封装了该操作的相关信息，应用程序需要知道的事件源上所发生事件的具体信息一般都是由 Event 对象来取得，如用户触摸屏幕的位置等。

（3）EventListener（事件监听器）：监听事件源发生的事件，并对被监听的事件做出相应的响应。一个事件监听器可包含多个事件处理器，每一个事件处理器就是一个事件处理方法。

基于监听的事件处理机制可以理解为一种委托式的事件处理机制。以上三个对象是按照一定方式进行分工协助的：UI 控件（事件源）将发生的事件委派给特定的对象（监听器）处理，再由监听器相应的事件处理器来具体处理这个事件，具体的流程如图 3-10 所示。

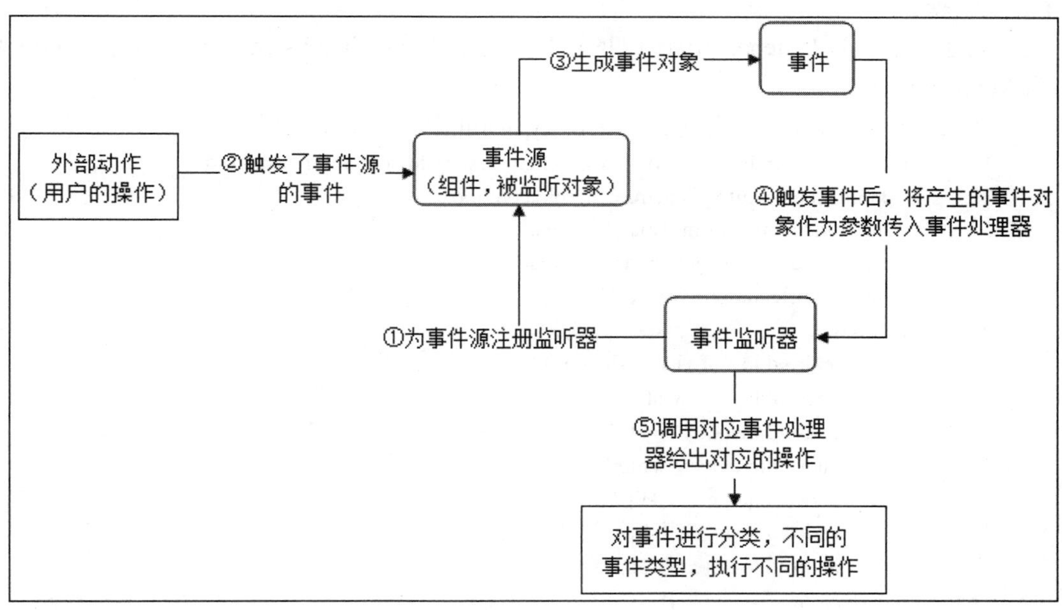


图 3-10　基于监听的事件处理过程

委托式的事件处理机制很容易理解，自己解决不了的棘手问题可以委托他人来帮助处理。这在生活中很常见，如工厂发生火灾时，自己并不能处理该事件，而是交给消防局处理。工厂需要原材料来加工产品，但自己不会生产原材料，就交给原材料供应商来处理该事件。可以这样来理解，我们自己是事件源，遇到的事情称为事件，处理这个事件的组织或公司叫事件监听器，具体解决这件事的人员就是事件处理器。

事件处理过程一般分为以下 3 个步骤：

（1）在应用程序中获取普通控件（事件源），即被监听的对象。

（2）实现事件监听器，它是一个特殊的 Java 类，要实现它的 XxxListener 接口中所有的方法。

（3）使用 setXxxListener()方法把事件监听器对象注册到事件源上，这样就完成了事件源与事件监听器的连接，在用户的触发事件发生后调用相应的方法。

在完成上述步骤的过程中，事件源一般通过 findViewById()方法来取得控件，接下来使用控件的 setXxxListener()方法中的参数传入一个具体的事件监听器对象。因为监听器是接口，要实现接口中的方法才算完成对象的实例化。在应用程序中实现监听器有以下 4 种形式：

- 匿名内部类实现：使用匿名内部类创建事件监听器。
- 外部类实现：对外部类实现事件监听器接口。
- 内部类实现：将事件监听器类定义到当前类中。
- 使用标签实现：直接在布局文件中指定标签绑定事件处理方法，主要是在布局文件中使用 android:onClick 来实现。

**【例 3.8】**实现一个简单的文字处理器，其可以改变文字背景、文字样式，且具有更改文字、擦除文字的功能。

**【说明】**使用匿名内部类、外部类、内部类、标签实现文字的处理。

**【开发步骤】**

（1）创建一个名为 Listener_Activity 的项目，包名为 com.hzu.listener_activity，Activity 组件名为 MainActivity。

（2）在 res/layout 目录下的布局文件 activity_main 中编写如下代码：

```
<LinearLayout xmlns:android="http://schemas.android.com/apk/res/android"
    xmlns:tools="http://schemas.android.com/tools"
    android:layout_width="match_parent"
    android:layout_height="match_parent"
    android:orientation="vertical">
<TextView
        android:id="@+id/tv_showmsg"
        android:layout_width="match_parent"
        android:layout_height="wrap_content"
        android:gravity="center"
        android:height="50dip"
        android:text="文字显示效果"
        android:textSize="25sp" />
<LinearLayout
        android:layout_width="wrap_content"
        android:layout_height="wrap_content"
        android:orientation="horizontal">
<TextView
            android:layout_width="wrap_content"
            android:layout_height="wrap_content"
            android:text="背景色" />
<Button
            android:id="@+id/bt_red"
            android:layout_width="wrap_content"
```

```
            android:layout_height="wrap_content"
            android:text="红色" />
<Button
            android:id="@+id/bt_green"
            android:layout_width="wrap_content"
            android:layout_height="wrap_content"
            android:text="绿色" />
<Button
            android:id="@+id/bt_yellow"
            android:layout_width="wrap_content"
            android:layout_height="wrap_content"
            android:text="黄色" />
</LinearLayout>
<LinearLayout
        android:layout_width="wrap_content"
        android:layout_height="wrap_content"
        android:orientation="horizontal">
<TextView
            android:layout_width="wrap_content"
            android:layout_height="wrap_content"
            android:text="样式" />
<Button
            android:id="@+id/bt_bold"
            android:layout_width="wrap_content"
            android:layout_height="wrap_content"
            android:text="加粗" />
<Button
            android:id="@+id/bt_italic"
            android:layout_width="wrap_content"
            android:layout_height="wrap_content"
            android:text="斜体" />
<Button
            android:id="@+id/bt_bold_italic"
            android:layout_width="wrap_content"
            android:layout_height="wrap_content"
            android:text="粗斜体" />
</LinearLayout>
<LinearLayout
        android:layout_width="wrap_content"
        android:layout_height="wrap_content"
        android:orientation="horizontal">
<TextView
            android:layout_width="wrap_content"
            android:layout_height="wrap_content"
            android:text="改变内容" />
<EditText
```

```
71                  android:id="@+id/et_content"
72                  android:layout_width="200dp"
73                  android:layout_height="wrap_content" />
74      <Button
75                  android:id="@+id/bt_commit"
76                  android:layout_width="wrap_content"
77                  android:layout_height="wrap_content"
78                  android:text="提交" />
79      </LinearLayout>
80      <LinearLayout
81              android:layout_width="wrap_content"
82              android:layout_height="wrap_content"
83              android:orientation="horizontal">
84      <TextView
85                  android:layout_width="wrap_content"
86                  android:layout_height="wrap_content"
87                  android:text="擦除内容" />
88      <Button
89                  android:id="@+id/bt_erase"
90                  android:layout_width="wrap_content"
91                  android:layout_height="wrap_content"
92                  android:text="擦除"
93                  android:onClick="eraseMethod"
94                  />
95      </LinearLayout>
96      </LinearLayout>
```

1）第 1～5 行声明了线性布局，方向为垂直方向。

2）第 6～13 行声明了一个 TextView 控件，其是处理文字的最终显示效果的地方。

3）第 14～37 行声明了一个子线性布局，方向为水平方向，在其线性布局内部声明了一个 TextView、三个 Button。第 38～61 行情况类似，不再叙述。

4）第 62～79 行声明了一个子线性布局，方向为水平方向，在其线性布局内部声明了一个 TextView、一个 EditText、一个 Button，用于处理用户输入的内容。

5）第 80～95 行声明了一个子线性布局，方向为水平方向，在其线性布局内部声明了一个 TextView、一个 Button，在 Button 中有一个属性 android:onClick，用于完成标签绑定事件的处理方法。

（3）编写逻辑代码。打开 src/com.hzu.listener_activity 中的 MainActivity.java 文件，根据要求编写代码。

为 TextView 中“文字显示效果”的背景色设置采用内部类的实现方式，其关键代码如下：

```
package com.hzu.listener_activity;
import com.hzu.listeseractivity.R;
import android.app.Activity;
import android.graphics.Color;
import android.graphics.Typeface;
import android.os.Bundle;
```

```
7	import android.view.View;
8	import android.view.View.OnClickListener;
9	import android.widget.Button;
10	import android.widget.EditText;
11	import android.widget.TextView;
12	import android.widget.Toast;
13	public class MainActivity extends Activity{
14	private EditText content;
15	private Button red, green, yellow;
16	protected void onCreate(Bundle savedInstanceState) {
17		super.onCreate(savedInstanceState);
18		setContentView(R.layout.activity_main);
19		showmsg = (TextView) findViewById(R.id.tv_showmsg);
20		red = (Button) findViewById(R.id.bt_red);
21		green = (Button) findViewById(R.id.bt_green);
22		yellow = (Button) findViewById(R.id.bt_yellow);
23		BackgroundChangeListener backgroundChangeListener = new
24			BackgroundChangeListener();
25		red.setOnClickListener(backgroundChangeListener);
26		green.setOnClickListener(backgroundChangeListener);
27		yellow.setOnClickListener(backgroundChangeListener);
28	 }
29	 private class BackgroundChangeListener implements OnClickListener {
30
31		public void onClick(View v) {
32		   switch (v.getId()) {
33		      case R.id.bt_red:
34			    showmsg.setBackgroundColor(Color.RED);
35			    break;
36		      case R.id.bt_green:
37			    showmsg.setBackgroundColor(Color.GREEN);
38			    break;
39		      case R.id.bt_yellow:
40			    showmsg.setBackgroundColor(Color.YELLOW);
41			    break;
42		   }
43		}
44	}
45 }
```

1）第 29～44 行在 MainActivity 上中实现了 BackgroundChangeListener 类。

2）第 31 行重写了 onClick ()方法。

为 TextView 中“文字显示效果”的文字样式设置采用外部类的实现方式，其关键代码如下：

```
public class MainActivity extends Activity implements OnClickListener {
  private TextView showmsg;
  private Button bold, italic, bold_italic;
```

```
4    protected void onCreate(Bundle savedInstanceState) {
5    super.onCreate(savedInstanceState);
6    setContentView(R.layout.activity_main);
7        showmsg = (TextView) findViewById(R.id.tv_showmsg);
8        bold = (Button) findViewById(R.id.bt_bold);
9        italic = (Button) findViewById(R.id.bt_italic);
10       bold.setOnClickListener(this);
11       italic.setOnClickListener(this);
12       bold_italic.setOnClickListener(this);
13   }
14   public void onClick(View v) {
15       switch (v.getId()) {
16         case R.id.bt_bold_italic:
17            showmsg.setTypeface(Typeface.MONOSPACE, Typeface.BOLD_ITALIC);
18            break;
19         case R.id.bt_italic:
20            showmsg.setTypeface(Typeface.MONOSPACE, Typeface.ITALIC);
21            break;
22         case R.id.bt_bold:
23            showmsg.setTypeface(Typeface.MONOSPACE, Typeface.BOLD);
24            break;
25       }
26   }
27 }
```

1）第 1 行在 MainActivity 上实现了 OnClickListener 接口。

2）第 14 行重写了 onClick()方法。

为 TextView 中“文字显示效果”的更改文字内容的设置采用匿名内部类的实现方式，其关键代码如下：

```
public class MainActivity extends Activity{
  private TextView showmsg;
  private EditText content;
  private Button commit;
  protected void onCreate(Bundle savedInstanceState) {
     super.onCreate(savedInstanceState);
        setContentView(R.layout.activity_main);
        showmsg = (TextView) findViewById(R.id.tv_showmsg);
        commit = (Button) findViewById(R.id.bt_commit);
        commit.setOnClickListener(new View.OnClickListener() {
           public void onClick(View v) {
              String ct=content.getText().toString();
              if(ct==null||ct.equals("")){
                 Toast.makeText(MainActivity.this, "内容不能为空",
                 Toast.LENGTH_LONG).show();
              }
```

```
17                        else{showmsg.setText(ct);}
18                    }
19                });
20        }
21    }
```

1）第 10 行在 commit 按钮上添加了匿名内部类。

2）第 11 行重写了 onClick()方法。

为 TextView 中“文字显示效果”的擦除文字的设置采用标签绑定的实现方式，其关键代码如下：

```
1    public class MainActivity extends Activity{
2      private TextView showmsg;
3      protected void onCreate(Bundle savedInstanceState) {
4                super.onCreate(savedInstanceState);
5                setContentView(R.layout.activity_main);
6                showmsg = (TextView) findViewById(R.id.tv_showmsg);
7      }
8      public void eraseMethod(View view){
9              showmsg.setText("");
10     }
11   }
```

1）在 res/layout 目录下的布局文件 activity_main 中第 93 行声明了方法名为 eraseMethod。

2）第 8～10 行实现了 eraseMethod()方法。

**【运行结果】**在 Eclipse 中启动 Android 模拟器，接着运行 Listener_Activity 项目。运行结果如图 3-11 所示。

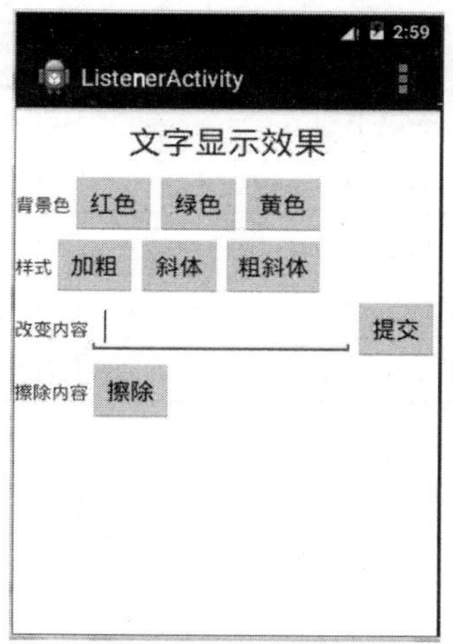


图 3-11　简单的文字处理器

在应用程序中实现监听器有以下 4 种形式，它们各有优缺点，在应用程序的编写过程中要根据具体情况来决定使用哪种监听器实现形式：

- 内部类形式：使用内部类可以在当前类中复用该监听器类，即多个事件源可以注册同一个监听器。使用内部类可以自由访问外部类的所有界面控件，内部类实质上是外部类的成员。内部类形式适用于多个事件源同时注册到同一事件监听器的情形。

- 外部类形式：这种事件监听器不能自由访问和创建 GUI 界面中的组件，且编程不够简洁。如果某个事件监听器确实需要被多个 GUI 界面共享，而且主要是为了完成某种业务逻辑的实现，则可以考虑使用外部类的形式来定义事件监听器类。
- 匿名内部类形式：这种事件处理器没有复用价值，大部分事件监听器只是临时使用一次，主要在类中监听事件少时使用。
- 使用标签形式：这种事件处理器涉及到布局文件与 Java 代码两个部分的处理，对于少量事件的监听还是可用的，灵活度不是很高。

常见的事件监听器接口与注册方法如表 3-7 与表 3-8 所示。

**表 3-7 常见事件监听器接口及其处理方法**

| 事件 | 接口 | 处理方法 | 描述 |
|---|---|---|---|
| 点击事件 | View.OnClickListener | abstract void onClick (View v) | 点击组件时触发 |
| 长按事件 | View.OnLongClickListener | abstract boolean onLongClick (View v) | 长按组件时触发 |
| 键盘事件 | View.OnKeyListener | abstract boolean onKey(View v, int keyCode, KeyEvent event) | 处理键盘事件 |
| 焦点事件 | View.OnFocusChangeListener | abstract void onFocusChange(View v, boolean hasFocus) | 当焦点发生改变时触发 |
| 触摸事件 | View.OnTouchListener | abstract boolean onTouch (View v, MotionEvent event) | 产生触摸事件 |

**表 3-8 View 类常见事件监听器注册方法**

| 方法 | 描述 |
|---|---|
| void setOnClickListener(View.OnClickListener l) | 注册点击事件 |
| void setOnLongClickListener(View.OnLongClickListener l) | 注册长按事件 |
| void setOnKeyListener(View.OnKeyListener l) | 注册键盘事件 |
| void setOnFocusChangeListener(View.OnFocusChangeListener l) | 注册焦点事件 |
| void setOnTouchListener(View.OnTouchListener l) | 注册触摸事件 |

## 3.4 本章小结

本章介绍了 Android 应用程序用户界面中各类控件的基本属性，并通过案例演示了各类控件的基本使用情况。基本上每个 Android 应用程序都会用到以上一些控件与事件处理机制，因此要求程序开发者掌握本章内容，为后面的学习奠定基础。

# 第 4 章　布局

（1）了解布局的基类。

（2）掌握各类布局的基本属性与使用方法。

（3）掌握嵌套布局的使用。

通过前面三章的学习，我们了解了几种简单的界面控件，并通过一些案例了解了几种常用控件的基本用法，可是应用程序的界面并不十分美观，控件排放杂乱无章。本章将介绍 Android 中几种常用的布局管理器来美化程序界面。

## 4.1　布局简介

当界面有多个控件时，需要按照需求进行合理摆放，布局管理器就是用来规范内部控件的摆放位置的。所有的布局管理器都是继承 ViewGroup 类的子类，都可以作为一个容器来使用。布局内部的控件被认为是布局的子控件，一个布局也可以成为另外一个布局的子控件，即一个布局管理器可以嵌套其他布局管理器。

## 4.2　常见布局

### 4.2.1　帧布局

帧布局（FrameLayout）是 Android 中最简单的一种布局。在这个布局中，整个界面被当成一块空白备用区域，所有的子元素都不能被指定放置的位置，它们都存放于这块区域的左上角，并且后面的子元素直接覆盖在前面的子元素之上，将前面的子元素全部遮挡。当然如果后一个子元素显示对象是透明的，则前一个对象会显示出来。

**【例 4.1】**设计出如图 4-1 所示的布局文件。

**【说明】**在帧布局中添加三个 TextView 控件，TextView1 是第一个添加的文本框，TextView2 是第二个添加的文本框，TextView3 是第三个添加的文本框，这三个文本框叠加显示在屏幕的左上角。

**【开发步骤】**

（1）创建一个名为 FrameLayout_Activtiy 的项目，包名为 com.hzu.framelayout_activtiy，Activity 组件名为 MainActivity。

（2）在 res/layout/activity_main.xml 文件中添加如下代码：

```
1 <FrameLayout xmlns:android="http://schemas.android.com/apk/res/android"
2   xmlns:tools="http://schemas.android.com/tools"
3   android:layout_width="match_parent"
4   android:layout_height="match_parent">
5     <TextView
6         android:layout_width="300dp"
7         android:layout_height="300dp"
8         android:background="#66ff1122"
9         android:gravity="bottom"
10        android:text="TextView1" />
11    <TextView
12        android:layout_width="200dp"
13        android:layout_height="200dp"
14        android:background="#55221122"
15        android:gravity="bottom"
16        android:text="TextView2" />
17    <TextView
18        android:layout_width="80dp"
19        android:layout_height="80dp"
20        android:background="#77441122"
21        android:gravity="bottom"
22        android:text="TextView3" />
23 </FrameLayout>
```

1）第 1 行声明一个 FrameLayout 布局。

2）第 5～22 行分别声明三个 TextView 控件，为了区分这三个控件，分别为它们设置不同的宽和高。

**【运行结果】**在 Eclipse 中启动 Android 模拟器，接着运行 FrameLayout_Activtiy 项目，显示效果如图 4-1 所示。

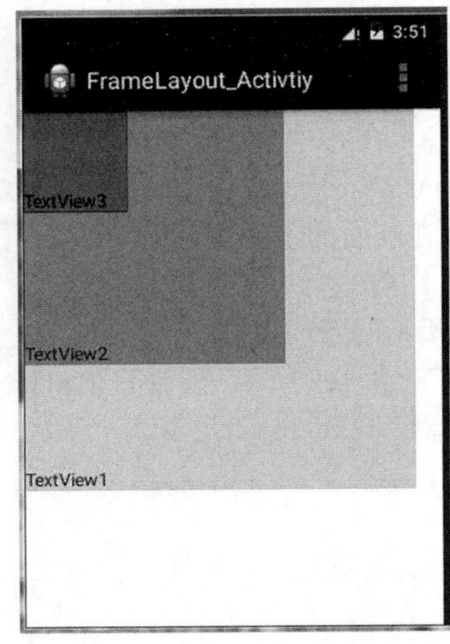


图 4-1　帧布局显示效果

### 4.2.2　线性布局

线性布局（LinearLayout）有两种类型：水平线性布局和垂直线性布局，LinearLayout 属性中 Android:orientation 为设置线性布局的方向，当其值为 vertical 时为垂直线性布局，当其值为 horizontal 时为水平线性布局，不管是水平还是垂直线性布局，它们的一行（列）都只能按照一个方向摆放子控件，如图 4-2 与图 4-3 所示。

图 4-2　水平线性布局

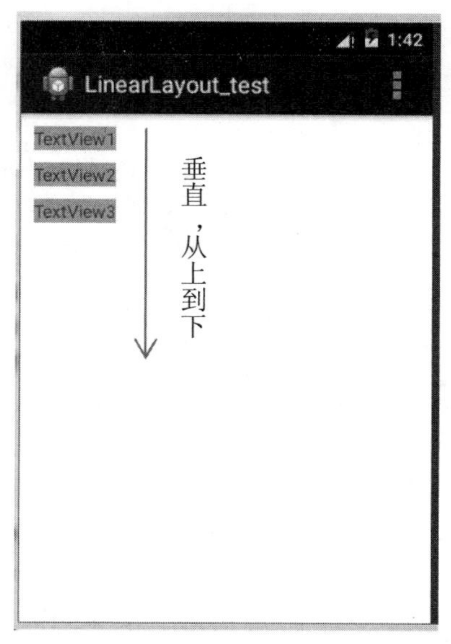


图 4-3　垂直线性布局

Android 在创建一个新的项目时，默认的布局是相对布局。如果需要使用线性布局，就要将其布局文件中的<RelativeLayout>属性改为<LinearLayout>，当然也可以在项目 layout 目录下新建 Android XML File。在编写线性布局过程中，要注意 Android 的线性布局不会自动换行，内部控件会按照相关方向一个挨着一个地排列，超出屏幕的控件将不会显示出来。

**【例 4.2】** 设计如图 4-3 所示的布局文件。

**【说明】** 图 4-3 有三个 TextView 控件，它们以垂直线性布局方式在屏幕左上方显示。

**【开发步骤】**

（1）创建一个名为 LinearLayout_test 的项目，包名为 com.hzu. linearlayout_test，Activity 组件名为 MainActivity。在 res/layout 目录下新建一个 vertical_main.xml 文件，将 MainActivity.java 文件中的 setContentView(R.layout.activity_main)改为 setContentView(R.layout.vertical_main)。其中编写的 vertical_main.xml 代码如下：

```
<LinearLayout xmlns:android="http://schemas.android.com/apk/res/android"
    xmlns:tools="http://schemas.android.com/tools"
    android:layout_width="match_parent"
    android:layout_height="match_parent"
    android:layout_marginLeft="10dp"
    android:layout_marginTop="10dp"
    android:orientation="vertical">
```

```
8        <TextView
9            android:layout_width="wrap_content"
10           android:layout_height="wrap_content"
11           android:background="#A6A6A6"
12           android:text="TextView1" />
13       <TextView
14           android:layout_width="wrap_content"
15           android:layout_height="wrap_content"
16           android:background="#A6A6A6"
17           android:layout_marginTop="10dp"
18           android:text="TextView2" />
19       <TextView
20           android:layout_width="wrap_content"
21           android:layout_height="wrap_content"
22           android:background="#A6A6A6"
23           android:layout_marginTop="10dp"
24           android:text="TextView3" />
25   </LinearLayout>
```

1）第 1～25 行声明了一个线性布局，第 5～6 行分别规定了线性布局与父容器左边和顶部的距离，第 7 行声明此布局为垂直方向。

2）第 8～12 行声明第一个 TextView 控件，第 9～10 行分别定义该控件的宽度和高度，第 11 行通过 android:background 设置 TextView 的背景色，第 12 行定义 TextView 控件默认显示的内容。

3）第 13～18 行声明第二个 TextView 控件，第 17 行通过 android:layout_marginTop 设置此 TextView 控件与上一个控件（TextView1）的距离。

4）第 19～24 行声明第三个 TextView 控件，与上一个控件的属性一致。

5）第 25 行为线性布局的结束标志。

开发者要想设计出漂亮的界面，对线性布局的相关属性就要比较清楚地了解，线性布局的常用属性如表 4-1 所示。

**表 4-1　线性布局的常用属性**

| 属性 | 说明 |
|---|---|
| android:orientation | 设置线性布局的方向，vertical 为垂直方向，horizontal 为水平方向 |
| android:background | 设置背景色 |
| android:gravity | 设置线性布局内部控件的对齐方式 |
| android:layout_width | 设置线性布局的宽度，match_parent 表示填充整个屏幕，wrap_content 表示根据对象上不同的文字宽度而确定显示对象的宽度 |
| android:layout_height | 设置线性布局的高度，属性值与 android:layout_width 相同 |
| android:layout_weight | 设置线性布局内部控件的重要度，按比例为其内部控件划分空间 |

### 4.2.3　表格布局

表格布局（TableLayout）是线性布局的特殊类，其采用行与列的形式来管理 UI 组件。表格布局没有明确指出有多少行与多少列，而是通过添加 TableRow 和其他控件来控制表格的行数与列数。当为 TableRow 对象时，可在 TableRow 下添加子控件，默认情况下每个子控件占据一列，而每列的宽度由表格各列中最宽的那个单元格来决定，整个表格布局的宽度取决于父容器的宽度。表格布局的常用属性如表 4-2 所示。

**表 4-2　表格布局的常用属性**

| 属性 | 说明 |
| --- | --- |
| android: stretchColumns | 设置允许被拉伸的列的宽度，用来填满表格中空闲的空间 |
| android:shrinkColumns | 设置指定可以进行收缩的列的宽度，使表格适用其父容器的大小 |
| android:collapseColumns | 设置需要被隐藏的列 |

**【例 4.3】**设计如图 4-4 所示的布局文件。

**【说明】**主要通过表格布局的相关属性来设置各个子控件在表格布局中的摆放。

**【开发步骤】**

（1）创建一个名为 TableLayout_test 的项目，包名为 com.hzu.tablelayout_test，Activity 组件名为 MainActivity。

（2）在 res/layout/activity_main.xml 文件中添加代码内容如下：

```
<?xml version="1.0" encoding="utf-8"?>
<LinearLayout xmlns:android="http://schemas.android.com/apk/res/android"
    android:layout_width="fill_parent"
    android:layout_height="fill_parent"
    android:orientation="vertical">
    <Button
        android:layout_width="wrap_content"
        android:layout_height="wrap_content"
        android:text="button1" />

     <TableLayout
         android:id="@+id/table1"
         android:layout_width="fill_parent"
         android:layout_height="wrap_content">
         <Button
             android:layout_width="wrap_content"
             android:layout_height="wrap_content"
             android:text="button2" />
     </TableLayout>

     <TableLayout
         android:id="@+id/table2"
         android:layout_width="fill_parent"
         android:layout_height="wrap_content">
         <TableRow>
             <Button
```

```
                android:layout_width="wrap_content"
                android:layout_height="wrap_content"
                android:text="button3" />
            <Button
                android:layout_width="wrap_content"
                android:layout_height="wrap_content"
                android:text="button4" />
            <Button
                android:layout_width="wrap_content"
                android:layout_height="wrap_content"
                android:text="button5" />
        </TableRow>
    </TableLayout>

    <TableLayout
        android:id="@+id/table3"
        android:layout_width="fill_parent"
        android:layout_height="wrap_content"
        android:stretchColumns="0,1,2">

        <TableRow>
            <Button
                android:layout_width="wrap_content"
                android:layout_height="wrap_content"
                android:text="button6" />
            <Button
                android:layout_width="wrap_content"
                android:layout_height="wrap_content"
                android:text="button7" />
            <Button
                android:layout_width="wrap_content"
                android:layout_height="wrap_content"
                android:text="button8" />
        </TableRow>
    </TableLayout>

    <TableLayout
        android:id="@+id/table4"
        android:layout_width="fill_parent"
        android:layout_height="wrap_content">
        <TableRow>
            <Button
                android:layout_width="wrap_content"
                android:layout_height="wrap_content"
                android:text="大家好，我是第 9 个按钮" />
            <Button
```

```
                    android:layout_width="wrap_content"
                    android:layout_height="wrap_content"
                    android:text="大家好，我是第 10 个按钮" />
                <Button
                    android:layout_width="wrap_content"
                    android:layout_height="wrap_content"
                    android:text="大家好，我是第 11 个按钮" />
            </TableRow>
        </TableLayout>

        <TableLayout
            android:id="@+id/table5"
            android:layout_width="fill_parent"
            android:layout_height="wrap_content"
            android:shrinkColumns="0,1,2">
            <TableRow>
                <Button
                    android:layout_width="wrap_content"
                    android:layout_height="wrap_content"
                    android:text="大家好，我是第 12 个按钮" />
                <Button
                    android:layout_width="wrap_content"
                    android:layout_height="wrap_content"
                    android:text="大家好，我是第 13 个按钮" />
                <Button
                    android:layout_width="wrap_content"
                    android:layout_height="wrap_content"
                    android:text="大家好，我是第 14 个按钮" />
            </TableRow>
        </TableLayout>

        <TableLayout
            android:id="@+id/table6"
            android:layout_width="fill_parent"
            android:layout_height="wrap_content"
            android:collapseColumns="1">
            <TableRow>
                <Button
                    android:layout_width="wrap_content"
                    android:layout_height="wrap_content"
                    android:text="button15" />
                <Button
                    android:layout_width="wrap_content"
                    android:layout_height="wrap_content"
                    android:text="button16" />
                <Button
```

```
119                     android:layout_width="wrap_content"
120                     android:layout_height="wrap_content"
121                     android:text="button17" />
122             </TableRow>
123         </TableLayout>
124
125     </LinearLayout>
```

1）第 1～5 行声明了一个垂直排列的线性布局控件，其宽度与高度都设置为 fill_parent 属性。

2）第 6～9 行声明了一个 Button，其中第 7～8 行定义宽度与高度都设置为 wrap_content 属性，第 9 行设置 Button 默认显示的内容。

3）第 11～19 行声明了一个 TableLayout，其中第 12 行声明其 id 为 table1，在其内部添加一个 Button，默认情况下铺满一行。

4）第 21～39 行声明了一个 TableLayout，在其内部添加了三个 Button，每个 Button 的宽度都是 wrap_content，因为三个 Button 的宽度之和没有超过屏幕宽度，所以可以正常显示 Button 上的文字内容。

5）第 41～61 行声明了一个 TableLayout，在其内部添加了三个 Button，因为第 45 行在表格布局中添加了 android:stretchColumns="0,1,2"属性，所以其内部三个 Button 在一行中平分宽度。

6）第 63～81 行声明了一个 TableLayout，在其内部添加了三个 Button，因为每个 Button 的文字内容都过长，所以看不到第三个 Button。

7）第 83～102 行声明了一个 TableLayout，在其内部添加了三个 Button，与上一个表格布局不同的是在此表格布局中添加了 android:shrinkColumns="0,1,2"，使得每一列都收缩排列。

8）第 104～123 行声明了一个 TableLayout，通过在表格布局中添加 android:collapseColumns="1"这个属性，使得第二个 Button（button16）消失。

**注意**：stretchColumns、shrinkColumns 与 collapseColumns 都是从第 0 个位置算起。

**【运行结果】**在 Eclipse 中启动 Android 模拟器，接着运行 FrameLayout_Activtiy 项目，显示效果如图 4-4 所示。

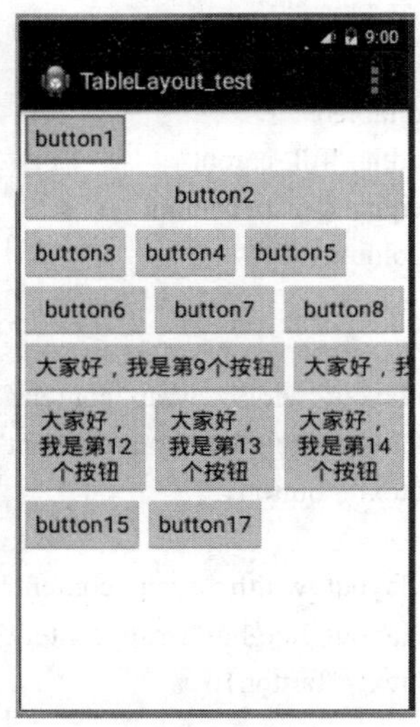


图 4-4　TableLayout 效果

### 4.2.4　相对布局

相对布局（RelativeLayout）允许子元素指定它们相对其父元素或兄弟元素的位置，是实际布局中最常用的布局方式之一。注意，如果 A 控件的位置是由 B 控件的位置来决定的，Android 要求先定义 B 控件，再定义 A 控件。相对布局灵活性较大，属性较多，操作难度较大，属性之间产生冲突的可能性也大，使用相对布局时要多做些测试。

相对布局的各类属性及说明详见表 4-3 至表 4-5。

表 4-3　相对布局取值为其他控件 id 的属性及说明

| 属性 | 说明 |
|---|---|
| android:layout_toRightOf | 控制该子控件位于给出 id 控件的右侧 |
| android:layout_toLeftOf | 控制该子控件位于给出 id 控件的左侧 |
| android:layout_above | 控制该子控件位于给出 id 控件的上方 |
| android:layout_below | 控制该子控件位于给出 id 控件的下方 |
| android:layout_alignTop | 控制该子控件位于给出 id 控件的上边界对齐 |
| android:layout_alignBottom | 控制该子控件位于给出 id 控件的下边界对齐 |
| android:layout_alignLeft | 控制该子控件位于给出 id 控件的左边界对齐 |
| android:layout_alignRight | 控制该子控件位于给出 id 控件的右边界对齐 |

表 4-4　相对布局取值为像素单位的属性及说明

| 属性 | 说明 |
|---|---|
| android:layout_marginBottom | 控制当前控件的下方的留白 |
| android:layout_marginTop | 控制当前控件的上方的留白 |
| android:layout_marginLeft | 控制当前控件的左侧的留白 |
| android:layout_marginRight | 控制当前控件的右侧的留白 |

表 4-5　相对布局取值为 boolean 值的属性及说明

| 属性 | 说明 |
|---|---|
| android:layout_centerHorizontal | 控制该子控件是否位于布局容器的水平居中位置 |
| android:layout_centerVertical | 控制该子控件是否位于布局容器的垂直居中位置 |
| android:layout_centerInparent | 控制该子控件是否位于布局容器的中央位置 |
| android:layout_alignParentBottom | 控制该子控件是否位于布局容器底端对齐 |
| android:layout_alignParentLeft | 控制该子控件是否位于布局容器左边对齐 |
| android:layout_alignParentRight | 控制该子控件是否位于布局容器右边对齐 |
| android:layout_alignParentTop | 控制该子控件是否位于布局容器顶端对齐 |

【例 4.4】设计如图 4-5 所示的布局文件。

【说明】使用相对布局的各类属性控制各个子控件的摆放。

【开发步骤】

（1）创建一个名为 RelativeLayout_test 的项目，包名为 com.hzu.relativelayout_test，Activity 组件名为 MainActivity。

（2）在 res/layout/activity_main.xml 文件中添加如下代码：

```
<RelativeLayout xmlns:android="http://schemas.android.com/apk/res/android"
    xmlns:tools="http://schemas.android.com/tools"
    android:layout_width="match_parent"
    android:layout_height="match_parent">
    <Button
        android:id="@+id/button1"
        android:layout_width="wrap_content"
        android:layout_height="wrap_content"
        android:text="1" />
    <Button
        android:id="@+id/button2"
        android:layout_width="wrap_content"
        android:layout_height="wrap_content"
        android:layout_alignParentBottom="true"
        android:text="2" />
    <Button
        android:id="@+id/button3"
        android:layout_width="wrap_content"
        android:layout_height="wrap_content"
        android:layout_alignParentRight="true"
        android:text="3" />
    <Button
        android:id="@+id/button4"
        android:layout_width="wrap_content"
        android:layout_height="wrap_content"
        android:layout_alignParentBottom="true"
        android:layout_alignParentRight="true"
        android:text="4" />
    <Button
        android:id="@+id/button5"
        android:layout_width="wrap_content"
        android:layout_height="wrap_content"
        android:layout_centerHorizontal="true"
        android:layout_centerVertical="true"
        android:text="5" />
    <Button
        android:id="@+id/button6"
        android:layout_width="wrap_content"
```

```
39            android:layout_height="wrap_content"
40            android:layout_alignTop="@id/button5"
41            android:layout_toLeftOf="@id/button5"
42            android:text="6" />
43        <Button
44            android:id="@+id/button7"
45            android:layout_width="wrap_content"
46            android:layout_height="wrap_content"
47            android:layout_alignLeft="@id/button5"
48            android:layout_below="@id/button5"
49            android:text="7" />
50        <Button
51            android:id="@+id/button8"
52            android:layout_width="wrap_content"
53            android:layout_height="wrap_content"
54            android:layout_alignTop="@id/button5"
55            android:layout_toRightOf="@id/button5"
56            android:text="8" />
57        <Button
58            android:id="@+id/button9"
59            android:layout_width="wrap_content"
60            android:layout_height="wrap_content"
61            android:layout_above="@id/button5"
62            android:layout_alignLeft="@id/button5"
63            android:text="9" />
64    </RelativeLayout>
```

1）第 1～64 行声明了一个相对布局控件。

2）第 5～9 行声明了第一个 Button，默认情况下这个 Button 摆放在屏幕左上角的位置。

3）第 10～15 行声明了第二个 Button，通过 android:layout_alignParentBottom 属性设置其在屏幕左下角显示。

4）第 16～21 行声明了第三个 Button，通过 android:layout_alignParentRight 属性设置其在屏幕右上角显示。

5）第 22～28 行声明了第四个 Button，通过 android:layout_ alignParentBottom 以及 android:layout_alignParentRight 属性设置其在屏幕右下角显示。

6）第 29～35 行声明了第五个 Button，通过 android:layout_ centerHorizontal 以及 android:layout_centerVertical 属性设置其居中显示。

7）第 36～42 行声明了第六个 Button，通过 android:layout_ alignTop="@id/button5"属性使其与 id 为 button5 的 Button 上边界对齐，通过 android:layout_toLeftOf="@id/button5"属性设置其在 id 为 button5 的 Button 左侧显示，后面第七、八、九个 Button 设置方式与第六个 Buton 类似。

**【运行结果】**在 Eclipse 中启动 Android 模拟器，接着运行 RelativeLayout_test 项目，显示效果如图 4-5 所示。

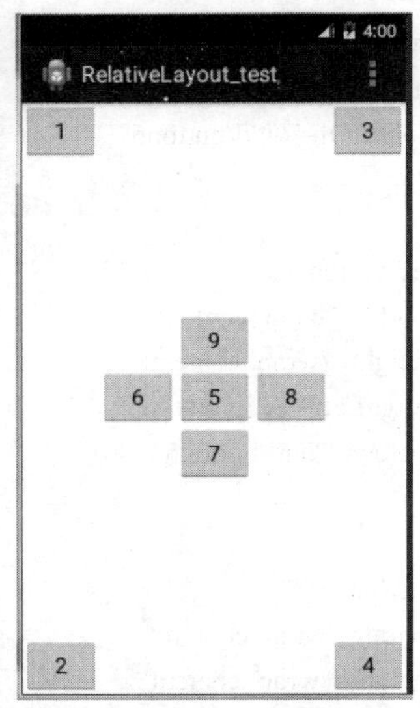


图 4-5 相对布局效果

### 4.2.5 绝对布局

绝对布局（AbsoluteLayout）以坐标形式来指定 View 对象的具体位置。当使用绝对布局时，布局管理器将不再管理子控件的位置和大小，所有这些都由开发者来处理。绝对布局最大的弊端在于位置是定死的，对于不同尺寸的手机，显示效果会出现不一致的情况，甚至可能出现有些控件溢出屏幕不可见的情况。

## 4.3 嵌套布局

在实现一些复杂界面时，使用单个布局并不能达到理想效果，需要使用多种布局或将某类布局嵌套使用，然后根据各类布局的属性，达到效果理想且实现起来简单的布局界面。从布局本身的角度来看，每个子布局都可理解为一个子控件，多个子控件组合就形成了嵌套布局。

**【例 4.5】**设计如图 4-6 所示的嵌套布局文件。

**【说明】**综合使用各种类型的布局管理器，控制各类子控件的摆放。

**【开发步骤】**

（1）创建一个名为 NestificationLayout_test 的项目，包名为 com.hzu.nestificationlayout_test，Activity 组件名为 MainActivity。

（2）将 bg_login.png 等四张图片复制到本项目的 res/ drawable-hdpi 目录下。

（3）在 res/ drawable-hdpi 目录下创建 bt_selector.xml 文件，其代码内容如下：

```
<?xml version="1.0" encoding="utf-8"?>
<selector xmlns:android="http://schemas.android.com/apk/res/android">
<item android:state_pressed="false">
<shape>
<corners android:radius="4dp" />
```

```
6 <solid android:color="#7700CDCD" />
7 </shape>
8 </item>
9
10 <item android:state_pressed="true">
11 <shape>
12 <corners android:radius="4dp" />
13 <solid android:color="#77009ACD" />
14 </shape>
15 </item>
16
17 </selector>
```

1）第 1～17 行为 activity_main.xml 的“注册”与“登录”两个按钮设置未按下与按下两种不同效果。

2）第 3～8 行设置未按下的效果，其中第 3 行表示按钮未按下的状态，第 4 行表示定义各种各样的形状，第 5 行表示按钮的边角弧度，第 6 行表示按钮的填充颜色。

（4）在 res/layout/activity_main.xml 文件中添加如下代码：

```
<RelativeLayout xmlns:android=http://schemas.android.com/apk/res/android
    xmlns:tools="http://schemas.android.com/tools"
    android:id="@+id/rl_first"
    android:layout_width="match_parent"
    android:layout_height="match_parent"
    android:background="@drawable/bg_login">

<LinearLayout
        android:id="@+id/ll_first"
        android:layout_width="match_parent"
        android:layout_height="wrap_content"
        android:layout_marginLeft="20dp"
        android:layout_marginRight="20dp"
        android:layout_marginTop="60dp"
        android:background="#883366cc"
        android:orientation="vertical">

<LinearLayout
            android:layout_width="match_parent"
            android:layout_height="50dp"
            android:orientation="horizontal">

<ImageView
                android:layout_width="30dp"
                android:layout_height="30dp"
                android:src="@drawable/ic_user">
</ImageView>
```

```

<EditText
        android:id="@+id/et_user"
        android:layout_width="0dp"
        android:layout_height="wrap_content"
        android:layout_weight="1"
        android:hint="邮箱/手机号"
        android:textColorHint="#acacac" />
</LinearLayout>

<LinearLayout
    android:layout_width="match_parent"
    android:layout_height="50dp"
    android:orientation="horizontal">

<ImageView
        android:layout_width="30dp"
        android:layout_height="30dp"
        android:src="@drawable/ic_password">
</ImageView>

<EditText
        android:id="@+id/et_pwd"
        android:layout_width="0dp"
        android:layout_height="wrap_content"
        android:layout_weight="1"
        android:ems="10"
        android:hint="密码"
        android:password="true"
        android:textColorHint="#acacac">
</EditText>
</LinearLayout>
</LinearLayout>

<LinearLayout
    android:id="@+id/ll_control"
    android:layout_width="match_parent"
    android:layout_height="wrap_content"
    android:layout_below="@id/ll_first"
    android:layout_marginLeft="26dp"
    android:layout_marginRight="26dp"
    android:layout_marginTop="40dp">

<Button
    android:id="@+id/bt_register"
```

```
73                  android:layout_width="0dp"
74                  android:layout_height="wrap_content"
75                  android:layout_weight="1"
76                  android:background="@drawable/bt_selector"
77                  android:gravity="center"
78                  android:text="注册" />
79
80   <Button
81                  android:id="@+id/bt_login"
82                  android:layout_width="0dp"
83                  android:layout_height="wrap_content"
84                  android:layout_marginLeft="8dp"
85                  android:layout_weight="1"
86                  android:background="@drawable/bt_selector"
87                  android:gravity="center"
88                  android:text="登录" />
89   </LinearLayout>
90
91   <TextView
92          android:layout_width="wrap_content"
93          android:layout_height="wrap_content"
94          android:layout_alignParentBottom="true"
95          android:layout_centerHorizontal="true"
96          android:text="贺州欢迎您"
97          android:textColor="#00B2EE" />
98
99   </RelativeLayout>
```

1）第 1～99 行声明了一个相对布局，内部分为三大部分，第一部分用于布局用户名与密码，第二部分用于布局登录与注册，第三部分用于布局“贺州欢迎您”标语。

2）第 8～60 行声明了一个线性布局，方向为垂直方向，内部又使用了两个线性布局，分别用于显示用户名图片与用户名输入框、密码图片与密码输入框。第 12～14 行分别设置最外层线性布局左边、右边和上边的边距。第 15 行设置最外层线性布局的背景色，第 31 行与第 33 行配合使用，使其 EditText 的宽度占据除去其他控件与左右边距宽度后的所有宽度。

3）第 62～89 行声明了一个线性布局，其内部再添加两个 Button，分别用于注册与登录，通过 android:layout_weight 属性来控制 Button 的宽度，第 76 行与第 86 行分别表示“注册”与“登录”按钮在用户点击与没有点击时的不同效果。

4）第 91～97 行声明了一个 TextView 控件，通过使用 alignParentBottom 与 centerHorizontal 属性使此控件位于屏幕底部中间的位置，第 96 行表示 TextView 默认显示的内容。第 97 行设置字体颜色。

**【运行结果】**在 Eclipse 中启动 Android 模拟器，接着运行 NestificationLayout_test 项目，显示效果如图 4-6 所示。

图 4-6　嵌套布局效果

## 4.4　本章小结

本章首先对 Android 应用程序布局文件的用途做了介绍，然后对帧布局、线性布局、表格布局、相对布局、绝对布局这五种常见布局做了详细介绍，最后介绍了嵌套布局，同时通过案例对各种布局进行了说明。

# 第 5 章　高级控件

（1）了解各类高级控件的使用场景。

（2）掌握 Spinner、ListView 控件的相关属性。

（3）掌握 TabHost、ViewPager 控件的使用方法。

通过第四章的学习，我们了解了基本布局的使用方法，然而当需要在界面上加载大量互联网数据时，基本布局就很难实现，这时使用高级控件便可以轻松解决此问题。

## 5.1　高级控件简介

通过前面四章的学习，我们学会了使用 Android 的一些常用基本控件，以及令这些基本控件按照程序的要求在界面上排列，从而设计出一些简单的界面。如何设计出一些复杂且功能强大的高级控件，如列表视图、滚动视图、进度条等，将在本章中进行详细介绍。

## 5.2　与适配器相关控件

Android 提供了一系列 AdapterView 的子类对象，这些对象可以称为 Adapter（适配器）控件。与其他的 View 控件相比，Adapter 控件通常包含多个格式相同的列表，对于这些列表而言，使用 setText()、setTextColor()等方法将几十到几百行格式相同的内容进行简单的设置是不可取的。所以事先把要加载的内容放入到一个列表中，然后把这个列表放到 Adapter 中对各项资源统一进行设置。这个存放 Adapter 控件的内容列表称为 Adapter（适配器）。Adapter 可以理解为一个显示器，它可以把复杂的数据按人们容易接受的方式进行显示。Android 提供的常用 Adapter 控件有 ListView、Spinner、GridView 等，开发者可以根据自己的需要继承 View 类自定义 Adapter 控件。Android 提供的常用 Adapter 对象有 ArrayAdapter、SimpleAdapter、SimpleCursorAdapter、BaseAdapter 等，开发者可以根据自己的需要继承 Adapter 类自定义 Adapter 的子类。

**注意：**BaseAdapter 是一个抽象类，这个类是不能被实例化的。开发者必须使用 BaseAdapter 类的子类，或者自己定义一个 Adapter。自定义的好处在于可以让数据按开发者自己定义的样式进行显示，缺点是编写代码量大，需要重写 Adapter 类中的各种方法。系统已经定义好了一些 BaseAdapter 子类，它们有着各自的特点。

- ArrayAdapter：它在默认情况下只显示文本信息，如果要显示其他的控件，一般需要重写 getView()方法。通常会将一个数组或集合放在 ArrayAdapter 中。

- SimpleAdapter：它可以将静态的数据关联到 XML 布局文件中的某个 View 控件上，也可以将 List 集合中多个对象包装成多个列表项。
- SimpleCursorAdapter：与 SimpleAdapter 类似，但只用于包装 Cursor 提供的数据。
- BaseAdapter：一般用于扩展，扩展 BaseAdapter 可以对各列表项进行最大限度的定制。

### 5.2.1 AutoCompleteTextView

AutoCompleteTextView控件继承自EditText控件，位于android.widget包下，它具有EditText的所有属性，可以让用户输入内容。它的特点是根据用户输入的少量内容，匹配指定的数据源时，就以列表的形式展示数据源中符合要求的数据内容供用户选择，减少用户的输入内容，提升程序的友好性。AutoCompleteTextView 控件有以下常用属性：

- android:completionThreshold：设置弹出列表时的最少字符个数，即用户在匹配数据源时至少要输入多少个字符才能弹出列表，默认是 2。
- android:completionHint：设置下拉列表下面的说明性文字。
- android:dropDownHeight：设置下拉列表高度。
- android:dropDownWidth：设置下拉列表宽度。
- android: popupBackground：设置下拉列表的背景。
- android:dropDownSelector：设置下拉列表被选中行的背景。
- android:dropDownHorizontalOffset：设置下拉列表与文本框之间的水平偏移像素，默认下拉列表与文本框左对齐。
- android:dropDownVerticalOffset：设置下拉列表与文本框之间的垂直偏移像素，默认下拉列表是紧跟着文本框的。

**【例 5.1】**设计如图 5-1 所示的显示效果。

**【说明】**在下拉列表的文本框中输入一个“贺”字，弹出所有开头为“贺”的选项。

**【开发步骤】**

（1）创建一个名为 AutoCompleteTextView_test 的项目，Activity 组件名为 MainActivity。在 res/layout 目录下的 activity_main.xml 文件内容如下：

```
1   <RelativeLayout xmlns:android="http://schemas.android.com/apk/res/android"
2       xmlns:tools="http://schemas.android.com/tools"
3       android:layout_width="match_parent"
4       android:layout_height="match_parent"
5       android:paddingBottom="@dimen/activity_vertical_margin"
6       android:paddingLeft="@dimen/activity_horizontal_margin"
7       android:paddingRight="@dimen/activity_horizontal_margin"
8       android:paddingTop="@dimen/activity_vertical_margin"
9       tools:context="com.example.autocompletetextview_test.MainActivity">
10  <AutoCompleteTextView
11              android:id="@+id/autotext"
12              android:layout_width="match_parent"
13              android:layout_height="wrap_content" />
14  </RelativeLayout>
```

第 10～13 行声明了一个 AutoCompleteTextView 控件，id 为 autotext。

（2）在 src 下包为 com.hzu.autocompletetextview_test 的 MainActivity.java 文件中编写如下内容：

```
package com.hzu.autocompletetextview_test;
import android.app.Activity;
import android.os.Bundle;
import android.widget.ArrayAdapter;
import android.widget.AutoCompleteTextView;
public class MainActivity extends Activity {
    private static final String[] str = new String[] { "贺州学院游泳馆", "贺州学院博物馆",
            "贺州学院艺术中心", "贺州学院图书馆" };
    @Override
    protected void onCreate(Bundle savedInstanceState) {
        super.onCreate(savedInstanceState);
        setContentView(R.layout.activity_main);
        ArrayAdapter<String> adapter = new ArrayAdapter<String>(this,
                android.R.layout.simple_dropdown_item_1line, str);
        AutoCompleteTextView autotext = (AutoCompleteTextView)
                                        findViewById(R.id.autotext);
        autotext.setAdapter(adapter);
        autotext.setThreshold(1);
    }
}
```

**【运行结果】**在 Eclipse 中启动 Android 模拟器，接着运行 AutoCompleteTextView_test 项目，显示效果如图 5-1 所示。

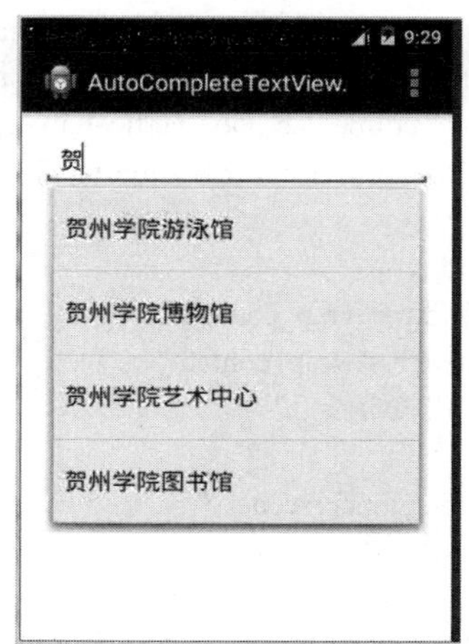


图 5-1　AutoCompleteTextView 效果

### 5.2.2　Spinner

Spinner（下拉列表）控件位于 android.widget 包下，类似于网页中常见的下拉列表框，它主要提供了一系列可供用户选择的列表项，可以减少用户的输入并帮助用户快速找到想要的结果。它的用法与 AutoCompleteTextView 非常相似，都需要指定一个数据源。Spinner 有两种形成数据源的方式：一种是在代码中使用数组或集合来形成数据源；另一种是使用 XML 文件中的<string-array>来形成数据源，然后为 Spinner 指定 android:entries 属性，不需要编写代码即可直接完成下拉列表的功能。Spinner 的常用方法及说明如表 5-1 所示。

表 5-1　Spinner 的常用方法及说明

| 方法 | 说明 |
|---|---|
| setPrompt(String) | 设置下拉列表的提示信息 |
| setSelection(int,boolean) | 设置 Spinner 在初始化自动调用一次 OnItemSelectedListener()事件时的下拉列表项，如果禁用首次调用，则使用 setSelection(0,true) |
| getSelectedItem() | 获取用户下拉列表时选择的数据 |
| getItemAtPosition(int) | 获取下拉列表中指定位置的数据 |

【例 5.2】设计如图 5-2 与图 5-3 所示的显示效果。

【说明】在代码中使用数据源的下拉列表显示效果。

【开发步骤】

（1）创建一个名为 Spinner_test 的项目，Activity 组件名为 MainActivity。在 res/layout 下的 activity_main.xml 文件内容如下：

```
1	<RelativeLayout xmlns:android=http://schemas.android.com/apk/res/android
2	    xmlns:tools=http://schemas.android.com/tools
3	    android:layout_width="match_parent"
4	    android:layout_height="match_parent"
5	    android:paddingBottom="@dimen/activity_vertical_margin"
6	    android:paddingLeft="@dimen/activity_horizontal_margin"
7	    android:paddingRight="@dimen/activity_horizontal_margin"
8	    android:paddingTop="@dimen/activity_vertical_margin"
9	    tools:context="com.hzu.spinner_test.MainActivity">
10	<TextView
11	        android:id="@+id/tv"
12	        android:layout_width="wrap_content"
13	        android:layout_height="wrap_content"
14	        android:text="贺州旅游景点" />
15	<Spinner
16	        android:id="@+id/spinner_scenery"
17	        android:layout_width="fill_parent"
18	        android:layout_height="wrap_content"
19	        android:spinnerMode="dialog"
20	        android:layout_below="@id/tv" />
21	</RelativeLayout>
```

1）第 10～14 声明了一个 TextView，其 id 为 tv。

2）第 15～20 声明了一个 Spinner，其 id 为 spinner_scenery，spinnerMode 负责设置下拉列表弹出样式。

（2）在 src 下包为 com.hzu.spinner_test 的 MainActivity.java 文件中编写如下内容：

```
package com.hzu.spinner_test;
```

```
2
3    import android.app.Activity;
4    import android.os.Bundle;
5    import android.view.View;
6    import android.widget.AdapterView;
7    import android.widget.ArrayAdapter;
8    import android.widget.Spinner;
9    import android.widget.Toast;
10
11   public class MainActivity extends Activity {
12       private static final String[] str = new String[] { "十八水原生态公园", "姑婆山国家森林公园",
13               "玉石林", "黄姚古镇", "客家围屋", "紫云仙境" };
14       private Spinner spinner;
15
16       @Override
17       protected void onCreate(Bundle savedInstanceState) {
18               super.onCreate(savedInstanceState);
19               setContentView(R.layout.activity_main);
20               spinner = (Spinner) findViewById(R.id.spinner_scenery);
21               ArrayAdapter<String> adapter = new ArrayAdapter<String>(this,
22                       android.R.layout.simple_spinner_item, str);
23               adapter.setDropDownViewResource(android.R.layout.simple_spinner_dropdown_item);
24               spinner.setAdapter(adapter);
25               spinner.setSelection(0, true);
26               spinner.setPrompt("请选择：");
27               spinner.setOnItemSelectedListener(new Spinner.OnItemSelectedListener() {
28
29               @Override
30               public void onItemSelected(AdapterView<?> parent, View view,
31                               int position, long id) {
32                               // TODO Auto-generated method stub
33                               Toast.makeText(MainActivity.this, "您选择的景点是：" + str[position],
34                                       Toast.LENGTH_SHORT).show();
35               }
36
37               @Override
38               public void onNothingSelected(AdapterView<?> parent) {
39                     // TODO Auto-generated method stub
40               }
41           });
42       }
43   }
```

1）第 3～9 行引入了一些类，在代码中使用了 Spinner、ArrayAdapter 等对象。

2）第 12 行定义了 Spinner 要加载的数据源。

3）第 17～42 行重写了 onCreate()方法。

4）第 21～22 行声明了一个名为 adapter 的 ArrayAdapter 对象。

5）第 23 行设置了 adapter 将要绑定的 Spinner 对象的下拉列表显示样式。

6）第 24 行为 Spinner 添加了适配器 adapter。

7）第 25 行保证项目第一次运行时不会加载 Spinner 的 OnItemSelectedListener()方法。

8）第 26 行为 Spinner 弹出的列表添加一个列表头。

9）第 27～41 行为 Spinner 添加了 OnItemSelectedListener()监听，重写了 onItemSelected()方法。

10）第 33～34 行为用户点击 Spinner 下拉列表中的某一项时，通过 Toast 显示出来。

**注意**：了解 onItemSelected(AdapterView<?> parent, View view,int position, long id)各个参数的说明。第一个参数 AdapterView 为适配器决定显示的视图类，<?>为适配器内容里的类型；第二个参数为适配器视图中被点击的对象；第三个参数为下拉列表中被选择的位置；第四个参数为用户点击的选项所在行的 id 号。

**【运行结果】**在 Eclipse 中启动 Android 模拟器，接着运行 Spinner_test 项目，显示效果如图 5-2 与图 5-3 所示。

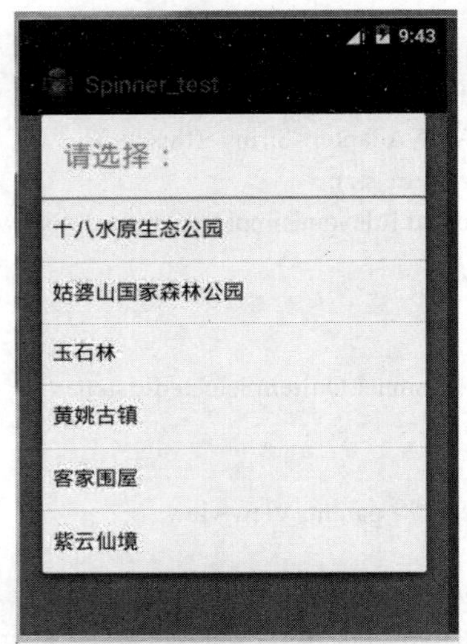


图 5-2　Spinner 效果 1

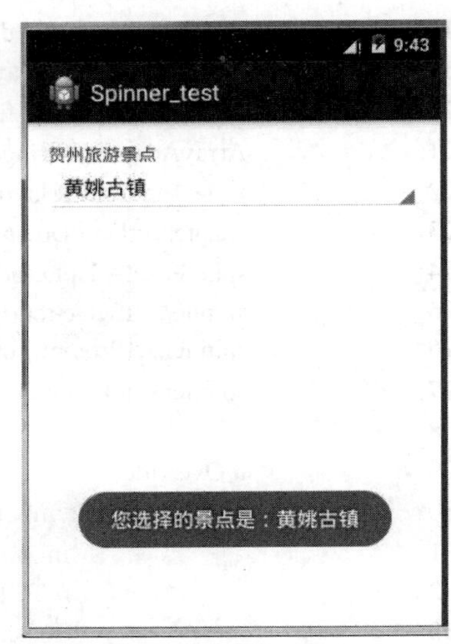


图 5-3　Spinner 效果 2

### 5.2.3　ListView

ListView 是手机系统中使用最广泛的一种控件，它以垂直列表的形式显示所有列表项，例如手机中经常看到的新闻客户端、应用商店等应用程序，都是一个页面能展示多项条目信息，并且每项条目信息的格式都是一样的。实现 ListView 的效果有两种形式：一种是让当前的 Activity 直接继承 ListActivity；另一种是在布局文件中添加一个 ListView，然后为 ListView 设置需要显示的内容（Adapter）。

ListView 是显示内容的地方，Adapter 规定显示内容的样式，Data Source 提供数据源，通过三者的结合可减少编写复杂显示样式的代码，三者关系如图 5-4 所示。

ListView 以列表的形式显示数据内容，并且可以根据数据的长度自适应屏幕来显示，接下来介绍 ListView 的常用属性，如表 5-2 所示。

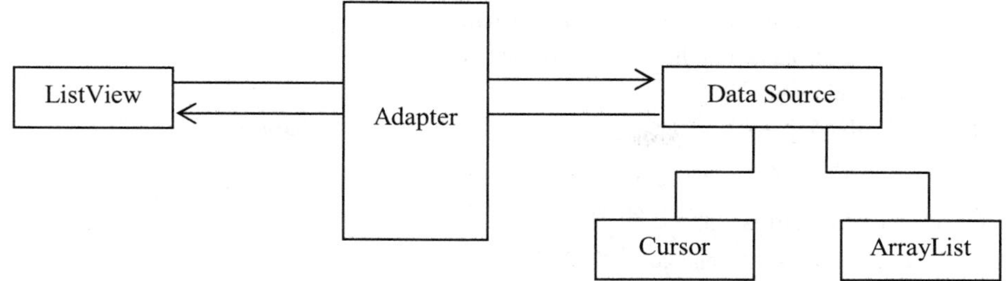


图 5-4　ListView、Adapter 与 Data Source 三者的关系

**表 5-2　ListView 的常用属性**

| 属性 | 说明 |
| --- | --- |
| android:divider | 设置分割线 |
| android:dividerHeight | 设置分割线高度 |
| android:listSelector | 设置 ListView item 选中时的颜色 |
| android:scrollbars | 设置 ListView 的滚动条 |
| android:cacheColorHint | 设置拖动时的背景色 |
| android:fadeScrollbars | 设置为 true 时实现滚动条的自动隐藏和显示 |

1. 使用当前的 Activity 直接继承 ListActivity

ListActivity 是一个自带的 ListView 部分方法的 Activity，在 ListView 控件实例化时不再需要使用 findViewById()方法，而是直接使用 getListView()方法来对 ListView 进行接下来的操作。在获取 ListView 时，布局文件中 ListView 的 id 必须为@android:id/list，其实现比较简单，在此不再详细介绍。

2. 使用布局文件添加一个 ListView

**【例 5.3】**显示贺州各个地区的地名，设计如图 5-5 所示的显示效果。

**【说明】**使用 ArrayAdapter 实现显示效果。

**【开发步骤】**

（1）创建一个名为 ListViewSimple_test 的项目，Activity 组件名为 MainActivity。在 res/layout 下的 activity_main.xml 文件内容如下：

```
<RelativeLayout xmlns:android=http://schemas.android.com/apk/res/android
    xmlns:tools="http://schemas.android.com/tools"
    android:layout_width="match_parent"
    android:layout_height="match_parent"
    android:paddingBottom="@dimen/activity_vertical_margin"
    android:paddingLeft="@dimen/activity_horizontal_margin"
    android:paddingRight="@dimen/activity_horizontal_margin"
    android:paddingTop="@dimen/activity_vertical_margin"
    tools:context="com.hzu.listviewsimple_test.MainActivity">
<ListView
        android:id="@+id/lv"
```

```
12            android:layout_width="wrap_content"
13            android:layout_height="wrap_content" />
14    </RelativeLayout>
```

1）第 1～14 行声明了一个相对布局。

2）第 10～13 行声明了一个 ListView，其 id 为 lv。

（2）在 res/values/strings.xml 文件中编写如下内容：

```
1    <?xml version="1.0" encoding="utf-8"?>
2    <resources>
3    <string name="app_name">ListViewSimple_test</string>
4    <string name="hello_world">Hello world!</string>
5    <string name="action_settings">Settings</string>
6    <string-array name="area">
7    <item>八步区</item>
8    <item>平桂区</item>
9    <item>昭平县</item>
10   <item>钟山县</item>
11   <item>富川瑶族自治县 </item>
12   </string-array>
13   </resources>
```

第 6～12 行声明了一个 string-array，其 name 为 area。

（3）在 res/layout 目录下创建文件 items.xml，其代码内容为：

```
1    <?xml version="1.0" encoding="utf-8"?>
2    <LinearLayout xmlns:android=http://schemas.android.com/apk/res/android
3        android:layout_width="match_parent"
4        android:layout_height="match_parent"
5        android:orientation="vertical">
6    <TextView
7            android:id="@+id/item_tv"
8            android:layout_width="wrap_content"
9            android:layout_height="wrap_content"
10           android:textSize="30sp" />
11   </LinearLayout>
```

1）第 2～11 行声明了一个线性布局。

2）第 6～10 行声明了一个 TextView，其 id 为 item_tv。

（4）在 src 目录下的 com.hzu.listviewsimple_test 包中，MainActivity.java 文件内容如下：

```
package com.hzu.listviewsimple_test;
import android.app.Activity;
import android.os.Bundle;
import android.widget.ArrayAdapter;
import android.widget.ListView;
public class MainActivity extends Activity {
    private String[] areas;
    private ArrayAdapter<String> adapter;
    private ListView mListView;

```

```
11        @Override
12        protected void onCreate(Bundle savedInstanceState) {
13            super.onCreate(savedInstanceState);
14            setContentView(R.layout.activity_main);
15            areas = getResources().getStringArray(R.array.area);
16            adapter = new ArrayAdapter<String>(MainActivity.this,
17                    R.layout.items,
18                    R.id.item_tv, areas);
19            mListView = (ListView) findViewById(R.id.lv);
20            mListView.setAdapter(adapter);
21        }
22    }
```

1）第 2～5 行引入了一些类，在代码中使用了 ArrayAdapter、ListView 等对象。

2）第 7 行声明了一个名为 areas 的数组。

3）第 8 行声明了一个名为 adapter 的适配器。

4）第 9 行声明了一个名为 mListView 的 ListView。

5）第 15 行进行了 areas 数组的赋值。

6）第 16～18 行进行 adapter 适配器的赋值。在 ArrayAdapter<String>(Context context, int resource, int textViewResourceId, String[]objects)中：第一个参数为当前 Activity 对象；第二个参数为实例化视图时规定的每一个条目的显示样式；第三个参数为一个资源 id，该资源代表一个 TextView 控件，该控件将作为 ArrayAdapter 中的一个列表项；第四个参数为 ArrayAdapter 提供的数据源。

7）第 19 行为 mListView 赋值。

8）第 20 行为 mListView 绑定适配器。

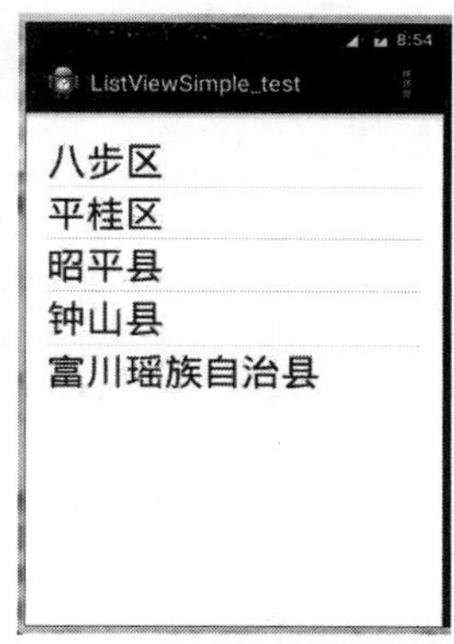


图 5-5　ListView 效果

**【运行结果】**在 Eclipse 中启动 Android 模拟器，接着运行 ListViewSimple_test 项目，显示效果如图 5-5 所示。

**【例 5.4】**设计如图 5-6 所示的贺州旅游新闻列表。

**【说明】**使用 BaseAdapter 实现显示效果。

**【开发步骤】**

（1）创建一个名为 ListViewComplex_test 的项目，Activity 组件名为 MainActivity。

（2）将图片资源复制到本项目的 res/drawable-hdpi 目录下，这组图片的文件名分别为 view01.png～view09.png。

（3）在 res/layout 下的 activity_main.xml 文件内容如下：

```
<RelativeLayout xmlns:android=http://schemas.android.com/apk/res/android
    xmlns:tools="http://schemas.android.com/tools"
    android:layout_width="match_parent"
    android:layout_height="match_parent">
<ListView
        android:id="@+id/listview_News"
        android:layout_width="match_parent"
        android:layout_height="match_parent">
```

```
9       </ListView>
10      </RelativeLayout>
```

第 5～9 行声明了一个 ListView，其 id 为 listview_News。

图 5-6　贺州旅游新闻列表

（4）在 res/layout 下创建一个 news_item.xml 文件，内容如下：

```
<?xml version="1.0" encoding="utf-8"?>
<RelativeLayout xmlns:android="http://schemas.android.com/apk/res/android"
    android:layout_width="match_parent"
    android:layout_height="match_parent">

<ImageView
        android:id="@+id/news_pic"
        android:layout_width="40dp"
        android:layout_height="40dp"
        android:paddingTop="7dp"
        android:src="@drawable/ic_launcher" />

<TextView
        android:id="@+id/news_title"
        android:layout_width="wrap_content"
        android:layout_height="wrap_content"
        android:layout_toRightOf="@id/news_pic"
        android:layout_marginTop="7dp"
        android:text="title"
        android:textSize="10sp" />

<TextView
        android:id="@+id/news_desc"
        android:layout_width="wrap_content"
```

```
25              android:layout_height="wrap_content"
26              android:layout_below="@id/news_title"
27              android:layout_toRightOf="@id/news_pic"
28              android:text="desc"
29              android:textSize="7sp" />
30
31      <TextView
32              android:id="@+id/news_time"
33              android:layout_width="wrap_content"
34              android:layout_height="wrap_content"
35              android:layout_alignParentRight="true"
36              android:paddingTop="7dp"
37              android:text="time"
38              android:textSize="10sp">
39      </TextView>
40      </RelativeLayout>
```

1）第 1～40 行声明了一个相对布局。

2）第 6～11 行声明了一个 ImageView，显示新闻左边的图片。

3）第 13～20 行声明了一个 TextView，其 id 为 news_title，显示新闻的标题。

4）第 22～29 行声明了一个 TextView，其 id 为 news_desc，显示新闻的基本概况。

5）第 31～39 行声明了一个 TextView，其 id 为 news_time，显示新闻发布的时间。

（5）在 src 下的 com.hzu.listviewcomplex_test 包中创建 News.java 文件，内容如下：

```
package com.hzu.listviewcomplex_test;

public class News {

    private int imageId;
    private String title;
    private String desc;
    private String time;

    public News() {
        super();
    }

   public News(int imageId, String title, String desc, String time) {
        super();
        this.imageId = imageId;
        this.title = title;
        this.desc = desc;
        this.time = time;

    }

    public int getImageId() {
        return imageId;
```

```
    }

    public void setImageId(int imageId) {
        this.imageId = imageId;
    }

    public String getTitle() {
        return title;
    }

    public void setTitle(String title) {
        this.title = title;
    }

    public String getDesc() {
        return desc;
    }

    public void setDesc(String desc) {
        this.desc = desc;
    }

    public String getTime() {
        return time;
    }

    public void setTime(String time) {
        this.time = time;
    }
}
```

把要显示的内容封装成一个 Java 类，为后面的处理做准备。

（6）在 src 下的 com.hzu.listviewcomplex_test 包中，MainActivity.java 文件内容如下：

```
package com.hzu.listviewcomplex_test;

import java.util.ArrayList;
import java.util.List;

import android.app.Activity;
import android.content.Context;
import android.os.Bundle;
import android.util.Log;
import android.view.LayoutInflater;
import android.view.View;
import android.view.ViewGroup;
import android.widget.AdapterView;
import android.widget.AdapterView.OnItemClickListener;
import android.widget.BaseAdapter;
```

```
import android.widget.ImageView;
import android.widget.ListView;
import android.widget.TextView;
import android.widget.Toast;

public class MainActivity extends Activity {
    private ListView listview_News;
    private NewsAdapter newsAdapter;
    private int[] imgs = new int[] { R.drawable.view01,
      R.drawable.view02, R.drawable.view03,
      R.drawable.view04, R.drawable.view05,
      R.drawable.view06, R.drawable.view07,
      R.drawable.view08,R.drawable.view09 };

     private String[] title = new String[] {"贺州市步头镇大桂山国家森林公园",
     "贺州市姑婆山大道 9 号","贺州市昭平县黄姚镇", "贺州市莲塘镇客家围屋景区",
     "贺州市平桂区黄田镇路花村十八水景区", "贺州市富川县朝东镇秀水状元村",
     "贺州市八步区黄田镇玉石林景区", "贺州市鹅塘镇栗木村紫云洞景区",
     "贺州市建设东路 180 号温泉" };

     private String[] desc = new String[] {"广西东部著名的旅游景区",
     "一处环境优美的天然大氧吧", "有着近千年历史的古镇",
     "客家围屋建于清乾隆末年", "众多落差、规模很大的瀑布最为著名",
     "出过一位状元和二十六位进士","由汉玉石柱、石笋组成",
     "石灰岩石山溶洞", "桂东南著名的旅游休闲胜地" };

     private String[] time = new String[] {
         "2017-02-23","2016-12-24", "2016-11-26",
         "2016-11-05", "2016-10-24", "2016-09-22",
         "2016-09-12", "2016-08-29", "2016-04-26" };

     @Override
     protected void onCreate(Bundle savedInstanceState) {
         super.onCreate(savedInstanceState);
         setContentView(R.layout.activity_main);
         listview_News = (ListView) findViewById(R.id.listview_News);
         newsAdapter = new NewsAdapter(imgs, title, desc, time, this);
         listview_News.setAdapter(newsAdapter);

         listview_News.setOnItemClickListener(new OnItemClickListener() {

           @Override
           public void onItemClick(AdapterView<?> parent, View view,
           int position, long id) {
           Toast.makeText(
                       MainActivity.this,
```

```
                        "您点击的新闻标题为:"
                        + newsAdapter.news.get(position).getTitle(),
                        Toast.LENGTH_LONG).show();
            }
        });
    }
}

class NewsAdapter extends BaseAdapter {

    public List<News> news;
    private LayoutInflater Inflater;

    public NewsAdapter(int[] images, String[] titles, String[] descs,
    String[] times, Context context) {
        super();
            news = new ArrayList<News>();
            Inflater = LayoutInflater.from(context);
        for (int i = 0; i < images.length; i++) {
            News all = new News(images[i], titles[i], descs[i], times[i]);
            news.add(all);
        }
    }

    @Override
    public int getCount() {
        if (null != news) {
            return news.size();
        } else {
            return 0;
        }
    }

    @Override
    public Object getItem(int position) {
        return news.get(position);
    }

    @Override
    public long getItemId(int position) {
        return position;
    }

    @Override
    public View getView(int position, View convertView, ViewGroup parent) {
        if (convertView == null) {
```

```
108                 convertView = Inflater.inflate(R.layout.news_item, null);
109             }
110
111             TextView news_title = (TextView) convertView.findViewById(R.id.news_title);
112             news_title.setText(news.get(position).getTitle());
113
114             TextView news_desc = (TextView) convertView .findViewById(R.id.news_desc);
115             news_desc.setText(news.get(position).getDesc());
116             TextView news_time = (TextView) convertView.findViewById(R.id.news_time);
117             news_time.setText(news.get(position).getTime());
118             ImageView news_pic = (ImageView) convertView.findViewById(R.id.news_pic);
119             news_pic.setImageResource(news.get(position).getImageId());
120             return convertView;
121         }
122     }
```

1）第 3～19 行引入了代码中需要使用的 Java 与 Android 的相关类。

2）第 21～68 行定义了 MainActivity 类。其中第 24～45 行定义了旅游新闻列表要显示的图片的 id 数组以及新闻标题、新闻概况、新闻发布时间的数组；第 48～67 行重写了 onCreate()方法；第 52 行实例化了 newsAdapter 适配器；第 53 行将 newsAdapter 适配器绑定到 listview_News 上；第 55 行为 listview_News 对象添加了 OnItemClickListener()监听，并重写了 onItemClick()方法。

3）第 70～122 行自定义了一个 NewsAdapter 适配器，让它继承 BaseAdapter。其中第 75～84 行定义了 NewsAdapter 的构造方法，将所有 News 对象封装到 List 中。此外 NewsAdapter 还重写了 getCount()、getItem()、getItemId()、getView()四个方法。

**【运行结果】**在 Eclipse 中启动 Android 模拟器，接着运行 ListViewComplex_test 项目，显示效果如图 5-7 所示，点击某条新闻的显示效果如图 5-8 所示。

图 5-7　运行后的结果

图 5-8　点击后的效果

### 5.2.4 GridView

GridView 是一个可以提供选择的二维选项网格控件，开发者可以控制网格列的宽度与数量，行的数量基于适配器提供的选项数，在保证有效显示的前提下动态确定。GridView 是一种常见的视图控件，一般用于显示多个图片的内容。GridView 的常用属性如表 5-3 所示。

表 5-3 GridView 的常用属性

| 属性 | 说明 |
| --- | --- |
| android:numColumn | 设置 GridView 的列数 |
| android:columnWidth | 设置 GridView 的列宽度 |
| android:stretchMode | 设置 GridView 的缩放模式 |
| android:verticalSpacing | 设置两行之间的间距 |
| android:horizontalSpacing | 设置两列之间的间距 |

**【例 5.5】**设计如图 5-9 所示的贺州景点，点击各个图片可以看到放大的图。

**【说明】**使用 SimpleAdapter 实现显示效果。

**【开发步骤】**

（1）创建一个名为 GridView_test 的项目，Activity 组件名为 MainActivity。

（2）将图片资源复制到本项目的 res/drawable-hdpi 目录下，这组图片的文件名分别为 view01.png～view09.png。

（3）在 res/layout 下的 activity_main.xml 文件内容如下：

```
<LinearLayout xmlns:android="http://schemas.android.com/apk/res/android"
    xmlns:tools="http://schemas.android.com/tools"
    android:layout_width="match_parent"
    android:layout_height="match_parent"
    android:orientation="vertical">

<GridView
        android:id="@+id/gridView"
        android:layout_width="match_parent"
        android:layout_height="wrap_content"
        android:numColumns="3"
        android:horizontalSpacing="1dp"
        android:verticalSpacing="1dp"
        android:gravity="center"
        />
<ImageView
        android:id="@+id/imageView"
        android:layout_marginTop="10dp"
        android:layout_width="250dp"
        android:layout_height="250dp"
        android:layout_gravity="center_horizontal"
```

```
22          />
23   </LinearLayout>
```

1）第 1～23 行声明了一个 LinearLayout 控件，为垂直方向排列。

2）第 7～15 行声明了一个 GridView 控件，其 id 为 gridView，并规定了 GridView 为三列，每行与每列之间的间距都为 1dp。

3）第 16～22 行声明了一个 ImageView 控件，其 id 为 imageView，并规定了与上边控件的间距为 10dp，宽与高分别为 250dp，水平居中对齐。

（4）在 res/layout 下创建一个 one.xml 文件，代码内容如下：

```
1    <?xml version="1.0" encoding="UTF-8"?>
2    <LinearLayout xmlns:android="http://schemas.android.com/apk/res/android"
3        android:orientation="horizontal"
4        android:layout_width="match_parent"
5        android:layout_height="match_parent"
6        android:gravity="center_horizontal"
7        android:padding="1dp">
8    <ImageView
9            android:id="@+id/image_one"
10           android:layout_width="50dp"
11           android:layout_height="50dp"
12           />
13   </LinearLayout>
```

1）第 1～13 行声明了一个线性布局，为水平方向，第 6 行设置内部元素（ImageView）水平居中，第 7 行设置内边距为 1dp。

2）第 8～11 行声明了一个 ImageView 控件，其 id 为 image_one，规定了每个网格中的图片宽与高都为 50dp。

（5）在 src 下的 com.hzu.gridview_test 包中，其 MainActivity.java 文件内容如下：

```
package com.hzu.gridview_test;
import android.app.Activity;
import android.os.Bundle;
import android.view.View;
import android.widget.AdapterView;
import android.widget.AdapterView.OnItemClickListener;
import android.widget.GridView;
import android.widget.ImageView;
import android.widget.SimpleAdapter;

import java.util.ArrayList;
import java.util.HashMap;
import java.util.List;
import java.util.Map;

public class MainActivity extends Activity {
    private GridView grid;
    private ImageView imageView;
    private int[] imgs = new int[] {
```

```
20                  R.drawable.view01, R.drawable.view02, R.drawable.view03,
21                  R.drawable.view04, R.drawable.view05, R.drawable.view06,
22                  R.drawable.view07, R.drawable.view08, R.drawable.view09
23          };
24
25      @Override
26      protected void onCreate(Bundle savedInstanceState) {
27          super.onCreate(savedInstanceState);
28          setContentView(R.layout.activity_main);
29          List<Map<String, Object>> items = new ArrayList<Map<String, Object>>();
30          for (int i = 0; i < imgs.length; i++) {
31              Map<String, Object> map = new HashMap<String, Object>();
32              map.put("image", imgs[i]);
33              items.add(map);
34          }
35
36          imageView = (ImageView) findViewById(R.id.imageView);
37          SimpleAdapter simpleAdapter = new SimpleAdapter(
38                  MainActivity.this,
39                  items,
40                  R.layout.one,
41                  new String[] { "image" },
42                  new int[] { R.id.image_one });
43          grid = (GridView) findViewById(R.id.gridView);
44          grid.setAdapter(simpleAdapter);
45          grid.setOnItemClickListener(new OnItemClickListener() {
46                  @Override
47                  public void onItemClick(AdapterView<?> parent, View view,
48                          int position, long id) {
49                          imageView.setImageResource(imgs[position]);
50                  }
51          });
52      }
53  }
```

1）第 2～14 行引入了代码中需要使用的 Java 与 Android 的相关类。

2）第 16～53 行定义了 MainActivity 类。其中第 19～23 行定义了网格中要显示的图片 id 数组；第 26～52 行重写了 onCreate()方法；第 29 行声明了一个 ArrayList 来存储 Map 对象；第 30～34 行通过 for 循环存储图片资源到 Map 对象。

3）第 37～42 行声明了一个 SimpleAdapter 对象。SimpleAdapter 中第一个参数为当前 Activity 对象；第二个参数是 ArrayList 类型的列表对象，用来向适配器中填充数据；第三个参数定义了每个网格中显示的 XML 文件；第四个参数为一个 String 类型的数组，该数组中的元素确定了 ArrayList 对象中的列；第五个参数为一个 int 类型的数组，该数组中的元素对应第三个参数所指的布局文件中的控件资源 id。

4）第 43 行将 grid 对象绑定到 SimpleAdapter 适配器。

5）第 45～51 行设置了对每个网格的监听，将用户点击的图片放大并在 imageView 上显示。

【运行结果】在 Eclipse 中启动 Android 模拟器，接着运行 GridView_test 项目，显示效果如图 5-9 所示。点击某张图片，显示放大的效果，如图 5-10 所示。

图 5-9　运行后的结果

图 5-10　点击后的效果

## 5.3　其他与视图相关的控件

在屏幕上显示一些复杂且每行格式相同的内容时要用适配器，在显示不复杂但一屏显示不完全的内容时可以使用一些视图控件来处理。

### 5.3.1　ScrollView

ScrollView 是一种可供用户滚动的层次结构布局容器，允许显示比实际多的内容。ScrollView 继承自 FrameLayout，是一种帧布局，这意味着需要在其中放置有自己滚动内容的子元素。子元素可以是一个复杂对象的布局管理器。通常用的子元素是垂直方向的 LinearLayout，显示在最上层垂直方向的内容可以让用户滚动查看。

### 5.3.2　TabHost

TabHost 继承自 FrameLayout，是一种帧布局。它是选项卡的封装类，用于创建选项卡窗口。使用 TabHost 时，Activity 界面的一部分是选项卡，点击某个选项卡就会切换到视图的另一部分并显示其他内容。

在构建选项卡视图时，要用到以下 3 个控件：

- TabHost：用于容纳选项卡标签和选项卡内容。
- TabWidget：用于容纳选项卡标签，每个标签由文本和可选的图标组成。
- FrameLayout：用于容纳选项卡的内容，每块内容都是 FrameLayout 的一个子类。

TabHost 的常用方法如下：

- setUp()：表示当 TabHost 实例不是通过 TabActivity 获取时调用此方法。

- addTab()：表示添加一个 tab 页面。
- newTabSpec()：表示获取一个新的 tab 页面。
- setCurrentTab()：表示设置当前要显示的 tab 页面。
- setOnTabChangedListener()：表示设置当 tab 页面发生改变时的监听。

TabSpec 有以下两个重要方法：

- setContent()：表示设置选项卡要包含的内容，一般要传入相应视图的 android:id。
- setIndicator()：表示设置选项卡按钮的标题。

使用 TabHost 要注意的几个地方如下：

- 在 XML 文件中使用 TabWidget 时，它的 android:id 要设置为@android:id/tabs。
- 如果使用 TabActivity，必须把 TabHost 的 android:id 设置为@android:id/tabhost。
- 如果不使用 TabActivity，那么在 Java 代码中要调用 TabHost 的 addTab()方法之前先调用 setUp()方法。

**【例 5.6】**通过 TabHost 设置新闻、咨询、地图三块内容。

**【说明】**点击选项卡时新闻、咨询、地图三块内容可自由切换。

**【开发步骤】**

（1）创建一个名为 TabHost_test 的项目，Activity 组件名为 MainActivity。

（2）在 res/layout 下的 activity_main.xml 文件内容如下：

```
<RelativeLayout xmlns:android="http://schemas.android.com/apk/res/android"
    xmlns:tools="http://schemas.android.com/tools"
    android:layout_width="match_parent"
    android:layout_height="match_parent"
>

<TabHost
        android:id="@android:id/tabhost"
        android:layout_width="match_parent"
        android:layout_height="match_parent"
        android:layout_alignParentLeft="true"
        android:layout_alignParentTop="true">

<LinearLayout
            android:layout_width="match_parent"
            android:layout_height="match_parent"
            android:orientation="vertical">

<TabWidget
                android:id="@android:id/tabs"
                android:layout_width="match_parent"
                android:layout_height="wrap_content">
</TabWidget>

<FrameLayout
                android:id="@android:id/tabcontent"
```

```
                    android:layout_width="match_parent"
                    android:layout_height="match_parent">

<LinearLayout
                        android:id="@+id/tab1"
                        android:layout_width="match_parent"
                        android:layout_height="match_parent"
                        android:orientation="vertical">

<TextView
                            android:layout_width="match_parent"
                            android:layout_height="match_parent"
                            android:gravity="center"
                            android:textSize="50sp"
                            android:text="旅游新闻"
                            />
</LinearLayout>

<LinearLayout
                        android:id="@+id/tab2"
                        android:layout_width="match_parent"
                        android:layout_height="match_parent"
                        android:orientation="vertical">
<TextView
                            android:layout_width="match_parent"
                            android:layout_height="match_parent"
                            android:gravity="center"
                            android:textSize="50sp"
                            android:text="旅游咨询"
                            />
</LinearLayout>

<LinearLayout
                        android:id="@+id/tab3"
                        android:layout_width="match_parent"
                        android:layout_height="match_parent"
                        android:orientation="vertical">
<TextView
                            android:layout_width="match_parent"
                            android:layout_height="match_parent"
                            android:gravity="center"
                            android:textSize="50sp"
                            android:text="旅游地图"
                            />
</LinearLayout>
</FrameLayout>
</LinearLayout>
```

```
74    </TabHost>
75    </RelativeLayout>
```

1）第 1～75 行声明了一个 RelativeLayout。

2）第 7～74 行声明了一个 TabHost，其 id 为 android:id/tabhost，此处为系统自定义的 id。

3）第 14～73 行声明了一个 LinearLayout，为垂直方向，在它的内部包含一个 TabWidget 与一个 FrameLayout。

4）第 19～23 行声明了一个 TabWidget，其 id 为 android:id/tabs，此处为系统自定义的 id，用于在 Java 代码中动态添加选项卡标签。

5）第 25～72 行声明了一个 FrameLayout，其 id 为 android:id/tabcontent，此处为系统自定义的 id。在此 FrameLayout 布局中添加三个选项卡内容。

（3）在 src 的 com.hzu.tabhost_test 包下 MainActivity.java 文件内容如下：

```
package com.hzu.tabhost_test;

import android.app.Activity;
import android.os.Bundle;
import android.widget.TabHost;
import android.widget.TabHost.OnTabChangeListener;
import android.widget.Toast;

public class MainActivity extends Activity {
    private TabHost myTabHost;

    @Override
    protected void onCreate(Bundle savedInstanceState) {
    super.onCreate(savedInstanceState);
    setContentView(R.layout.activity_main);
    myTabHost = (TabHost) findViewById(android.R.id.tabhost);
    myTabHost.setup();
    myTabHost.addTab(myTabHost.newTabSpec("tab1").setIndicator("新闻").setContent(R.id.tab1));
    myTabHost.addTab(myTabHost.newTabSpec("tab2").setIndicator("咨询").setContent(R.id.tab2));
    myTabHost.addTab(myTabHost.newTabSpec("tab3").setIndicator("地图").setContent(R.id.tab3));
    myTabHost.setOnTabChangedListener(new OnTabChangeListener() {

            @Override
            public void onTabChanged(String tabId) {
                if (tabId == "tab1") {
Toast.makeText(MainActivity.this, "您点击了新闻选项卡", Toast.LENGTH_SHORT).show();
                } else if (tabId == "tab2") {
Toast.makeText(MainActivity.this, "您点击了咨询选项卡",Toast.LENGTH_SHORT).show();
                } else {
Toast.makeText(MainActivity.this, "您点击了地图选项卡",Toast.LENGTH_SHORT).show();
                }
            }
    });
  }
}
```

1）第 3～7 行引入了代码中需要使用的 Java 和 Android 的相关类。

2）第 9～35 行定义了 MainActivity 类，重写了 onCreate()方法，其中第 16 行获取了 TabHost 对象，第 17 行的 setup()方法用于初始化 TabHost 容器。

3）第 18 行在 myTabHost 中添加了第一个名为“新闻”的选项卡，并设置此 Tab 页显示的布局控件。

4）第 19～20 行设置了另外两个选项卡。

5）第 21～33 行为 myTabHost 添加监听事件，当点击不同的选项卡时触发事件，通过 Toast 显示相关信息。

**【运行结果】**在 Eclipse 中启动 Android 模拟器，接着运行 TabHost_test 项目，显示效果如图 5-11 所示。点击顶部“咨询”选项卡时的显示效果如图 5-12 所示。

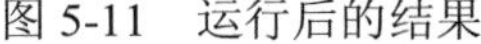

图 5-11　运行后的结果

图 5-12　点击后的效果

### 5.3.3　ViewPager

ViewPager 能够实现最基本的页面左右滑动功能，使用它时和选择控件一样需要适配器。与 TabHost 不同的是 ViewPager 每个页面都可以单独设置，这样就提升了用户体验。

ViewPager 的使用步骤如下：

（1）在布局文件中添加 ViewPager 控件。

（2）初始化要显示的页面。

（3）创建 ViewPager 对象。

（4）将适配器添加到 ViewPager 中。

ViewPager 的常用方法有：

- setCurrentItem()：表示设置当前显示的页面。
- setAdapter()：表示添加适配器。
- setOnPageChangeListener()：表示添加页面切换的监听。

**【例 5.7】** 通过 ViewPager 设置可以滑动的 5 张图片内容。

**【说明】** 滑动时，在靠近底部的位置显示当前是滑动到了第几张图片。

**【开发步骤】**

（1）创建一个名为 ViewPager_test 的项目，Activity 组件名为 MainActivity。

（2）将图片资源复制到本项目的 res/drawable-hdpi 目录下，这组图片的文件名分别为 pic1.png～pic5.png，还有一个蓝色小圆点图片 bule.png 和灰色小圆点图片 gray.png。

（3）在 res/layout 下的 activity_main.xml 文件内容如下：

```
1   <FrameLayout xmlns:android="http://schemas.android.com/apk/res/android"
2       xmlns:tools="http://schemas.android.com/tools"
3       android:layout_width="match_parent"
4       android:layout_height="match_parent">
5
6   <android.support.v4.view.ViewPager
7           android:id="@+id/vp1"
8           android:layout_width="match_parent"
9           android:layout_height="match_parent">
10  </android.support.v4.view.ViewPager>
11
12  <LinearLayout
13          android:id="@+id/tipsBox"
14          android:layout_width="match_parent"
15          android:layout_height="match_parent"
16          android:layout_alignParentBottom="true"
17          android:layout_centerHorizontal="true"
18          android:layout_marginBottom="36dp"
19          android:orientation="horizontal"
20          android:gravity="bottom|center"
21  >
22  </LinearLayout>
23
24  </FrameLayout>
```

1）第 1～24 行声明了一个 FrameLayout 布局。

2）第 6～10 行添加了 ViewPager，因为 ViewPager 是 android 扩展包（v4 包）中的类，所以引入时使用了 android.support.v4.view.ViewPager。

3）第 12～22 行添加了一个线性布局，为切换每一张图片时显示一个小圆点做准备。

（4）在 res/layout 下创建一个 one.xml 文件，内容为：

```
<?xml version="1.0" encoding="utf-8"?>
<LinearLayout xmlns:android="http://schemas.android.com/apk/res/android"
    android:layout_width="match_parent"
    android:layout_height="match_parent"
    android:orientation="vertical">

<ImageView
        android:id="@+id/iv1"
```

```
9              android:layout_width="match_parent"
10             android:layout_height="match_parent"
11             android:src="@drawable/pic1"
12             android:scaleType="centerCrop"/>
13
14     </LinearLayout>
```

1）第 1～14 行声明了一个 LinearLayout 布局，为垂直方向。

2）第 7～12 行声明了一个 ImageView，id 为 iv1，默认显示 pic1.png。通过将 ImageView 的 scaleType 属性设置为 centerCrop 使得图片布满整个屏幕。

（5）在 src 的 com.hzu.viewpager_test 包下 MainActivity.java 文件内容如下：

```
package com.hzu.viewpager_test;

import android.app.Activity;
import android.os.Bundle;
import android.support.v4.view.PagerAdapter;
import android.support.v4.view.ViewPager;
import android.support.v4.view.ViewPager.OnPageChangeListener;
import android.util.Log;
import android.view.View;
import android.view.ViewGroup;
import android.view.ViewGroup.LayoutParams;
import android.widget.ImageView;
import android.widget.LinearLayout;

public class MainActivity extends Activity {
    private int[] images = new int[] { R.drawable.pic1, R.drawable.pic2,
            R.drawable.pic3, R.drawable.pic4, R.drawable.pic5, };
    private PagerAdapter pagerAdapter;
    private ViewPager viewPager;
    public ImageView[] tips;
    private int currentPage = 0;

    @Override
    protected void onCreate(Bundle savedInstanceState) {
        super.onCreate(savedInstanceState);
        setContentView(R.layout.activity_main);
        viewPager = (ViewPager) findViewById(R.id.vp1);
        LinearLayout tipsBox = (LinearLayout) findViewById(R.id.tipsBox);
        tips = new ImageView[5];

     for (int i = 0; i < tips.length; i++) {
            ImageView img = new ImageView(this);        //实例化一个小圆点
            img.setLayoutParams(new LayoutParams(5, 5));
            tips[i] = img;
            if (i == 0) {
                img.setBackgroundResource(R.drawable.bule);     //蓝色背景
```

```
            } else {
                img.setBackgroundResource(R.drawable.gray);      //灰色背景
            }

            LinearLayout.LayoutParams params = new LinearLayout.LayoutParams(
                    new ViewGroup.LayoutParams(LayoutParams.WRAP_CONTENT,
                            LayoutParams.WRAP_CONTENT));
            params.leftMargin = 5;
            params.rightMargin = 5;
            tipsBox.addView(img, params);      //将小圆点添加到容器中
        }
        pagerAdapter = new PagerAdapter() {
            @Override
            public int getCount() {
                return images.length;
            }

            @Override
            public boolean isViewFromObject(View arg0, Object arg1) {
                return arg0 == arg1;
            }

            @Override
            public Object instantiateItem(ViewGroup container, int position) {
                View view = getLayoutInflater().inflate(R.layout.one, null);
                ImageView imageView = (ImageView) view.findViewById(R.id.iv1);
                imageView.setImageResource(images[position]);
                container.addView(view);
                return view;
            }

            @Override
            public void destroyItem(ViewGroup container, int position,
                    Object object) {
                container.removeView((View) object);
            }
        };
        viewPager.setAdapter(pagerAdapter);
        viewPager.setOnPageChangeListener(new OnPageChangeListener() {

            @Override
            public void onPageScrollStateChanged(int arg0) {
                // TODO Auto-generated method stub
            }

            @Override
            public void onPageScrolled(int arg0, float arg1, int arg2) {
```

```
84                  // TODO Auto-generated method stub
85              }
86
87              @Override
88              public void onPageSelected(int arg0) {
89                  Log.i("MainActivity", "您当前选择的是第" + arg0 + "页面");
90                  tips[currentPage].setBackgroundResource(R.drawable.gray);
91                  currentPage = arg0;
92                  tips[arg0].setBackgroundResource(R.drawable.bule);
93              }
94          });
95      }
96  }
```

1）第 3～13 行引入了代码中需要使用的 Java 与 Android 的相关类。

2）第 15～96 行定义了 MainActivity 类，其中第 16～17 行定义要显示的图片的 id 数组。

3）第 24～95 行重写了 onCreate()，其中第 27 行实例化了 viewPager，第 28 行实例化了 tipsBox。

**【运行结果】**在 Eclipse 中启动 Android 模拟器，接着运行 ViewPager_test 项目，显示效果如图 5-13 所示，滑动后的效果如图 5-14 所示。

图 5-13　运行后的结果

图 5-14　滑动后的效果

## 5.4　进度条与滑动块

1．ProgressBar

ProgressBar（进度条）是一种向用户显示进度的展现形式，在 Android 中的样式有多种，例如有长条状的，也有圆形转动的。它展现给用户的进度是不可以被用户随意更改的。ProgressBar 的常用属性如表 5-4 所示。

表 5-4 ProgressBar 的常用属性

| 属性 | 说明 |
|---|---|
| android:max | 设置进度条的最大值 |
| android:progress | 设置第一层进度条的初始值 |
| android:secondaryProgress | 设置第二层进度条的初始值 |

进度条的常用方法：

- getMax()：获取进度条的最大值。
- getProgress()：返回进度条当前的进度。
- getSecondProgress()：返回进度条当前的次要进度。
- incrementProgressBy()：指定增加的进度，即每次推进的程序。

**【例 5.8】**通过点击按钮模拟完成下载效果。

**【说明】**主要通过多线程以及消息处理机制 Handler 来完成下载效果。

**【开发步骤】**

（1）创建一个名为 ProgressBar_test 的项目，Activity 组件名为 MainActivity。

（2）在 res/layout 下的 activity_main.xml 文件内容如下：

```
<RelativeLayout xmlns:android="http://schemas.android.com/apk/res/android"
    xmlns:tools="http://schemas.android.com/tools"
    android:layout_width="match_parent"
    android:layout_height="match_parent"
    android:paddingBottom="@dimen/activity_vertical_margin"
    android:paddingLeft="@dimen/activity_horizontal_margin"
    android:paddingRight="@dimen/activity_horizontal_margin"
    android:paddingTop="@dimen/activity_vertical_margin"
    tools:context="com.hzu.progressbar_test.MainActivity">

<TextView
        android:id="@+id/tv1"
        android:layout_width="wrap_content"
        android:layout_height="wrap_content"
        android:text="当前进度：" />

<ProgressBar
        android:id="@+id/progressBar1"
        style="@android:style/Widget.ProgressBar.Horizontal"
        android:layout_width="wrap_content"
        android:layout_height="wrap_content"
        android:layout_alignLeft="@+id/tv1"
        android:layout_alignParentRight="true"
        android:layout_below="@+id/tv1"
        android:max="100" />

<Button
```

```
28            android:id="@+id/button1"
29            android:layout_width="wrap_content"
30            android:layout_height="wrap_content"
31            android:layout_alignParentLeft="true"
32            android:layout_alignRight="@+id/progressBar1"
33            android:layout_below="@+id/progressBar1"
34            android:text="开始" />
35
36    </RelativeLayout>
```

1）第 11～15 行声明了一个 TextView，其 id 为 tv1。

2）第 17～25 行声明了一个 ProgressBar，其 id 为 progressBar1。其中第 19 行表示使用 Android 自带的主题样式，第 25 行表示设置 ProgressBar 的最大值为 100。

3）第 27～34 行声明了一个 Button，其 id 为 button1，其中第 34 行表示 Button 上默认显示的内容。

（3）在 src 的 com.hzu.progressbar_test 包下 MainActivity.java 文件内容如下：

```
package com.hzu.progressbar_test;

import android.app.Activity;
import android.os.Bundle;
import android.os.Handler;
import android.os.Message;
import android.view.View;
import android.widget.Button;
import android.widget.ProgressBar;
import android.widget.TextView;

public class MainActivity extends Activity {
    private ProgressBar progressBar;
    private Button button;
    private TextView textView;
    private Handler handler;

    @Override
    protected void onCreate(Bundle savedInstanceState) {
        super.onCreate(savedInstanceState);
        setContentView(R.layout.activity_main);
        progressBar = (ProgressBar) findViewById(R.id.progressBar1);
        button = (Button) findViewById(R.id.button1);
        textView = (TextView) findViewById(R.id.tv1);
        handler = new Handler() {
            @Override
            public void handleMessage(Message msg) {
                if (msg.what == 0x0002) {
                    button.setEnabled(true);
                } else if (msg.what == 0x0001) {
```

```
31                         textView.setText("当前进度是：" + progressBar.getProgress() + "%");
32                     }
33                 }
34             };
35
36             button.setOnClickListener(new View.OnClickListener() {
37                 @Override
38                 public void onClick(View v) {
39                     new myThread().start();
40                     button.setEnabled(false);
41                 }
42             });
43
44         }
45
46         private class myThread extends Thread {
47             @Override
48             public void run() {
49                 for (int i = 0; i < 100; i++) {
50                     progressBar.setProgress(i + 1);
51                     handler.sendEmptyMessage(0x0001);
52                     try {
53                         Thread.sleep(300);
54                     } catch (Exception e) {
55                     }
56                 }
57                 handler.sendEmptyMessage(0x0002);
58             }
59         }
60     }
```

1）第 19～44 行重写了 onCreate()方法，其中第 22 行获取了 ProgressBar 对象，第 23 行获取了 Button 对象，第 24 行获取了 TextView 对象，第 25～34 行创建了 Handler 对象，第 28～29 行表示进度条到达 100 后才可以点击“开始”按钮进行操作，第 30～31 行设置了显示进度条的前进速度。

2）第 36～42 行设置了“开始”按钮的监听，其中第 39 行表示启动线程，第 40 行表示设置“开始”按钮不可用。

3）第 46～59 行表示声明了一个 myThread 类，其继承自 Thread 类，其中第 49～56 行表示每 300 毫秒发送一次消息，并使进度条每次增加 1，第 57 行表示进度条加载完成 100%后再发一次消息。

**【运行结果】**在 Eclipse 中启动 Android 模拟器，接着运行 ProgressBar_test 项目，显示效果如图 5-15 所示，点击“开始”按钮后的效果如图 5-16 所示。

2. SeekBar

SeekBar 是 ProgressBar 的扩展，在其基础上增加了一个可拖动的滑块，即允许用户控制进度，一般用于调节音量和亮度等。

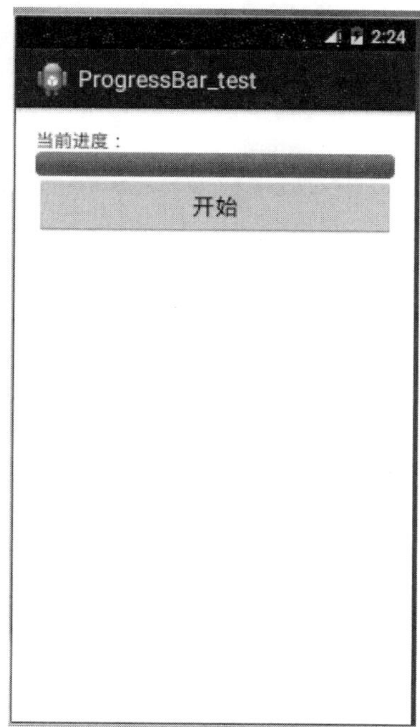


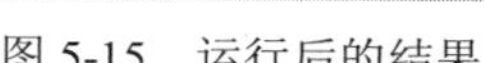
图 5-15 运行后的结果

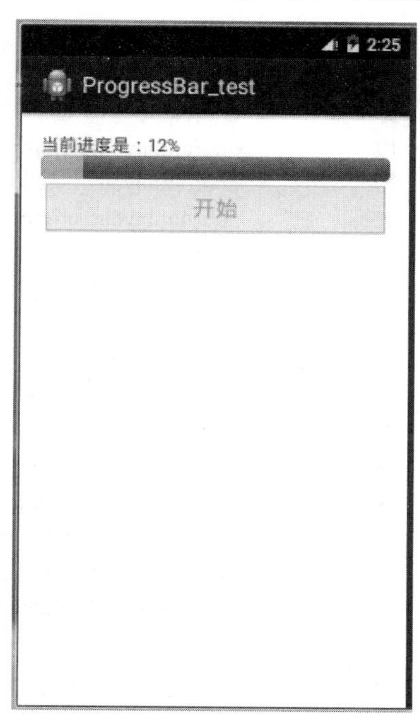


图 5-16 点击后的效果

3. RatingBar

RatingBar 是基于 ProgressBar 和 SeekBar 的扩展，用星形来显示等级评定。RatingBar 与 SeekBar 在用法和功能上都非常相似，它们都允许用户控制进度。

**【例 5.9】**分别通过滑动 SeekBar 和点击 RatingBar 来显示当前进度值。

**【说明】**分别使用 SeekBar 的 setOnSeekBarChangeListener()监听方法与 RatingBar 的 setOnRatingBarChangeListener()监听方法实时获取用户动作对应的值。

**【开发步骤】**

（1）创建一个名为 SeekBar_test 的项目，Activity 组件名为 MainActivity。

（2）在 res/layout 下的 activity_main.xml 文件内容如下：

```
<RelativeLayout xmlns:android="http://schemas.android.com/apk/res/android"
     xmlns:tools="http://schemas.android.com/tools"
     android:layout_width="match_parent"
     android:layout_height="match_parent"
     android:paddingBottom="@dimen/activity_vertical_margin"
     android:paddingLeft="@dimen/activity_horizontal_margin"
     android:paddingRight="@dimen/activity_horizontal_margin"
     android:paddingTop="@dimen/activity_vertical_margin"
     tools:context="com.hzu.seekbar_test.MainActivity">

<TextView
          android:id="@+id/tv1"
          android:layout_width="wrap_content"
          android:layout_height="wrap_content"
          android:text="SeekBar 当前取值为：" />
```

```
16
17  <SeekBar
18          android:id="@+id/sb1"
19          android:layout_width="match_parent"
20          android:layout_height="wrap_content"
21          android:layout_alignLeft="@+id/tv1"
22          android:layout_below="@+id/tv1"
23          android:layout_marginTop="17dp" />
24
25  <TextView
26          android:id="@+id/tv2"
27          android:layout_width="wrap_content"
28          android:layout_height="wrap_content"
29          android:layout_alignLeft="@+id/sb1"
30          android:layout_below="@+id/sb1"
31          android:layout_marginTop="22dp"
32          android:text="RatingBar 当前取值为：" />
33
34  <RatingBar
35          android:id="@+id/rb1"
36          android:layout_width="wrap_content"
37          android:layout_height="wrap_content"
38          android:layout_alignLeft="@+id/tv2"
39          android:layout_below="@+id/tv2"
40          android:max="5"
41          android:rating="1" />
42
43  </RelativeLayout>
```

1）第 1～43 行声明了一个相对布局，其内部有两个 TextView、一个 SeekBar、一个 RatingBar。

2）第 34～41 行声明了一个 RatingBar，其中第 40 行表示评分条最大值，第 41 行表示默认评分值。

（3）在 src 的 com.hzu.seekbar_test 包下 MainActivity.java 文件内容如下：

```
package com.hzu.seekbar_test;

import android.app.Activity;
import android.os.Bundle;
import android.widget.RatingBar;
import android.widget.RatingBar.OnRatingBarChangeListener;
import android.widget.SeekBar;
import android.widget.SeekBar.OnSeekBarChangeListener;
import android.widget.TextView;

public class MainActivity extends Activity {
        private TextView textView1;
        private TextView textView2;
```

```
14 	private SeekBar seekBar;
15 	private RatingBar ratingBar;
16 
17 	@Override
18 	protected void onCreate(Bundle savedInstanceState) {
19 		super.onCreate(savedInstanceState);
20 		setContentView(R.layout.activity_main);
21 		textView1 = (TextView) findViewById(R.id.tv1);
22 		textView2 = (TextView) findViewById(R.id.tv2);
23 		seekBar = (SeekBar) findViewById(R.id.sb1);
24 		ratingBar = (RatingBar) findViewById(R.id.rb1);
25 		seekBar.setOnSeekBarChangeListener(new OnSeekBarChangeListener() {
26 
27 			@Override
28 			public void onProgressChanged(SeekBar seekBar, int progress,
29 					boolean fromUser) {
30 				textView1.setText("当前取值为：" + progress);
31 			}
32 
33 			@Override
34 			public void onStartTrackingTouch(SeekBar seekBar) {
35 				// TODO Auto-generated method stub
36 			}
37 
38 			@Override
39 			public void onStopTrackingTouch(SeekBar seekBar) {
40 				// TODO Auto-generated method stub
41 			}
42 		});
43 
44 		ratingBar.setOnRatingBarChangeListener(new OnRatingBarChangeListener() {
45 			@Override
46 			public void onRatingChanged(RatingBar ratingBar, float rating,
47 					boolean fromUser) {
48 				textView2.setText("当前取值为：" + rating);
49 			}
50 		});
51 	}
52 
53 }
```

1）第 25～42 行为 SeekBar 对象添加了监听器，重写了 onProgressChanged()、onStartTrackingTouch()和 onStopTrackingTouch()三个方法。

2）第 44～50 行为 RatingBar 对象添加了监听器，重写了 onRatingChanged()方法。

**【运行结果】**在 Eclipse 中启动 Android 模拟器，接着运行 SeekBar_test 项目，显示效果如图 5-17 所示，滑动 SeekBar 与点击 RatingBar 的效果如图 5-18 所示。

图 5-17　运行后的结果

图 5-18　用户滑动和点击后的效果

## 5.5　本章小结

本章首先讲解了与适配器相关的高级控件 AutoCompleteTextView、Spinner、ListView、GridView，然后对与之相关的适配器的构建和使用做了说明，最后介绍了 ScrollView、TabHost、ViewPager 控件，并给出了相关实例。

# 第 6 章　菜单与相关控件

（1）了解菜单的基本使用方法。

（2）理解 ActionBar 的基本属性。

（3）掌握 Dialog、Toast 的使用场景。

常见的 Android 手机下面都会有三个触摸功能键，分别是“返回”键、Home 键和“菜单”键，其中当用户触摸到“菜单”键时，就会弹出当前应用程序的菜单。菜单可分为选项菜单、子菜单和上下文菜单。ActionBar 位于应用程序显示界面的顶部，常被用来显示程序的标题、LOGO、搜索功能等。菜单和 ActionBar 提高了应用程序的界面空间使用率。通过本章的学习，读者将会掌握菜单、ActionBar 等应用方法。

## 6.1　菜单

### 6.1.1　菜单简介

程序菜单为 Android 应用程序提供了十分人性化的人机交互界面。程序菜单将不同的应用功能进行分类管理，极大地提高了程序的可操作性，优化了用户体验。本节将讨论菜单的三种类型，分别是选项菜单（OptionMenu）、子菜单（SubMenu）和上下文菜单（ContextMenu）。

### 6.1.2　选项菜单

当前智能手机的底端一般都有三个触摸键，其中有一个触摸键是“菜单”键，虽然不同的手机制造商生产的手机“菜单”键位置不同，但它们的功能是一样的。当用户触摸到“菜单”键时，拥有菜单选项功能的程序会在程序底部弹出菜单项以供用户选择，这弹出的菜单就是选项菜单（OptionMenu）。

1．选项菜单中常用的回调方法和说明

（1）public boolean onCreateOptionsMenu(Menu menu)。它是菜单在初始化的时候所使用的方法。菜单项的添加操作均可在此方法里面实现。其中参数 menu 是一个菜单（Menu）对象，这个对象可以添加多个菜单项（MenuItem）。

（2）public boolean onOptionItemSelected(MenuItem item)。当菜单中的某一项被选中时将调用此方法。

（3）public boolean onPrepareOptionsMenu(Menu menu)。每次显示选项菜单都会调用此方法。在这个方法中可以实现菜单项的修改，或者定义菜单是否可用。

（4）public void onOptionsMenuClosed(Menu menu)。当用户选择了某个菜单项或点击了“返回”键，亦或菜单被关闭的时候，将调用此方法。

2. 选项菜单常用的方法和说明

MenuItem add(int groupId,int itemId,int order,CharSequence title)，其功能是向菜单中添加一个菜单项，返回一个菜单项对象。参数说明：groupId 表示菜单项所在组的 ID；itemId 表示菜单项 ID；order 表示菜单项顺序；title 表示菜单项的标题，其中菜单项的标题也可以用文本资源符方式表示，若使用文本资源符则需重写 MenuItem add(int groupId,int itemId,int order,int titleRes)方法。需要注意的是，添加菜单项操作既可以使用此方法来实现，也可以在 res/menu 文件夹下的 main.xml 文件里面添加菜单项，并在 onCreateOptionsMenu()方法中采用“getMenuInflater().inflate(R.menu.main, menu);”语句来实现。

**【例 6.1】**开发一款关于贺州景点导游的手机 App。假定这款软件在运行过程中，如果用户点击手机“菜单”键，则可以弹出“景区”“住宿”“美食”“交通”“设置”等菜单项供游客使用。当点击这些菜单项之后，文本控件显示用户选择的内容。

**【说明】**实现这些功能需重写 Activity 中的 onCreateOptionMenu(menu)方法；要在菜单中添加五个菜单项，则需要调用菜单中的 add(groupId,int itemId,int order,CharSequence title)方法；如果实现菜单项的点击，可以给每个菜单项添加一个 OnMenuItemClickListener 监听器，当然也可以不使用这个监听器，而重写 onOptionsItemSelected()方法。

**【开发步骤】**

（1）创建一个名为 Sample_OptionMenu 的项目，Activity 组件名为 OptionMenuActivity。

（2）前台布局文件。在应用程序的目录结构中定位到 res/layout 目录，编辑该目录下的 activity_option_menu.xml 文件，文件里面有系统初始的代码，可有选择地使用或修改。

```
1    <RelativeLayout xmlns:android="http://schemas.android.com/apk/res/android"
2        xmlns:tools="http://schemas.android.com/tools"
3        android:layout_width="match_parent"
4        android:layout_height="match_parent"
5        android:paddingBottom="@dimen/activity_vertical_margin"
6        android:paddingLeft="@dimen/activity_horizontal_margin"
7        android:paddingRight="@dimen/activity_horizontal_margin"
8        android:paddingTop="@dimen/activity_vertical_margin"
9        tools:context="cn.gx.hzu.optionmenu.OptionMenuActivity">
10
11   <TextView
12           android:id="@+id/textView1"
13           android:layout_width="wrap_content"
14           android:layout_height="wrap_content"
15           android:text="这里显示选择的内容..." />
16
17   </RelativeLayout>
```

第 11～15 行是一个文本控件，它用来显示用户选中菜单项的内容。

（3）后台代码编写。在应用程序的目录结构中定位到 cn.gx.hzu.optionmenu 包，打开 OptionMenuActivity.java 文件，并根据其中的初始代码完成如下编码工作：

```
package cn.gx.hzu.optionmenu;
```

```

import android.app.Activity;
import android.os.Bundle;
import android.view.Menu;
import android.view.MenuItem;
import android.view.MenuItem.OnMenuItemClickListener;
import android.widget.TextView;

public class OptionMenuActivity extends Activity {

    TextView tv;
    //给每一个菜单项设置一个 ID 常量
    private static final int ITEM1 = Menu.FIRST;
    private static final int ITEM2 = Menu.FIRST+1;
    private static final int ITEM3 = Menu.FIRST+2;
    private static final int ITEM4 = Menu.FIRST+3;
    private static final int ITEM5 = Menu.FIRST+4;

    @Override
    protected void onCreate(Bundle savedInstanceState) {
        super.onCreate(savedInstanceState);
        setContentView(R.layout.activity_option_menu);
        tv = (TextView)findViewById(R.id.textView1);
    }

    @Override
    public boolean onCreateOptionsMenu(Menu menu) {
        //添加五个菜单项
        MenuItem spotitem = menu.add(0, ITEM1, 0, "景区");
        MenuItem hotelitem = menu.add(0, ITEM2, 0, "住宿");
        MenuItem fooditem = menu.add(0, ITEM3, 0, "美食");
        MenuItem trafficitem = menu.add(0, ITEM4, 0, "交通");
        MenuItem settingitem = menu.add(0, ITEM5, 0, "设置");

        OnMenuItemClickListener lsn = new OnMenuItemClickListener(){
            @Override
            public boolean onMenuItemClick(MenuItem arg0) {
                // TODO Auto-generated method stub
                switch(arg0.getItemId()){
                case ITEM1:tv.setText("你点击了景区");break;
                case ITEM2:tv.setText("你点击了住宿");break;
                case ITEM3:tv.setText("你点击了美食");break;
                case ITEM4:tv.setText("你点击了交通");break;
                case ITEM5:tv.setText("你点击了设置");break;
                }
                return true;
```

```
48                    }
49                };
50                spotitem.setOnMenuItemClickListener(lsn);
51                hotelitem.setOnMenuItemClickListener(lsn);
52                fooditem.setOnMenuItemClickListener(lsn);
53                trafficitem.setOnMenuItemClickListener(lsn);
54                settingitem.setOnMenuItemClickListener(lsn);
55                return true;
56        }
57    }
```

1）第 1 行是包名。

2）第 3～8 行引入相关类。

3）第 14～18 行定义菜单项的 ID。

4）第 28～56 行重写了 onCreateOptionsMenu()方法，在这个方法里面添加了五个菜单项。由于本例暂未考虑到分组和排序的问题，所以暂将 groupId 和 order 参数值都设置为 0。其中第 36～49 行定义了 onMenuItemClickListener 监听器，第 50～54 行把监听器安装到每一个菜单项中。

**【运行结果】**项目运行效果如图 6-1 所示。

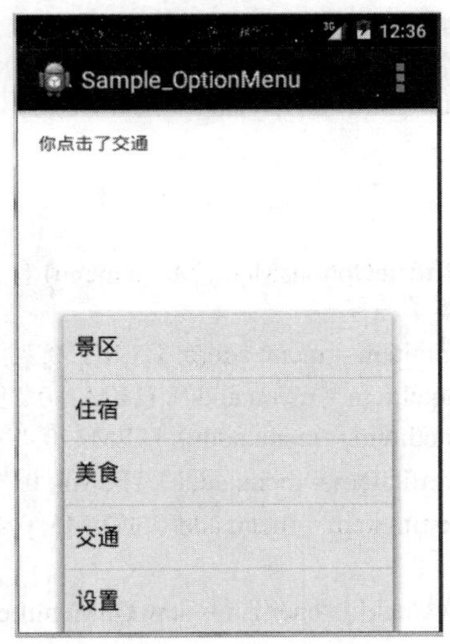


图 6-1　选项菜单

### 6.1.3　子菜单

子菜单（SubMenu）通常和选项菜单一起使用。子菜单可以看作是选项菜单的下一级操作功能。子菜单的添加和菜单项相比既有相同的地方也有不同的地方。

下面我们介绍如何添加子菜单。子菜单（SubMenu）是菜单类（Menu）的子类。所以子菜单（SubMenu）类继承了菜单（Menu）类中所有方法。

添加一个子菜单与添加一个菜单项的方法类似。添加子菜单的方法为

SubMenu addSubMenu(int groupId,int itemId,int order,int titleRes)

或者

SubMenu addSubMenu(int groupId,int itemId,int order,ChardSequences title)

添加子菜单的方法参数与添加菜单项的方法参数是一样的，可参考 6.1.2 小节中的内容。

**【例 6.2】**现在我们开发一款关于贺州景点导游的手机 App。假定该项目在运行过程中，如果用户点击手机菜单键，则可以弹出“景区”“住宿”“美食”等子菜单，当点击子菜单之后就可以弹出子菜单下的菜单项，例如点击景区子菜单，则弹出“姑婆山”“十八水”和“黄瑶古镇”等菜单项供游客使用。当点击这些菜单项之后，文本控件显示用户选择的内容。

**【说明】**实现这些功能需重写 Activity 里面的 onCreateOptionMenu(menu)方法，如果要在菜单中添加三个子菜单，则需要调用添加子菜单的方法 addSubMenu(groupId,int itemId,int order, CharSequence title)；如果还要实现菜单项点击事件可以给每个菜单项添加一个 OnMenuItem ClickListener 监听器，当然也可以不使用这个监听器，而是重写 onOptionsItemSelected()方法。为了便于跟例 6.1 形成比较，这里使用重写 onOptionsItemSelected()方法来实现。

**【开发步骤】**

（1）创建一个名为 Sample_SubMenu 的项目，Activity 组件名为 MainActivity。

（2）前台布局文件。在应用程序的目录结构中定位到 res/layout 目录，编辑目录下面的 activity_main.xml 文件，文件里面有系统初始的代码，可有选择性地使用或修改。

```
1   <RelativeLayout xmlns:android="http://schemas.android.com/apk/res/android"
2       xmlns:tools="http://schemas.android.com/tools"
3       android:layout_width="match_parent"
4       android:layout_height="match_parent"
5       android:paddingBottom="@dimen/activity_vertical_margin"
6       android:paddingLeft="@dimen/activity_horizontal_margin"
7       android:paddingRight="@dimen/activity_horizontal_margin"
8       android:paddingTop="@dimen/activity_vertical_margin"
9       tools:context="cn.gx.hzu_submenu.MainActivity">
10
11  <TextView
12          android:id="@+id/tv_Show"
13          android:layout_width="wrap_content"
14          android:layout_height="wrap_content"
15          android:text="显示选择项......" />
16
17  </RelativeLayout>
```

第 11～15 行是一个文本控件，它用来显示用户选中的菜单项内容。

（3）后台代码编写。在应用程序的目录结构中定位到 cn.gx.hzu_submenu 包下，接着打开 MainActivity.java 文件，并根据其中的初始代码完成如下编码工作：

```
package cn.gx.hzu_submenu;

import android.app.Activity;
import android.os.Bundle;
import android.view.Menu;
import android.view.MenuItem;
import android.view.SubMenu;
```

```
import android.widget.TextView;

public class MainActivity extends Activity {
    //定义组的 ID
    final int GROUP_FIRST=100;      //菜单总组的 ID
    final int GROUP_SPOT=0;         //景点组 ID
    final int GROUP_HOTEL=1;        //酒店组 ID
    final int GROUP_FOOD=2;         //美食组 ID
    //定义菜单项的 ID
    final int ITEM_SPOT1=3;         //景点 1ID
    final int ITEM_SPOT2=4;         //景点 2ID
    final int ITEM_SPOT3=5;         //景点 3ID
    final int ITEM_HOTEL1=6;        //酒店 1ID
    final int ITEM_HOTEL2=7;        //酒店 2ID
    final int ITEM_HOTEL3=8;        //酒店 3ID
    final int ITEM_FOOD1=9;         //美食 1ID
    final int ITEM_FOOD2=10;        //美食 2ID
    final int ITEM_FOOD3=11;        //美食 3ID

    TextView tv_Show;
    @Override
    protected void onCreate(Bundle savedInstanceState) {
        super.onCreate(savedInstanceState);
        setContentView(R.layout.activity_main);
        tv_Show = (TextView)findViewById(R.id.tv_Show);
    }

    @Override
    public boolean onCreateOptionsMenu(Menu menu) {
    SubMenu subMenuSpot,subMenuHotel,subMenuFood;             //声明三个子菜单
    MenuItem menuItemSpot1,menuItemSpot2,menuItemSpot3;       //声明景区菜单项
    MenuItem menuItemHotel1,menuItemHotel2,menuItemHotel3;    //声明酒店菜单项
    MenuItem menuItemFood1,menuItemFood2,menuItemFood3;       //声明美食菜单项

 subMenuSpot = menu.addSubMenu(GROUP_FIRST,GROUP_SPOT,0,"景区");     //景区子菜单
 subMenuHotel = menu.addSubMenu(GROUP_FIRST,GROUP_HOTEL,1,"酒店"); //酒店子菜单
 subMenuFood = menu.addSubMenu(GROUP_FIRST,GROUP_FOOD,2,"美食");    //美食子菜单

    //初始化景区菜单项
    menuItemSpot1 = subMenuSpot.add(GROUP_SPOT, ITEM_SPOT1, 0, "姑婆山");
    menuItemSpot2 = subMenuSpot.add(GROUP_SPOT, ITEM_SPOT2, 1, "十八水");
    menuItemSpot3 = subMenuSpot.add(GROUP_SPOT, ITEM_SPOT3, 2, "黄瑶古镇");

    //初始化酒店菜单项
menuItemHotel1 = subMenuHotel.add(GROUP_HOTEL, ITEM_HOTEL1, 0, "贺州国际大酒店");
menuItemHotel2 = subMenuHotel.add(GROUP_HOTEL, ITEM_HOTEL2, 1, "贺州维也纳大酒店");
```

```
54  menuItemHotel3 = subMenuHotel.add(GROUP_HOTEL, ITEM_HOTEL3, 2, "利源酒店");
55
56          //初始化美食菜单项
57          menuItemFood1 = subMenuFood.add(GROUP_FOOD, ITEM_FOOD1, 0, "羊肠酸");
58          menuItemFood2 = subMenuFood.add(GROUP_FOOD, ITEM_FOOD2, 1, "信都红瓜子");
59          menuItemFood3 = subMenuFood.add(GROUP_FOOD, ITEM_FOOD3, 2, "黄田扣肉");
60
61                  return true;
62              }
63
64              @Override
65              public boolean onOptionsItemSelected(MenuItem item) {
66                  tv_Show.setText(item.getTitle());
67                  return true;
68              }
69
70  }
```

1）第 12～15 行定义了菜单组的 ID。本例使用了三个子菜单，所以对应定义了三个菜单组。但是所有的子菜单又同属一个总菜单，所以也对应定义了一个总菜单 ID。

2）第 36～62 行重写了 onCreateOptionsMenu 方法。在这个方法里面声明了三个子菜单，也声明了九个菜单项。每个子菜单下面有三个菜单项。

3）第 65～68 行重写了 onOptionsItemSelected 方法，定义了点击某个菜单项的响应事件。

**【运行结果】**运行项目显示效果如图 6-2 所示，点击图 6-2 中的“景区”菜单的显示效果如图 6-3 所示。

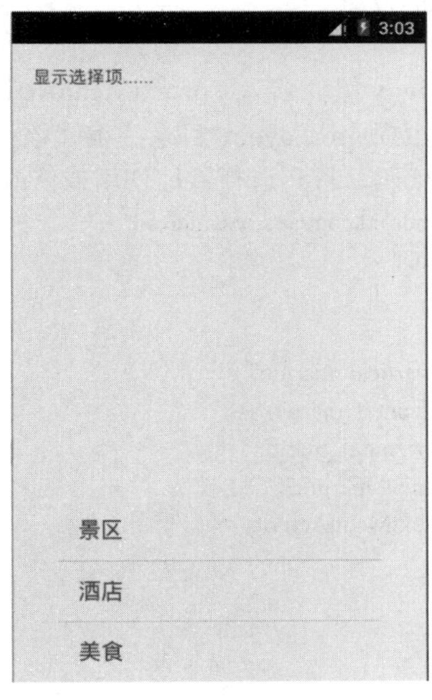

图 6-2　子菜单

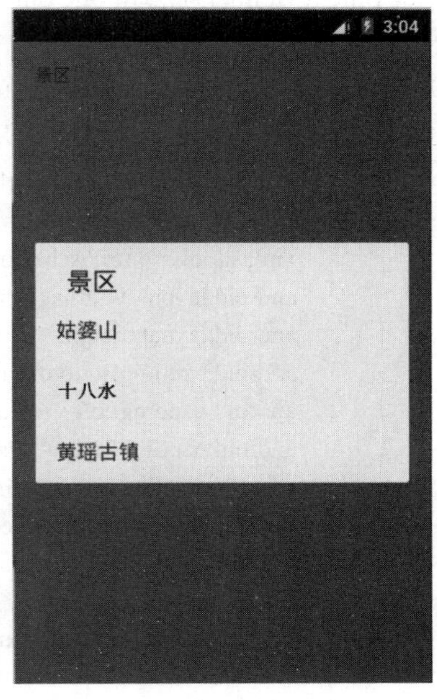


图 6-3　子菜单里面的菜单项

### 6.1.4 上下文菜单

上下文菜单（ContextMenu）也是Menu的一个子类。与选项菜单不同的是，上下文菜单不具有独立的操作功能，它必须依托于某一个视图（View）控件。当用户长按这个视图控件的时候，才会弹出上下文菜单。上下文菜单的创建离不开onCreateContextMenu()方法，也在该方法中完成对上下文菜单中的菜单项初始化操作。

上下文菜单被创建后，还需要把它注册到一个具体的视图控件上面，才能让上下文菜单生效。当用户点击一个上下文菜单中的菜单项时，将调用onContextItemSelected()方法来响应用户的点击事件。与上下文菜单相关的方法有以下3个：

- public void onCreateContextMenu(ContextMenu menu, View v, ContextMenuInfo menuInfo)：menu表示我们当前所创建的上下文菜单；v表示被上下文菜单注册到的视图控件；menuInfo表示上下文菜单的相关显示信息。
- public boolean onContextItemSelected(MenuItem item)：item表示被用户选中的上下文菜单中的菜单项。当用户选中上下文菜单中的某个菜单项时，则调用这个方法。
- public void registerForContextMenu(View view)：view表示上下文菜单需要注册到这个视图控件。只有将上下文菜单注册到该视图控件上，才能使用户长按这个控件时弹出上下文菜单。

**【例6.3】**假定我们开发一张有关图片浏览的应用程序。长按页面上的图片，程序弹出“收藏”和“分享”两个菜单项，分别点击这两个菜单项之后，系统会提示“收藏成功”或“分享成功”的信息。

**【说明】**这个功能需要用到上下文菜单（ContextMenu），实现这个功能需重写Activity中的onCreateContextMenu()方法和onContextItemSelected()方法。

**【开发步骤】**

（1）创建一个Sample_ContextMenu的项目，Activity组件名为ContextMenuActivity。

（2）前台布局文件。在应用程序的目录结构中定位到res/layout目录，编辑该目录下的activity_context_menu.xml文件，文件中有系统初始的代码，可有选择性地使用或修改。

```
<RelativeLayout xmlns:android="http://schemas.android.com/apk/res/android"
        xmlns:tools="http://schemas.android.com/tools"
        android:layout_width="match_parent"
        android:layout_height="match_parent"
        android:paddingBottom="@dimen/activity_vertical_margin"
        android:paddingLeft="@dimen/activity_horizontal_margin"
        android:paddingRight="@dimen/activity_horizontal_margin"
        android:paddingTop="@dimen/activity_vertical_margin"
        tools:context="cn.gx.hzu_contextmenu.ContextMenuActivity">

<TextView
            android:id="@+id/textView1"
            android:layout_width="wrap_content"
            android:layout_height="wrap_content"
            android:text="精彩瞬间" />
```

```
16
17  <ImageView
18          android:id="@+id/imageView1"
19          android:layout_width="wrap_content"
20          android:layout_height="wrap_content"
21          android:layout_below="@+id/textView1"
22          android:layout_marginTop="20dp"
23          android:src="@drawable/pic"
24          android:contentDescription="@string/app_name"/>
25
26  </RelativeLayout>
```

第 17～24 行是一个图像视图控件，上下文菜单就是要依托这个控件。

（3）后台代码编写。在应用程序的目录结构中定位到 cn.gx.hzu_contextmenu 包，打开包中的 ContextMenuActivity.java 文件，并根据文件的初始代码完成如下编码工作：

```
package cn.gx.hzu_contextmenu;

import android.app.Activity;
import android.os.Bundle;
import android.view.ContextMenu;
import android.view.ContextMenu.ContextMenuInfo;
import android.view.Menu;
import android.view.MenuItem;
import android.view.View;
import android.widget.ImageView;
import android.widget.Toast;

public class ContextMenuActivity extends Activity {

    ImageView iv;
    final int MENU1 = 1;
    final int MENU2 = 2;
    @Override
    protected void onCreate(Bundle savedInstanceState) {
        super.onCreate(savedInstanceState);
        setContentView(R.layout.activity_context_menu);
        iv =(ImageView)findViewById(R.id.imageView1);
        this.registerForContextMenu(iv);
    }

    @Override
    public void onCreateContextMenu(ContextMenu menu, View v,
            ContextMenuInfo menuInfo) {
        // TODO Auto-generated method stub
        super.onCreateContextMenu(menu, v, menuInfo);
        if(v==iv){
            menu.add(0,MENU1,0,"收藏");
```

```
33                  menu.add(0,MENU2,0,"分享");
34              }
35          }
36
37          @Override
38          public boolean onContextItemSelected(MenuItem item) {
39              // TODO Auto-generated method stub
40              if(item.getItemId()==MENU1){
41                  Toast.makeText(this, "你已经将这个图片收藏到收藏夹了",
42                  Toast.LENGTH_LONG).show();
43              }else if(item.getItemId()==MENU2){
44                  Toast.makeText(this, "你已经将这个图片分享到朋友圈了",
45                  Toast.LENGTH_LONG).show();
46              }
47              return true;
48          }
49      }
```

1）第 16～17 行定义菜单项的 ID。

2）第 22 行根据配置文件初始化图像视图控件。

3）第 23 行将上下文菜单注册到图像视图控件。

4）第 27～35 行重写了 onCreateContextMenu()方法。该方法上添加了“收藏”和“分享”两个菜单项。如第 31 行所示，在添加这两个菜单项之前需要判断当前长按的控件是不是我们想要的图像控件，即比较两者 ID。在添加菜单项时同样使用了“add(groupId,itemId,order,title);”语句，本例暂没有考虑分组和排序问题，所以暂将这两个参数置为 0。

5）第 38～48 行重写了 onContextItemSelected()方法。该方法参数是一个菜单项对象，在响应事件之前需要判断当前点击的菜单项是哪一项。如第 40 行所示，如果当前点击的是第一个菜单项，则提示已经收藏成功。如第 43 行所示，如果当前点击的是第二个菜单项，则提示已经分享成功。

【运行结果】运行项目显示效果如图 6-4 所示。

图 6-4　上下文菜单

# 6.2　ActionBar

## 6.2.1　ActionBar 简介

ActionBar 的中文名称是动作栏，它为用户提供了一种导航模式，标识应用程序的位置。大多数的手机应用程序在运行的过程中，其界面最上面都会有一块区域来显示程序的图标、标题和登录的用户头像等，而这一块区域正是动作栏（ActionBar）显示区。动作栏在应用程序不同的操作界面切换过程中具有一定的稳定性，正是这种显示稳定的优点，使其可用来标示应用程序的属性和用户位置。开发人员可以直接将应用程序的菜单项作为动作项，显示在动作栏中，也可将动作视图放置在动作栏中，比如常用的搜索等功能。与动作项无关的其他选项菜单都被折叠在 overflow 中。只有点击菜单键或动作栏中的 overflow 按钮才会显示被折叠的选项菜单。

## 6.2.2　ActionBar 的创建与使用

我们多数的 Activity 的风格都默认继承了 Theme.Holo 主题，而从 Android 5.0 开始 Theme.Holo 主题已经把 ActionBar 包含在其中，即我们新建 Android 项目时已经默认拥有 ActionBar，因此 ActionBar 不需要开发人员手动创建。既然 ActionBar 是 Android 项目默认有的，在一些特定情况下，怎样才能把 ActionBar 移除或者不显示 ActionBar 呢？有下面两种常用方法来解决此问题。

第一种方法是在 AndroidManiffest.xml 中修改对应 Activity 的主题。把默认的主题修改为：<activity android:theme:"@android:style/Theme.Holo.NoActionBar">。因为各个主题的显示效果不同，在 Theme.Holo.NoActionBar 这种主题中并没有包括 ActionBar。打开 theme_holo.xml 文件可以看到关于 Theme.Holo.NoActionBar 的描述。

```
<style name="Theme.Holo.NoActionBar">
<item name="windowActionBar">false</item>
<item name="windowNoTitle">true</item>
</style>
```

第二种方法是在后台 Activity 代码中的 onCreate()方法中将动作栏隐藏，即在 onCreate()方法中添加如下代码：getActionBard().hide()，这样即可达到效果。

## 6.2.3　ActionBar 的不同样式

### 1. 把选项菜单项作为动作项添加到 ActionBar 中

如果用户访问某个菜单项时需要点击菜单键，系统就可弹出选项菜单的菜单项，或者点击动作栏中的 overflow 按钮，系统也可弹出选项菜单的菜单项。有时候用户想直接访问菜单项，这时可以考虑将菜单项作为动作项添加到动作栏中。6.1 节中讲过选项菜单的菜单项有两种添加方法。

一种是在 onCreateOptionMenu()方法中调用 Menu 对象中的 add()方法。在这种情况下，如果要将菜单项作为动作项添加到动作栏中，需要调用 MenuItem 对象的 setShowAsAction()方法。

setShowAsAction(int actionEnum)方法用于设置是否将该菜单栏显示在 ActionBar 上，作为动作项。参数 actionEnum 是菜单项（MenuItem）中的静态属性。actionEnum 的可选值有以下几种类型：

- SHOW_AS_ACTION_ALWAYS：总是将该 MenuItem 显示在 ActionBar 上。
- SHOW_AS_ACTION_COLLAPSE_ACTION_VIEW：将该 Action View 折叠成普通菜单项。
- SHOW_AS_ACTION_IF_ROOM：当 ActionBar 位置有足够空间时才显示该 MenuItem。
- SHOW_AS_ACTION_NEVER：不将该 MenuItem 显示在 ActionBar 上。
- SHOW_AS_ACTION_WITH_TEXT：将该 MenuItem 显示在 ActionBar 上且显示该菜单项文本。

另外一种方法是在 res/menu 目录下的 main.xml 文件中添加菜单项，并且在 onCreateOptionsMenu()方法中采用“getMenuInflater().inflate(R.menu.main, menu)”语句来调用。而这种方法是 Android 开发平台新建项目时默认的，该方法就需要添加配置文件菜单项的属性。

```
<item
    android:id="@+id/action_settings"
    android:orderInCategory="100"
    android:showAsAction="ifRoom"
    android:title="@string/action_settings"/>
```

上面代码中“android:showAsAction="ifRoom"”语句表示将菜单项作为动作项显示在动作栏中。

**【例 6.4】**在例 6.1 的 Activity 中有“景区”“住宿”“美食”“交通”“设置”五个选项菜单。当用户点击动作栏中的 overflow 按钮时，就会弹出这五个选项菜单。假如现在把除“设置”之外的四个菜单项作为动作项置于动作栏中，请完成这个任务。

**【说明】**将菜单项作为动作项置于动作栏中，那就必须把菜单项转化为动作项，有两种转化方法：一种是使用 setShowAsAction()方法，另一种是修改 res/menu 目录下的 main.xml 文件。

**【开发步骤】**

（1）创建一个名为 Sample_ActionItem 的项目，Activity 组件名为 ActionItemActivity。

（2）前台布局文件。在应用程序的目录结构中定位到 res/layout 目录，编辑该目录下面的 activity_action_item.xml 文件，文件中有系统初始的代码，可有选择性地使用或修改。

```
<RelativeLayout xmlns:android="http://schemas.android.com/apk/res/android"
    xmlns:tools="http://schemas.android.com/tools"
    android:layout_width="match_parent"
    android:layout_height="match_parent"
    android:paddingBottom="@dimen/activity_vertical_margin"
    android:paddingLeft="@dimen/activity_horizontal_margin"
    android:paddingRight="@dimen/activity_horizontal_margin"
    android:paddingTop="@dimen/activity_vertical_margin"
    tools:context=" cn.gx.hzu_actionitem.ActionItemActivity ">

<TextView
        android:id="@+id/textView1"
        android:layout_width="wrap_content"
```

```
            android:layout_height="wrap_content"
            android:text="这里显示选择的内容..." />

</RelativeLayout>
```

上面的代码与例 6.1 中的代码类似。

（3）菜单布局文件。使用第二种方法时要用到菜单布局文件。在应用程序的目录结构中定位到 res/menu 目录，编辑目录下面的 action_item.xml 文件，文件中有系统初始的代码，可有选择性地使用或修改。

```
1   <menu xmlns:android="http://schemas.android.com/apk/res/android"
2         xmlns:tools="http://schemas.android.com/tools"
3         tools:context="cn.gx.hzu_actionitem.ActionItemActivity">
4
5   <item
6             android:id="@+id/new_settings"
7             android:orderInCategory="100"
8             android:showAsAction="ifRoom"
9             android:title="设置"/>
10
11  </menu>
```

1）第 1～11 行是菜单布局文件的内容。前后有成对的<menu></menu>标签。

2）第 5～9 行是菜单中的一个菜单项，前后有成对的<item></item>标签。第 8 行则是将菜单项设置为动作栏，可置于动作栏里。

（4）后台代码编写。在应用程序的目录结构中定位到 cn.gx.hzu_actionitem 包，打开包中 ActionItemActivity.java 文件，并根据该文件中的初始代码完成如下编码工作：

```
package cn.gx.hzu_actionitem;

import android.app.Activity;
import android.os.Bundle;
import android.view.Menu;
import android.view.MenuItem;
import android.view.MenuItem.OnMenuItemClickListener;
import android.widget.TextView;

public class ActionItemActivity extends Activity {

      TextView tv;
      //给每一个菜单项设置一个 ID 常量
      private static final int ITEM1 = Menu.FIRST;
      private static final int ITEM2 = Menu.FIRST+1;
      private static final int ITEM3 = Menu.FIRST+2;
      private static final int ITEM4 = Menu.FIRST+3;
      private static final int ITEM5 = Menu.FIRST+4;
      @Override
      protected void onCreate(Bundle savedInstanceState) {
            super.onCreate(savedInstanceState);
```

```
22              setContentView(R.layout.activity_action_item);
23              tv = (TextView)findViewById(R.id.textView1);
24          }
25
26          @Override
27          public boolean onCreateOptionsMenu(Menu menu) {
28          //第一种方法来实现
29          getMenuInflater().inflate(R.menu.action_item, menu);//利用布局文件来生成菜单项
30          MenuItem settingitem=menu.findItem(R.id.new_settings);
31          //添加菜单项
32              MenuItem spotitem = menu.add(0, ITEM1, 0, "景区");
33              MenuItem hotelitem = menu.add(0, ITEM2, 0, "住宿");
34              MenuItem fooditem = menu.add(0, ITEM3, 0, "美食");
35              MenuItem trafficitem = menu.add(0, ITEM4, 0, "交通");
36
37              //第二种实现方法
38              //MenuItem settingitem = menu.add(0, ITEM5, 0, "设置");
39              //settingitem.setShowAsAction(MenuItem.SHOW_AS_ACTION_IF_ROOM);
40
41              OnMenuItemClickListener lsn = new OnMenuItemClickListener(){
42                  @Override
43                  public boolean onMenuItemClick(MenuItem arg0) {
44                      // TODO Auto-generated method stub
45                      switch(arg0.getItemId()){
46                      case ITEM1:tv.setText("你点击了景区");break;
47                      case ITEM2:tv.setText("你点击了住宿");break;
48                      case ITEM3:tv.setText("你点击了美食");break;
49                      case ITEM4:tv.setText("你点击了交通");break;
50                      //case ITEM5:tv.setText("你点击了设置");break;
51                      case R.id.new_settings:tv.setText("你点击了设置");break;
52                      }
53                      return true;
54                  }
55              };
56              spotitem.setOnMenuItemClickListener(lsn);
57              hotelitem.setOnMenuItemClickListener(lsn);
58              fooditem.setOnMenuItemClickListener(lsn);
59              trafficitem.setOnMenuItemClickListener(lsn);
60              settingitem.setOnMenuItemClickListener(lsn);
61              return true;
62          }
63      }
```

1）第 29～30 行表示使用第一种方法将菜单项转为动作项，这种方法需要与菜单布局文件 action_item.xml 相配合。第 29 行是采用加载布局文件的方式来初始化菜单。第 30 行是通过调用 Menu 类对象中的 findItem()方法来获取菜单项对象，为菜单项添加监听事件做准备。

2）第 38～39 行是使用第二种方法将菜单项转为动作项。SHOW_AS_ACTION_IF_ROOM

是 MenuItem 类中的一个静态属性，表示如果动作栏中有空间则显示，这种方法不需要与菜单布局文件 action_item.xml 相配合。

3）第 45 行，在点击事件响应处理方法中进行菜单项判断，需要使用的菜单项 ID 是菜单布局文件中菜单项所设置的 ID。

【运行结果】运行项目显示效果如图 6-5 所示。

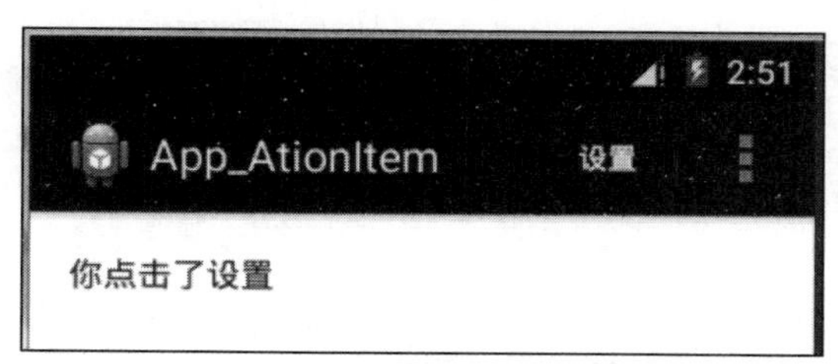


图 6-5 添加 Action Item 项

2. *在动作栏中添加动作视图*（Action View）

可以在动作栏中添加动作视图，动作视图是在菜单布局文件中的菜单项中进行定义。在菜单项中添加一个 actionViewClass 属性。下面的例子中，将在动作栏中添加一个“搜索”动作视图，用户在动作栏中就可以实现对数据的搜索。

【例 6.5】假定有一个关于旅游景区的信息列表，现在动作栏里添加一个搜索视图按钮，当点击这个搜索按钮时，可弹出搜索框，在该搜索框中输入搜索内容，就可以从景区列表中进行搜索，将搜索结果显示出来。

【说明】在动作栏中添加动作视图，需要修改菜单布局文件，即 res/menu 目录下的 main.xml。在菜单布局文件中添加一个新项，在项属性里面添加 actionViewClass 属性，值为“android.widget.SearchView”，从而将该项作为一个动作视图。对于动作视图的响应操作，采用了 onQueryTextSubmit()方法，即提交搜索之后响应该事件。还可采用 onQueryTextChange()方法，即搜索框内的文字发生改变时，响应该方法。

【开发步骤】

（1）创建一个名为 Sample_ActionView 的项目，Activity 组件名为 MainActivity。

（2）前台布局文件。在应用程序的目录结构中定位到 res/layout 目录，编辑该目录下面的 activity_action_item.xml 文件，文件中有系统初始的代码，可有选择地使用或修改。

```
<RelativeLayout xmlns:android="http://schemas.android.com/apk/res/android"
        xmlns:tools="http://schemas.android.com/tools"
        android:layout_width="match_parent"
        android:layout_height="match_parent"
        android:paddingBottom="@dimen/activity_vertical_margin"
        android:paddingLeft="@dimen/activity_horizontal_margin"
        android:paddingRight="@dimen/activity_horizontal_margin"
        android:paddingTop="@dimen/activity_vertical_margin"
        tools:context="com.example.app_actionview.MainActivity">

<ListView
            android:id="@+id/listView1"
```

```
13                  android:layout_width="match_parent"
14                  android:layout_height="wrap_content">
15      </ListView>
16
17      </RelativeLayout>
```

第 11～15 行是新建的一个 ListView 控件，在 ListView 中存放待搜索的内容项。

（3）在 res/menu 目录下的 main.xml 中添加的内容如下：

```
1       <menu xmlns:android="http://schemas.android.com/apk/res/android"
2           xmlns:tools="http://schemas.android.com/tools"
3           tools:context="com.example.app_actionview.MainActivity">
4
5       <item
6               android:id="@+id/action_search"
7               android:orderInCategory="100"
8               android:actionViewClass="android.widget.SearchView"
9               android:showAsAction="ifRoom"
10              android:title="搜索"/>
11
12      </menu>
```

第 5～10 行是一个菜单项。该菜单项作为动作视图放置于动作栏中，此动作视图的作用是搜索。第 8 行添加一个 actionViewClass 属性，该属性设置值为 android.widget.SearchView。通过该属性设置，使得此动作视图作为搜索的功能出现。

（4）在 com.example.sample_actionview 包下的 MainActivity.java 文件中，其代码内容如下：

```
package com.example. sample_actionview;

import android.app.Activity;
import android.os.Bundle;
import android.view.Menu;
import android.view.MenuItem;
import android.widget.ArrayAdapter;
import android.widget.ListView;
import android.widget.SearchView;
import android.widget.Toast;

public class MainActivity extends Activity {
    ListView lv;
    static final String[] viewSpot= new String[]{
        "姑婆山","黄瑶","十八水","大桂山","紫云洞","客家围屋","玉石林"
    };
    ArrayAdapter<String>adapter;
    @Override
    protected void onCreate(Bundle savedInstanceState) {
        super.onCreate(savedInstanceState);
        setContentView(R.layout.activity_main);
```

```
        lv = (ListView)findViewById(R.id.listView1);
        adapter = new ArrayAdapter<String>(
                this,
                android.R.layout.simple_list_item_1,
                viewSpot);
        lv.setAdapter(adapter);
    }

    @Override
    protected void onResume() {
        // TODO Auto-generated method stub
        super.onResume();

    }

    @Override
    public boolean onCreateOptionsMenu(Menu menu) {
    // Inflate the menu; this adds items to the action bar if it is present.
        getMenuInflater().inflate(R.menu.main, menu);

        SearchView searchView = (SearchView)menu.findItem
                    (R.id.action_search).getActionView();
        searchView.setOnQueryTextListener(new SearchView.OnQueryTextListener() {

            @Override
            public boolean onQueryTextSubmit(String query) {
                // TODO Auto-generated method stub
                boolean flag = false;
                for(String tempStr : viewSpot){
                    if(tempStr.equals(query)){
                        flag=true;
                    }
                }
                if(flag){
                    lv.setAdapter(null);
                    ArrayAdapter<String> adapter2 = new ArrayAdapter<String>(
                            MainActivity.this,
                            android.R.layout.simple_list_item_1,
                            new String[]{query});
                    lv.setAdapter(adapter2);
                }
                return true;
            }

            @Override
            public boolean onQueryTextChange(String newText) {
```

```
68                  // TODO Auto-generated method stub
69                  if(newText.equals("")){
70                      ArrayAdapter<String> adapter3 = new ArrayAdapter<String>(
71                              MainActivity.this,
72                              android.R.layout.simple_list_item_1,
73                              viewSpot);
74                      lv.setAdapter(adapter3);
75
76                  }else{
77                      lv.setAdapter(null);
78                  }
79
80                  return false;
81              }
82          });
83
84          return true;
85      }
86
87      @Override
88      public boolean onOptionsItemSelected(MenuItem item) {
89          Toast.makeText(this, "你选择了"+item.getTitle(),Toast.LENGTH_LONG);
90          return true;
91      }
92  }
```

1）第 13 行声明了一个 ListView 对象。

2）第 22 行是从布局文件中读取 ListView 控件来完成对 ListView 对象的初始化。

3）第 14～16 行新建了一个字符串数组，这个数组内容是 ListView 控件里显示的内容，也将作为搜索动作视图的搜索内容。

4）第 23～27 行是将 viewSpot 数组内容通过适配器绑定到 ListView 控件上。

5）第 40 行，从布局文件中读取菜单项，该搜索菜单项作为动作视图添加到动作栏中。

6）第 42 行，通过 Menu 对象的 findItem()方法将菜单项转换为动作视图，并完成了视图对象的初始化。

7）第 44～82 行表示为动作视图添加一个监听器。该监听器用来监听搜索框文本的变化，监听器是一个 OnQueryTextListener 对象。第 46～47 行重写了 OnQueryTextListener 类的 onQueryTextSubmit 方法，即当递交搜索框内容时响应事件，此处是完成搜索任务。第 49 行定义了一个 flag 布尔变量，来判断有没有查到内容，默认值是 false。第 50～54 行是进行 for 循环，在循环中进行比较，如果查询到搜索框里的文本，则 flag 返回 true。第 55～62 行判断 flag 的值是否为 true，即是否搜索到内容，如果搜索到内容，则将原来的 ListView 置空，将搜索结果显示在 ListView 上。第 66～81 行重写了 OnQueryTextListener 类的 onQueryTextChange()方法，即当搜索框文本内容发生改变时响应该事件，此处是正在输入内容。正输入文本内容时 ListView 的内容为空，当搜索框的内容被清空时，ListView 的内容又恢复为初始景区列表值。

**【运行结果】**运行项目显示效果如图 6-6 所示。

图 6-6　动作栏中的动作视图

## 6.3　对话框

### 6.3.1　Dialog

Dialog（对话框）是系统进行人机对话最常见的方式之一，对话框的表示形式常常是一个小窗口，当对话框弹出之后，Activity 界面暂时失去焦点，此刻就由对话框负责与用户进行交互。对话框有提示、可供选择、可供设置和显示进度等功能。其中提示功能是对话框中最常用的功能。接下来我们以提示对话框为例讲解对话框的使用方法。

Android 系统提供了 AlertDialog 类实现提示对话框。使用 AlertDialog 之前需导入 android.app.AlertDialog 类。

Activity 类中最常见的和对话框相关的方法如下：

- public void showDialog(int id)，参数说明：整数类型的参数 id 是对话框的唯一标识，这个参数将传递给 onCreateDialog()方法。方法说明：该方法用来显示一个对话框，当被首次调用时，系统将调用 onCreateDialog()方法与之对应，来生成对话框。
- public Dialog onCreateDialog(int id)，参数说明：整型参数 id 是对话框的唯一标识，它与 showDialog()方法中的 id 是对应的。方法说明：该方法用来创建对话框，只有首次显示对话框时调用，在该方法内定义对话框具体的显示样式和相关事件。

AlertDialog.Builder 类中最常用的方法如下：

- setTitle(charSequence title)，参数说明：字符序列的参数 title 表示对话框的标题。方法说明：这个方法用来为对话框设置标题。
- setTitle(int titleId)，参数说明：整型的参数 titleId 表示对话框标题的文本资源 ID。方法说明：与上一个方法相同。
- setIcon(Drawable icon)，参数说明：Drawable 类型的参数 icon 表示要设置的对话框图标。方法说明：这个方法用来为对话框设置图标。

- setIcon(int iconId)，参数说明：整型的参数 iconId 表示要设置对话框图标所对应的资源 ID。方法说明：与上一个方法相同。
- setPositiveButton(CharSequence text,DialogInterface.OnClickListener listener)，参数说明：字符序列类型的参数 text 是 YES 按钮上要显示的文本内容，OnClickListener 类型的参数 listener 是一个监听器，它监听用户是否点击了 YES 按钮。方法说明：该方法用来为对话设置一个 YES 按钮，即确认性质的按钮。
- setPositiveButton(int textId,DialogInterface.OnClickListener listener)，参数说明：整型的参数 textId 是 YES 按钮上要显示的文本的资源 ID。方法说明：与上一个方法相同。
- setMessage(CharSequence message)，参数说明：显示主体信息的文本内容。方法说明：该方法用来设置对话框主体信息的内容。
- setMessage(int messageId)，参数说明：显示主体信息的文本资源 ID。方法说明：与上一个方法相同。
- create()，该方法用来生成对话框。

**【例 6.6】**假定有一个关于旅游景区的信息列表，每个景区名称的右边有一个“详情”按钮，点击“详情”按钮，则弹出一个介绍景区详细信息的提示对话框，请选择合适的控件和方法来实现该功能。

**【说明】**这里要用到 AlertDialog 对话框，要实现点击“详情”按钮，就可以弹出对话框，需要在“详情”按钮上添加监听事件，如监听到有点击事件则调用 showDialog()方法。要响应并生成对话框，则需要在 Activity 中重写 onCreateDialog()方法，在该方法中设置对话框的显示样式，如标题、图标、信息内容等。但需要注意的是本例是一个关于旅游景区的信息列表，ListView 中有很多项，每一项中有一个“详情”按钮。点击不同项的按钮，显示的信息内容也是不同的。如何才能做到这一点呢？这就需要用到信息传递。具体实现方法可在本例中细细体会。

**【开发步骤】**

（1）创建一个名为 Sample_AlertDialog 的项目，Activity 组件名为 AlertDialogActivity。

（2）前台布局文件。在应用程序的目录结构中定位到 res/layout 目录，编辑该目录下的 activity_alert_dialog.xml 文件，文件中有系统初始的代码，可有选择性地使用或修改。

```
<RelativeLayout xmlns:android="http://schemas.android.com/apk/res/android"
    xmlns:tools="http://schemas.android.com/tools"
    android:layout_width="match_parent"
    android:layout_height="match_parent"
    android:paddingBottom="@dimen/activity_vertical_margin"
    android:paddingLeft="@dimen/activity_horizontal_margin"
    android:paddingRight="@dimen/activity_horizontal_margin"
    android:paddingTop="@dimen/activity_vertical_margin"
    tools:context="com.example.sample_alertdialog.AlertDialogActivity">

<ListView
        android:id="@+id/listView1"
        android:layout_width="match_parent"
        android:layout_height="wrap_content"
        android:layout_alignParentLeft="true"
```

```
16                 android:layout_alignParentTop="true">
17     </ListView>
18
19     </RelativeLayout>
```

第 11～17 行，添加了一个 ListView 控件。

（3）准备字符串资源，修改 res/values 目录下的文件 string.xml，其内容如下：

```
1     <?xml version="1.0" encoding="utf-8"?>
2     <resources>
3
4     <string name="app_name">App_AlertDialog</string>
5     <string name="hello_world">Hello world!</string>
6     <string name="action_settings">Settings</string>
7     <string name="guposhan">姑婆山</string>
8     <string name="huangyao">黄姚古镇</string>
9     <string name="shibashui">十八水</string>
10    <string name="daguishan">大桂山</string>
11    <string name="gpsdetail">姑婆山详情：
12                姑婆山国家森林公园位于广西东北部，湘、桂、粤三省（区）交界处的萌渚岭
13                南端广西贺州市八步区境内，距市区（八步）中心仅 26 公里，姑婆山方圆 80 公
14                里是天然动植物王国，境内海拔 1000 米以上的山峰有 25 座，最高峰海拔 1844
15                米是桂东第一高峰。</string>
16    <string name="hydetail">黄姚古镇详情：
17                黄姚古镇方圆 3.6 公里，属喀斯特地貌。发祥于宋朝年间，有着近 1000 年历史。
18                自然景观有八大景二十四小景，保存有寺观庙祠 20 多座，亭台楼阁 10 多处，多
19                为明清建筑。</string>
20    <string name="sbsdetail">十八水详情：
21                广西贺州十八水原生态景区位于广西贺州市境内，距市中心仅 21 公里，景区
22                总面积 18 平方公里，景区共有二十多个景点。</string>
23    <string name="dgsdetail">大桂山详情：
24                桂山国家森林公园，始建于 1993 年，是广西第一个国家级森林公园。位于湘、
25                桂、粤三省交界处，距贺州市区约 30 公里，是桂东、桂林两地区唯一拥有成
26                片森林的旅游度假胜地。</string>
27
28    </resources>
```

第 7～26 行声明了 8 个字符串资源，分别对应四个景区的名称字符串和四个景区的介绍字符串。

（4）创建数组资源。创建并编写 res/values 目录下的数组描述文件 arrs.xml，其代码如下：

```
<?xml version="1.0" encoding="utf-8"?>
<resources>

<string-array name="viewspots">
<item>guposhan</item>
<item>huangyao</item>
<item>shibashui</item>
<item>daguishan</item>
```

```
9    </string-array>
10   <string-array name="viewdetails">
11   <item>gpsdetail</item>
12   <item>hydetail</item>
13   <item>sbsdetail</item>
14   <item>dgsdetail</item>
15   </string-array>
16
17   </resources>
```

1）第 4～9 行声明了一组内容，景区名称的字符串数组 viewspots。

2）第 10～15 行声明了一组内容，景区详细信息的字符串数组 viewdetails。

（5）开发逻辑代码。打开 src/com.example.sample_alertdialog 包下的 AlertDialogActivity 文件，其代码内容如下：

```
package com.example.sample_alertdialog;

import android.app.Activity;
import android.app.AlertDialog;
import android.app.AlertDialog.Builder;
import android.app.Dialog;
import android.os.Bundle;
import android.view.Gravity;
import android.view.Menu;
import android.view.MenuItem;
import android.view.View;
import android.view.ViewGroup;
import android.view.ViewGroup.LayoutParams;
import android.widget.*;

public class AlertDialogActivity extends Activity {
    ListView lv;
    //final int ALERT_DIALOG = 0;
    String[] temparrs ={"","","",""};
    @Override
    protected void onCreate(Bundle savedInstanceState) {
        super.onCreate(savedInstanceState);
        setContentView(R.layout.activity_alert_dialog);
        lv=(ListView)findViewById(R.id.listView1);

        BaseAdapter ba = new BaseAdapter(){

            String[] viewspots = getResources().getStringArray(R.array.viewspots);
            String[] viewdetails = getResources().getStringArray(R.array.viewdetails);
            @Override
            public int getCount() {
                return 4;
            }
```

```
            @Override
            public Object getItem(int position) {
                return null;
            }

            @Override
            public long getItemId(int position) {
                return 0;
            }

            @Override
            public View getView(int position, View convertView,ViewGroup parent) {
                LinearLayout ll = new LinearLayout(AlertDialogActivity.this);
                ll.setOrientation(LinearLayout.HORIZONTAL);

                TextView tv = new TextView(AlertDialogActivity.this);
                tv.setText(getResources().getText(getResources().
                    getIdentifier(viewspots[position],
                    "string", getPackageName())));
                tv.setLayoutParams(new LinearLayout.
                    LayoutParams(LayoutParams.WRAP_CONTENT,
                    LayoutParams.WRAP_CONTENT));
                tv.setTextSize(22);

                LinearLayout ll1 = new LinearLayout(AlertDialogActivity.this);
                ll1.setOrientation(LinearLayout.HORIZONTAL);
                ll1.setGravity(Gravity.RIGHT);
                ll1.setPadding(3, 0, 3, 0);
                ll1.setLayoutParams(new LinearLayout.LayoutParams
                    (LayoutParams.FILL_PARENT,LayoutParams.WRAP_CONTENT));

                Button btn = new Button(AlertDialogActivity.this);
                btn.setLayoutParams(new LinearLayout.LayoutParams(100,
                    LayoutParams.WRAP_CONTENT));
                btn.setId(position);
                btn.setOnClickListener(new View.OnClickListener() {

                    @Override
                    public void onClick(View v) {

                        temparrs[v.getId()]=getResources()
                                .getText(getResources()
                                .getIdentifier(viewdetails[v.getId()],
                        "string", getPackageName())).toString();
                        showDialog(v.getId());
```

```
80                              }
81                          });
82                          btn.setText("详情");
83                          ll1.addView(btn);
84                          ll.addView(tv);
85                          ll.addView(ll1);
86                          return ll;
87                  }};
88                  lv.setAdapter(ba);
89          }
90          @Override
91          @Deprecated
92          protected Dialog onCreateDialog(int id) {
93
94                  Dialog dialog = null;
95                  Builder b = new AlertDialog.Builder(this);
96                  b.setTitle("景区详情");
97                  b.setPositiveButton("确定", null);
98                  switch(id){
99                  case 0:b.setMessage(temparrs[0]);break;
100                 case 1:b.setMessage(temparrs[1]);break;
101                 case 2:b.setMessage(temparrs[2]);break;
102                 case 3:b.setMessage(temparrs[3]);break;
103                 }
104                 dialog = b.create();
105                 return dialog;
106         }
107
108 }
```

1）第 17 行声明了一个 ListView 对象。

2）第 19 行新建了一个字符串数组，用来存放四个景区详情信息的字符串。在该案例中用户点击“详情”按钮，则弹出景区详情的提示框，景区详情的内容正是存储在该字符串数组中。

3）第 24 行完成了 ListView 对象的初始化。

4）第 26～87 行表示新建一个 BaseAdapter 对象，并且重写 BaseAdapter 类中的四个方法。从而实现界面数据的列表布局。

5）第 28 行表示获取资源目录下的字符串数组 viewspots 中的内容，并将这些内容赋值给字符串数组 viewspots。

6）第 29 行表示获取当地资源目录下的字符串数组 viewdetails 的内容，并将这些内容赋值给字符串数组 viewdetails。

7）第 32 行，此处返回值为 4，说明本例景区的个数假定为 4。

8）第 47～48 行新建了一个水平布局。

9）第 50～57 行新建了一个文本视图，用来存放每一个景区的名称。

10）第 59～64 行新建了一个水平布局，这个水平布局是为了容纳后面的“详情”按钮。并使“详情”按钮能够在水平方向上靠右显示。

11）第 66～82 行新建了一个按钮视图，即“详情”按钮。

12）第 69 行是将当前行的值赋值为按钮 ID，为了传值的需要，否则系统将不能确定当前点击的按钮是哪一行的“详情”按钮，因为每一行都有一个“详情”按钮。

13）第 70 行是给按钮添加一个点击监听器，使用匿名类的方式来实现的。

14）第 75～79 行是监听器的响应事件，即当用户点击“详情”按钮之后响应的操作。请注意，在该 onClick()方法中无法获取 AlertDialogActivity 类中的变量值，因为此时它们已经不属于同一类别。那该如何确定是点击的哪一行按钮呢？这时第 69 行给行号组按钮赋值 ID 的方法就起到了作用。onClick()方法中的参数 View 对象就是被点击的按钮对象，这时使用 v.getID()方法将会得到按钮的 ID，即当前的行号，这样就确定了当点响应的是哪一行的“详情”按钮。第 75 行是在确定哪一行的“详情”按钮之后，根据该行的行号，找出 viewdetails 数组对应的字符串，即这一行对应的景区详情字符串内容，并把这个内容放到与索引相对应的 temparrs 数组中。为什么这样做呢？是因为字符串数组 temparrs 是一个全局变量，目的是为了在主 Activity 中与对话框之间进行传值。而字符串数组 viewdetails 并不是全局变量，无法实现传值。第 79 行，即打开一个提示框（AlertDialog），此处只表明需要打开一个提示框，但提示框里的具体初始化内容，将在重写 onCreateDialog()方法时进行实现。需要注意的是，showDialog()方法中的参数是 v.getID()，即把当前的行号做为参数传递出去。

15）第 92～106 行表示重写 onCreateDialog()方法，完成对提示对话框的初始化工作。第 94 行声明一个空对话框对象。第 95 行新建并初始化 Builder 对象。第 96 行给这个提示对话框定义标题。第 97 行给这个提示对话框定义确定按钮。第 98～103 行使用 switch 语句来确定提示对话框内的主体信息到底是显示哪一行所对应的景区详情字符串。switch 语句的判断条件是当前 ID 值，即第 79 行 showDialog()方法所传递过来的行值。第 104 行使用 Builder 对象来创建对话框。第 105 行返回对话框对象。

**【运行结果】**运行项目显示效果如图 6-7 所示。

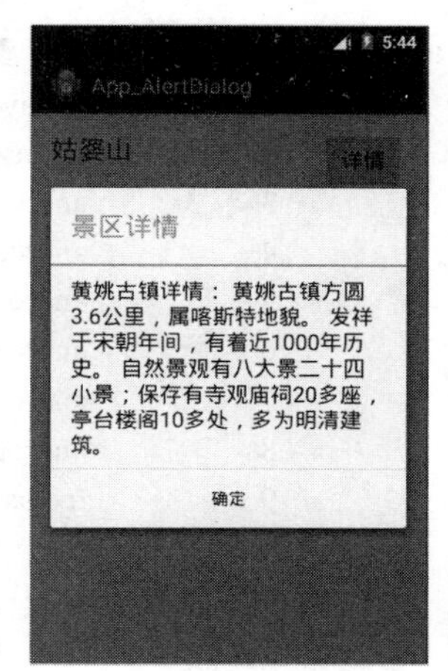


图 6-7 提示对话框

### 6.3.2 Toast

Toast 的中文名称是烤面包，它这里指的是友好的消息提示机制。Toast 与 AlertDialog 都能给用户提供提示信息，但两者有着很大差别。AlertDialog 作为对话框是有焦点的，而且当对话框弹出之后，Activity 则失去焦点。在用户与对话框对话完毕之后，对话框才会消失。而 Toast 则不一样，首先 Toast 没有焦点，它也不会使 Activity 失去焦点。其次它提供的是一种快速的即时消息，也就是说这个消息所停留的时间不会很长，然后自行消失。

常见的与 Toast 相关的方法有以下几个：

- makeText(Context context,String message,int duration)，参数说明：Context 类型的参数 context 表示当前的上下文，字符串类型的参数 message 表示提示信息的内容，整型的

参数 duration 表示提示信息在屏幕中的持续时间，这个取值为 Toast.LENGTH_LONG（表示时间较长）和 Toast.LENGTH_SHORT（表示时间较短）两个值。方法说明：该方法是 Toast 类的一个静态方法，使用该方法可以完成 Toast 对象的创建和初始化。

- makeText(Context context,int resId,int duration)，参数说明：整型的参数 resId 表示提示信息文本对应的资源 ID 值。方法说明：与上一个方法相同。
- setGravity(int gravity, int xOffset, int yOffset)，参数说明：整型参数 gravity 是设置 Toast 在屏幕中显示的位置，若居中值为 Gravity.CENTER，整型参数 xOffset 是设置相对于第一个参数设置 Toast 位置的横向 X 轴的偏移量，正数向右偏移，负数向左偏移，整型参数 yOffset 同第二个参数原理一样。方法说明：设置提示信息的位置。
- show()，将 Toast 对象的消息提示内容显示在屏幕上。

**【例 6.7】**假定有一个存放景区信息的列表，列表旁边有一个“刷新”按钮，点击“刷新”按钮完成列表内容的刷新，并且使用 Toast 对象来显示“刷新成功”的提示信息。

**【说明】**使用 ListView 控件来实现景区信息的存放。本例还需要使用按钮事件的监听。使用 Toast 对象来显示“刷新成功”的提示信息，那么本例也需要创建 Toast 对象来显示提示信息，具体实现方法可在本例中细细体会。

**【开发步骤】**

（1）创建一个名为 Sample_Toast 的项目，Activity 组件名为 ToastActivity。

（2）前台布局文件。在应用程序的目录结构中定位到 res/layout 目录，编辑该目录下的 activity_toast.xml 文件，文件中有系统初始的代码，可有选择地使用或修改。

```
<RelativeLayout xmlns:android="http://schemas.android.com/apk/res/android"
        xmlns:tools="http://schemas.android.com/tools"
        android:layout_width="match_parent"
        android:layout_height="match_parent"
        android:paddingBottom="@dimen/activity_vertical_margin"
        android:paddingLeft="@dimen/activity_horizontal_margin"
        android:paddingRight="@dimen/activity_horizontal_margin"
        android:paddingTop="@dimen/activity_vertical_margin"
        tools:context="cn.gx.hzu.sample_toast.ToastActivity">

<Button
            android:id="@+id/btn_Reflash"
            android:layout_width="wrap_content"
            android:layout_height="wrap_content"
            android:layout_alignParentLeft="true"
            android:layout_alignParentTop="true"
            android:text="刷新" />

<ListView
            android:id="@+id/listView1"
            android:layout_width="match_parent"
            android:layout_height="wrap_content"
            android:layout_alignParentBottom="true"
            android:layout_alignParentLeft="true"
```

```
25                    android:layout_below="@id/btn_Reflash">
26    </ListView>
27
28    </RelativeLayout>
```

1）第 11～17 行表示新建一个“刷新”按钮控件。

2）第 19～26 行表示新建一个 ListView 控件。第 25 行说明 ListView 控件是在按钮控件的下方。

（3）开发逻辑代码。打开 src/cn.gx.hzu.sample_toast 包下的 ToastActivity 文件，其代码内容如下：

```
package cn.gx.hzu.sample_toast;

import java.util.ArrayList;
import java.util.List;

import android.app.Activity;
import android.os.Bundle;
import android.view.Gravity;
import android.view.Menu;
import android.view.MenuItem;
import android.view.View;
import android.widget.*;

public class ToastActivity extends Activity implements View.OnClickListener {

        ListView lv;
        Button btn;
        List<String> strName;
        ArrayAdapter adapter;

        @Override
        protected void onCreate(Bundle savedInstanceState) {
                super.onCreate(savedInstanceState);
                setContentView(R.layout.activity_toast);
                lv=(ListView)findViewById(R.id.listView1);
                btn = (Button)findViewById(R.id.btn_Reflash);
                strName = new ArrayList<String>();
                adapter= new ArrayAdapter<String>(this,android.R.layout.simple_list_item_1, strName);
                lv.setAdapter(adapter);
                strName.add("姑婆山");
                strName.add("黄瑶古镇");
                strName.add("十八水");
                strName.add("大桂山");
                adapter.notifyDataSetChanged();
                btn.setOnClickListener(this);
        }
```

```
37
38          @Override
39          public void onClick(View v) {
40                adapter.notifyDataSetChanged();
41                Toast toast = Toast.makeText(this, "刷新成功", Toast.LENGTH_LONG);
42                toast.setGravity(Gravity.CENTER, 0, 0);
43                toast.show();
44          }
45    }
```

1）第 14 行 ToastActivity 继承 View.OnClickListener 接口。

2）第 16 行声明 ListView 对象是用来存放景区名称信息列表。

3）第 17 行声明 Button，该 Button 是刷新按钮。

4）第 18 行声明 List<String>，把景区的名称字符串都添加到 List 中，它也是 ListView 对象的适配器中的数据来源。

5）第 19 行声明数组列表适配器。

6）第 25～28 行分别将上面的 ListView 对象、Button 对象、List 对象和适配器对象实例化。

7）第 29 行把适配器绑定在 ListView 对象上。

8）第 30～33 行表示向 List 中添加字符串。

9）第 34 行表示刷新适配器中的数据源。

10）第 35 行将监听器安装到 Button 上。请注意我们这里使用的是外部类实现事件监听器接口方法，所以监听器的对象使用的是 this。

11）第 38～44 行定义点击事件响应。第 40 行为适配器数据的刷新，即刷新 ListView 中的数据。第 41 行为新建并实例化 Toast 对象，定义提示文本是“刷新成功”，定义消息的显示时间较长。第 42 行定义 Toast 的显示位置。第 43 行将 Toast 显示出来。值得注意的是，此处的按钮点击监听，使用外部类实现事件监听接口的方法。为什么要这样做呢？因为这些方法可以将 onClick()方法写在 Activity 中，那么就可以在 onClick()方法中使用适配器对象，实现数据的刷新。如果不使用该方法，而使用匿名类和内部类的方法，就不能直接调用 Activity 中的适配器对象，因为它们已经不属于同一类。

**【运行结果】**运行项目显示效果如图 6-8 所示。

图 6-8　Toast 消息

### 6.3.3　其他 Dialog

日期及时间选择对话框可以为用户提供自由选择和设定日期与时间的功能，它们都要用到 Calendar 类。

日期选择对话框（DatePickerDialog）主要用到 DatePicker 类，它主要针对日期相关类的开发，它是提示对话框（AlertDialog）的子类。涉及到日期选择对话框的相关方法如下：

- DatePickerDialog(Context context, DatePickerDialog.OnDateSetListener callback,int year, int monthOfYear,int dayOfMonth)，参数说明：Context 类型参数 context 代表当前上下

文，OnDateSetListener 类型的参数 callback 表示日期设置监听器，整型参数 year 表示年份，整型参数 monthOfYear 表示月份，整型参数 dayOfMonth 表示天数。方法说明：该方法用来初始化 DatePickerDialog 类对象。

- void onDateSet(DatePicker view,int year,int month,int day)，参数说明：DatePicker 类型参数 view 代表要设置的 DatePicker，整型参数 year 表示年份，整型参数 month 表示月份，整型参数 day 表示天数。方法说明：该方法用来设置指定的 TimePicker 时间。

时间选择对话框（TimePickerDialog）主要用到 TimePicker 类，涉及到时间选择对话框的相关方法如下：

- TimePickerDialog(Context context, TimePickerDialog.OnTimeSetListener callback,int hourOfDay, int minute,boolean is24HourView)，参数说明：Context 类型参数 context 代表当前上下文，OnTimeSetListener 类型的参数 callback 指时间设置监听器，整型参数 hourOfDay 表示小时，整型参数 minute 表示分钟，boolean 参数 is24HourView 表示是否以 24 小时显示格式。方法说明：该方法用来初始化 TimePickerDialog 类对象。
- void onTimeSet(TimePicker view,int hourOfDay,int minute)，参数说明：TimePicker 类型参数 view 代表要设置的 TimePicker，整型参数 hourOfDay 代表时间中的小时，整型参数 minute 代表时间中的分钟。方法说明：该方法用来设置指定的 TimePicker 时间。

跟 Calendar 对象相关的方法有以下几个：

- Calendar getInstance()，构造一个日历的新实例。
- int get(int field)，参数说明：整型参数 field 是日历的字段值，通常情况下取值为 Calendar.YEAR、Calendar.MONTH、Calendar.DAY_OF_MONTH、Calendar.HOUR_OF_DAY、Calendar.MINUTE，分别表示年、月、日、时、分。方法说明：该方法返回日历的字段值，如果参数值是年字段，则返回要传入的年份。如果参数值是月字段，则返回要传入的月份。如果参数值是日字段，则返回要传入的天数。

**【例 6.8】**假定设计一个关于旅行日程计划的程序，要求日程计划中的日期和时间设置需要使用时间对话框和日期对话框。

**【说明】**使用日期对话框和时间对话框来设定旅行日程的时间和日期，那么就必须要用到对话框中的日期对话框（DatePickerDialog）和时间对话框（TimePickerDialog）。这两种对话框与提示对话框相比，有相同的地方也有不同的地方，具体的实例如下所示。

**【开发步骤】**

（1）创建一个名为 Sample_DatePicker_TimePicker 的项目，Activity 组件名称为 Date_Time_Activity。

（2）前台布局文件。在应用程序的目录结构中定位到 res/layout 目录，编辑该目录下的 activity_date_time.xml 文件，文件中有系统初始的代码，可有选择性地使用或修改。

```
<RelativeLayout xmlns:android="http://schemas.android.com/apk/res/android"
        xmlns:tools="http://schemas.android.com/tools"
        android:layout_width="match_parent"
        android:layout_height="match_parent"
        android:paddingBottom="@dimen/activity_vertical_margin"
        android:paddingLeft="@dimen/activity_horizontal_margin"
```

```
        android:paddingRight="@dimen/activity_horizontal_margin"
        android:paddingTop="@dimen/activity_vertical_margin"
        tools:context="cn.gx.hzu_datepicker_timepicker.Date_Time_Activity">

    <LinearLayout
            android:id="@+id/linearLayout1"
            android:layout_width="match_parent"
            android:layout_height="200sp"
            android:background="#DEDEDE"
            android:orientation="vertical">
    <TextView
                android:id="@+id/tv_travel"
                android:layout_width="wrap_content"
                android:layout_height="wrap_content"
                android:text="" />
    </LinearLayout>

    <LinearLayout
            android:id="@+id/linearLayout2"
            android:layout_width="match_parent"
            android:layout_height="wrap_content"
            android:layout_below="@+id/linearLayout1"
            android:orientation="vertical">

    <LinearLayout
                    android:layout_width="wrap_content"
                    android:layout_height="wrap_content"
                    android:orientation="horizontal">

    <TextView
                        android:id="@+id/tv_date"
                        android:layout_width="wrap_content"
                        android:layout_height="wrap_content"
                        android:text="2017.9.5" />

    <Button
                        android:id="@+id/btn_date"
                        android:layout_width="wrap_content"
                        android:layout_height="wrap_content"
                        android:text="选择日期" />
    </LinearLayout>

    <LinearLayout
                  android:layout_width="wrap_content"
                  android:layout_height="wrap_content"
                  android:orientation="horizontal">
```

```
53
54  <TextView
55                      android:id="@+id/tv_time"
56                      android:layout_width="wrap_content"
57                      android:layout_height="wrap_content"
58                      android:text="12:00:00" />
59
60  <Button
61                      android:id="@+id/btn_time"
62                      android:layout_width="wrap_content"
63                      android:layout_height="wrap_content"
64                      android:text="选择时间" />
65  </LinearLayout>
66
67  <LinearLayout
68                  android:layout_width="wrap_content"
69                  android:layout_height="wrap_content"
70                  android:orientation="horizontal">
71
72  <EditText
73                      android:id="@+id/et_content"
74                      android:layout_width="wrap_content"
75                      android:layout_height="wrap_content"
76                      android:ems="10"
77                      android:text="点击此处添加行程"
78                      android:textColor="#969696" />
79
80  <Button
81                    android:id="@+id/btn_add"
82                    android:layout_width="wrap_content"
83                    android:layout_height="wrap_content"
84                    android:text="添加" />
85  </LinearLayout>
86  </LinearLayout>
87
88  </RelativeLayout>
```

1）第 11～22 行声明了 id 值为 linearLayout1 的垂直线性布局。第 14 行为线性布局 linearLayout1 定义高度。第 15 行为线性布局 linearLayout1 定义背景色。第 17～21 行在 linearLayout1 线性布局里面声明 id 为 tv_travel 的文本对象。

2）第 24～86 行声明 id 值为 linearLayout2 的垂直线性布局。第 28 行定义线性布局 linearLayout2 在线性布局 linearLayout1 的下方。第 31～47 行声明一个水平线性布局。第 36～40 行，在该线性布局里声明一个文本对象，用来显示新设置的日期。第 42～46 行在该线性布局里声明一个按钮对象，用来打开日期对话框。第 49～65 行声明一个水平线性布局。第 54～58 行在该线性布局里声明一个文本对象，用来显示新设置的时间。第 60～64 行在该线性布局里声明一个按钮对象，用来打开时间对话框。第 67～85 行声明一个水平线性布局。第 72～78 行在该线性布局里声明一个编辑文本框对象，用来设置日程地点。第 80～84 行在该线性布局里声明一个按钮对象，用来将日程添加到日程文本中。

（3）开发逻辑代码。打开 src/cn.gx.hzu_datepicker_timepicker 包下的 Date_Time_Activity 文件，其代码内容如下：

```
package cn.gx.hzu_datepicker_timepicker;

import java.util.Calendar;
import android.app.Activity;
import android.app.DatePickerDialog;
import android.app.Dialog;
import android.app.TimePickerDialog;
import android.os.Bundle;
import android.view.*;
import android.widget.*;

public class Date_Time_Activity extends Activity {

        final int DATE_DIALOG =1;
        final int TIME_DIALOG =2;
        TextView tv_travel,tv_date,tv_time;
        EditText et_content;
        Button btn_add,btn_date,btn_time;
        Calendar c = null;

        @Override
        protected void onCreate(Bundle savedInstanceState) {
                super.onCreate(savedInstanceState);
                setContentView(R.layout.activity_date_time);
                tv_travel = (TextView)findViewById(R.id.tv_travel);
                tv_date = (TextView)findViewById(R.id.tv_date);
                tv_time = (TextView)findViewById(R.id.tv_time);
                et_content = (EditText)findViewById(R.id.et_content);
                btn_add = (Button)findViewById(R.id.btn_add);
                btn_date = (Button)findViewById(R.id.btn_date);
                btn_time = (Button)findViewById(R.id.btn_time);

                btn_add.setOnClickListener(new View.OnClickListener(){

                        @Override
                        public void onClick(View v) {
                         String my_travel = "你将于";
                         my_travel+=tv_date.getText();
                         my_travel+=tv_time.getText();
                         my_travel+="去"+et_content.getText()+"游玩";

                         tv_travel.append(my_travel+"\n");
```

```
            }
        });

        btn_date.setOnClickListener(new View.OnClickListener(){

            @Override
            public void onClick(View v) {
                showDialog(DATE_DIALOG);
            }
        });

        btn_time.setOnClickListener(new View.OnClickListener(){
            @Override
            public void onClick(View v) {
                showDialog(TIME_DIALOG);
            }
        });
    }

    @Override
    protected Dialog onCreateDialog(int id) {
        Dialog dialog = null;
        c = Calendar.getInstance();
        switch(id){
          case DATE_DIALOG:
            dialog = new DatePickerDialog(
                this,
                new DatePickerDialog.OnDateSetListener(){
                    @Override
                    public void onDateSet(DatePicker dp,int year,int month,int day){
                    month=month+1;tv_date.setText(year+"年"+month+"月"+day+"日");
                    }
                },
                c.get(Calendar.YEAR),
                c.get(Calendar.MONTH),
                c.get(Calendar.DAY_OF_MONTH)
            );
            break;

          case TIME_DIALOG:
            dialog = new TimePickerDialog(this, new TimePickerDialog.OnTimeSetListener(){
                        @Override
                        public void onTimeSet(TimePicker tp,int hour,int minute){
                            tv_time.setText(","+hour+"时"+minute+"分");
                        }
                    },
                    c.get(Calendar.HOUR_OF_DAY),
```

```
90                              c.get(Calendar.MINUTE),
91                              false
92                          );
93                      break;
94                  }
95                  return dialog;
96          }
97
98  }
```

1）第 3～10 行导入相关类。

2）第 14 行定义日期对话框的 ID。因为这个 ID 值是整型，为了增强程序的可读性，避免混淆，因此将为其定义一个全局的 final 类型的整型变量。

3）第 15 行定义时间对话框 ID。

4）第 16 行分别声明了日程文本对象，日期文本对象，时间文本对象。

5）第 17 行声明了日程编辑框对象。

6）第 18 行分别声明了添加日程按钮、打开日期对话框按钮，打开时间对话框按钮。

7）第 19 行声明日历对象，通过这个日期对象可以获取和设置关于年、月、日、时、分、秒的数据。

8）第 25～31 行分别完成对日程文本对象、日期文本对象、时间文本对象、日程编辑框对象、添加日程按钮、打开日期对话框按钮，打开时间对话框按钮的初始化。

9）第 33～44 行定义了“添加日程按钮”的点击事件监听器。

10）第 37 行新建 String 类型的变量 my_travel，这个变量用来存放单次日程的内容。

11）第 38 行将日期文本附加到 my_travel 字符串上。

12）第 39 行将时间文本附加到 my_travel 字符串上。

13）第 40 行将日程的地点文本附加到 my_travel 字符串上。

14）第 42 行将 my_travel 这个记载了单次日程的字符串附加到整个日程文本框中。

15）第 46～52 行定义了“打开日期对话框按钮”的点击事件监听器。

16）第 54～59 行定义了“打开时间对话框按钮”的点击事件监听器。

17）第 63～96 行重写了 onCreateDialog()方法，在这个方法中完成对话框的创建和初始化。

18）第 66～94 行是一个 switch 判断语句，即判断对话框的 ID。当该 ID 值是 DATE_DIALOG 时，则将对话框对象创建为日期对话框。当该 ID 值是 TIME_DIALOG 时，则将对话框对象创建为时间对话框。

**【运行结果】**运行项目显示效果如图 6-9 所示。

除了前面介绍的几种对话框之外，还有一种使用起来稍微复杂的对话框，即进度对话框，与前几种对话框静态显示数据不同的是进度对话框需要动态显示数据的变化。因此在使用进度对话框的时候，还需要使用到线程操作和 Handler 通信机制。进度对话框广泛运用于复杂计算、数据获取等耗时的操作应用中。下面具体介绍进度对话框在使用过程中经常使用的一些方法。

- protected void onPrepareDialog(int id, Dialog dialog)，参数说明：整型参数 id 表示要打开的对话框 ID，Dialog 对象 dialog 是第一次调用 onCreateDialog()方法返回的 Dialog

对象。方法说明：假如要在每一次对话框被打开时改变它的任何属性，可以重写该方法，该方法在每次打开对话框时被调用。

- void setMax(int max)，参数说明：整型参数 max 表示当前进度对话框设定的最大值。方法说明：设置进度对话框中进度条的最大值。
- void setProgressStyle(int style)，参数说明：整型参数 style 表示进度对话框中对话条的样式。方法说明：设置进度对话框中进度条的样式。
- void setTitle(CharSequence title)，参数说明：字符序列参数 title 表示对话框标题内容。方法说明：为对话框设置标题。
- void setMessage(CharSequence message)，参数说明：字符序列参数 message 表示对话框消息内容。方法说明：为对话框设置消息内容。

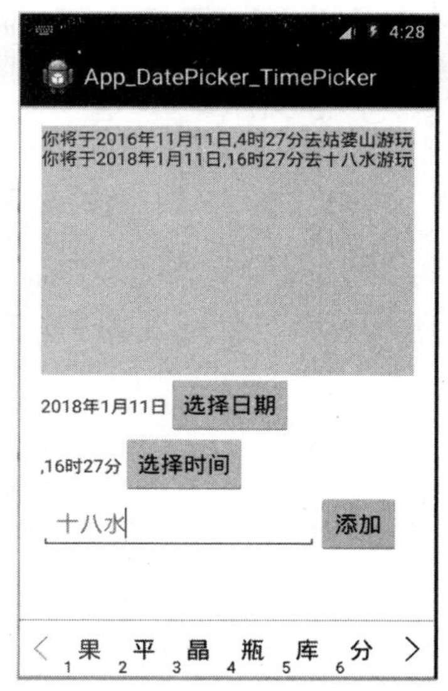


图 6-9　日期对话框和时间对话框

- void setCancelable(boolean cancelable)，设置按“返回”键时，对话框是否会退出。
- void setOnDismissListener(DialogInterface.OnCancelListener callBack)，参数说明：DialogInterface.OnCancelListener 类型的参数 callBack 表示监听对话框消失的监听器。方法说明：为对话框设置监听器，监听对话框消失事件。
- void incrementProgressBy(int data)，参数说明：整型参数 data 表示进度条每次增加的数量级。方法说明：为进度对话框的进度条设置每次增加的数量级。
- int getProgress()，获取进度条的当前值。

**【例 6.9】**设计一个进度对话框来实现 1+2+3+…+100 这个计算过程。

**【说明】**这个程序需要使用到进度对话框，加法的操作需要与进度对话框的进度相配合。同时为了实现进度对话框的动态更新，这个程序还要用到 Handler 机制和线程操作。

**【开发步骤】**

（1）创建一个名为 Sample_ProgressDialog 的项目，Activity 组件名称为 ProgressDialogActivity。

（2）前台布局文件。在应用程序的目录结构中定位到 res/layout 目录，编辑该目录下的 activity_progress_dialog.xml 文件，文件中有系统初始的代码，可有选择地使用或修改。

```
<RelativeLayout xmlns:android="http://schemas.android.com/apk/res/android"
        xmlns:tools="http://schemas.android.com/tools"
        android:layout_width="match_parent"
        android:layout_height="match_parent"
        android:paddingBottom="@dimen/activity_vertical_margin"
        android:paddingLeft="@dimen/activity_horizontal_margin"
        android:paddingRight="@dimen/activity_horizontal_margin"
        android:paddingTop="@dimen/activity_vertical_margin"
        tools:context="cn.gx.hzu_progressdialog.ProgressDialogActivity">

<LinearLayout
            android:id="@+id/ll01"
```

```
13              android:layout_width="wrap_content"
14              android:layout_height="wrap_content">
15
16      <TextView
17                  android:id="@+id/tv_gs"
18                  android:layout_width="wrap_content"
19                  android:layout_height="wrap_content"
20                  android:text="1+2+3+..+100=" />
21
22      <TextView
23                  android:id="@+id/tv_result"
24                  android:layout_width="wrap_content"
25                  android:layout_height="wrap_content"
26                  android:text="@string/hello_world" />
27      </LinearLayout>
28
29      <Button
30              android:id="@+id/button1"
31              android:layout_width="wrap_content"
32              android:layout_height="wrap_content"
33              android:layout_below="@+id/ll01"
34              android:text="点击开始计算" />
35
36      </RelativeLayout>
```

1）第 16～20 行文件控件用来显示前面的式子。

2）第 22～26 行文本控件用来显示最终的计算结果。

3）第 29～34 行按钮控件用来打开进度对话框。

（3）开发逻辑代码。打开 src/cn.gx.hzu_progressdialog 包下的 ProgressDialogActivity.java 文件，其代码内容如下：

```
package cn.gx.hzu_progressdialog;

import android.app.Activity;
import android.app.Dialog;
import android.app.ProgressDialog;
import android.content.DialogInterface;
import android.os.Bundle;
import android.os.Handler;
import android.os.Message;
import android.renderscript.Sampler.Value;
import android.view.Menu;
import android.view.MenuItem;
import android.view.View;
import android.widget.*;

public class ProgressDialogActivity extends Activity {
```

```
    ProgressDialog pd;
    Button btn;
    TextView tv;
    Handler myHandler;
    final int PROGRESSDIALOG = 1;
    final int INCREASE = 2;
    int sum = 0;

    @Override
    protected void onCreate(Bundle savedInstanceState) {
        super.onCreate(savedInstanceState);
        setContentView(R.layout.activity_progress_dialog);
        btn = (Button) findViewById(R.id.button1);
        btn.setOnClickListener(new View.OnClickListener() {

            @Override
            public void onClick(View v) {
                showDialog(PROGRESSDIALOG);
            }
        });
        myHandler = new Handler() {

            @Override
            public void handleMessage(Message msg) {
                switch (msg.what) {
                    case INCREASE:
                        pd.incrementProgressBy(1);
                        pd.setMessage("同步计算结果:" +String.valueOf(sum));
                        if (pd.getProgress() >= 100){
                            tv= (TextView) findViewById(R.id.tv_result);
                            tv.setText(String.valueOf(sum));
                        }
                        break;
                }
                super.handleMessage(msg);
            }
        };
    }

    @Override
    @Deprecated
    protected Dialog onCreateDialog(int id) {
        // TODO Auto-generated method stub
        Dialog dialog = null;
        switch (id) {
        case PROGRESSDIALOG:
            pd = new ProgressDialog(this);
```

```
                pd.setMax(100);
                pd.setProgressStyle(ProgressDialog.STYLE_HORIZONTAL);
                pd.setTitle("亲，正在努力计算哟！");
                pd.setMessage("请稍等...");
                pd.setCancelable(true);
                pd.setOnDismissListener(new DialogInterface
                    .OnDismissListener() {
                    @Override
                    public void onDismiss(DialogInterface dialog) {
                        sum=0;
                    }
                });
                dialog = pd;
                break;
            }
            return dialog;
        }

        @Override
        @Deprecated
        protected void onPrepareDialog(int id, Dialog dialog) {
            super.onPrepareDialog(id, dialog);
            switch (id) {
            case PROGRESSDIALOG:
                pd.incrementProgressBy(-pd.getProgress());
                new Thread() {

                    @Override
                    public void run() {
                        // TODO Auto-generated method stub
                        for (int i = 1; i <= 100; i++) {
                            sum += i;
                            myHandler.sendEmptyMessage(INCREASE);
                            try {
                                Thread.sleep(100);
                            } catch (Exception e) {
                                e.printStackTrace();
                            }
                        }
                    }
                }.start();
                break;
            }
        }
    }
```

1）第 17 行声明进度对话框，这个对话框来显示计算的进度。

2）第 18 行声明一个按钮，点击这个按钮打开进度对话框。

3）第 19 行声明一个文本控件，这个控件用来显示最终的计算结果。

4）第 20 行声明 Handler 对象，来实现线程的通信。

5）第 21 行声明整型的变量 PROGRESSDIALOG，来作为进度对话框的 ID。

6）第 22 行声明整型的变量 INCREASE，来作为消息内容。

7）第 23 行声明整型变量 sum，用来临时存放最终的计算结果。

8）第 29 行初始化按钮控件。

9）第 30 行给按钮添加监听器，当点击按钮时打开进度对话框。

10）第 37～53 行创建 Handler 对象，完成了线程通信。

11）第 40 行 handMessage()方法是 Handler 对象处理接收消息的方法。Message 类型参数 msg 表示接收到的消息。

12）第 41 行使用 switch 语句对接收到的消息进行判断。

13）第 42～48 行表示如果接收到的消息是整型变量 INCREASE，首先将进度对话框的进度加 1（43 行）；其次，在进度对话框的消息内容位置同步暂时的计算结果（44 行）；再次判断进度对话框有没有达到最大值，如果达到了最大值，说明已经计算结束，这时获取结果文件控件对象，将结果文本对象的值设置为最终计算的值（46 行和 47 行）。

14）第 58～80 行重写 onCreateDialog()方法，在这个方法中实现对话框的创建。

15）第 61～62 行使用 switch 判断语句，如果要创建的是进度对话框，则创建进度对话框。

16）第 63 行创建进度对话框。

17）第 64 行给进度对话框设置最大值。

18）第 65 行设置进度对话框的风格为水平条状。

19）第 66 行为进度对话框设置标题。

20）第 67 行为进度对话框设置消息内容。

21）第 68 行设置进度对话框是否可取消，如果方法参数是 true，说明是可取消的，当点击对话框之后的区域，对话框将取消。如果方法参数是 false，说明是不可取消的，必须等进度达到最大值之后，才能决定对话框是否消失。

22）第 69～75 行给对话框添加监听器，当对话框消失时，让 sum 的值为 0。为什么要这样呢？因为 sum 用来暂存每一次的计算结果，如果进度对话框消失时，不将 sum 重置为 0 的话，那么第二次再打开进度对话框进行计算的时候，计算的结果将会从上一次计算的结果中进行累加。

23）第 84～107 行重写 onPrepareDialog()方法，每次打开进度对话框的时候都用到这个方法。这个方法很重要，如果不重写这个方法，进度对话框的进度是没有办法更新的，进度对话框只能保留第一次打开的样子。

24）第 89～104 行新建一个线程，并启动这个线程，在线程里完成对进度对话框的更新。

25）第 94～102 行使用 for 循环来实现从 1 到 100 相加的操作，每完成一次加法，就利用 Handler 对象发送一次消息，Handler 对象接收到这个消息之后，就将进度的值加 1。

26）第 98 行表示每一次相加操作后都让线性休眠一会儿。因为计算机的加法操作是非常快的，这样大家看不到进度对话框变化的过程，为了让大家看到这个变化过程，每次让线程休眠一会儿。

【运行结果】程序的运行结果如图 6-10 所示。

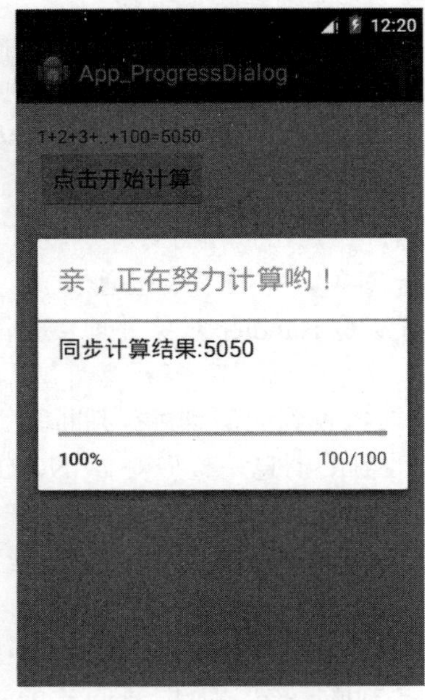


图 6-10　进度对话框

## 6.4　本章小结

本章首先讲解了菜单的各种类型和使用场景，接着介绍了 ActionBar 的基本属性与方法，最后讲解了对话框 Dialog、Toast 以及其他 Dialog 控件，并给出了相应案例。

# 第 7 章　Activity

（1）了解 Activity 的四种状态。

（2）掌握 Activity 的生命周期。

（3）掌握 Intent 与 Bundle 的相关属性与使用方法。

在 Android 的应用程序中，Activity 用于提供可视化的用户界面，一个 Android 应用程序由一个或多个 Activity 组成。从一个 Activity 跳转到另一个 Activity 需要使用 Intent 对象，当在跳转时如果需要传递数据，还要用到 Bundle 对象。接下来将以上内容进行详细介绍。

## 7.1　Activity 简介

Android 程序是由 Activity、Service、Content Provider 和 BroadcastReceiver 等组件组成，其中 Activity 是最基本、最常用的组件。Activity 中所有操作都与用户密切相关，它负责与用户直接进行交互，通过使用 setContentView(View)来显示具体内容。每个 Activity 都有相对应的生命周期，通过使用生命周期可以精准地控制程序流程，提升用户体验。

## 7.2　Activity 的四种状态

栈是一个特殊线性表，它遵循着后进先出的原则。栈有两个动作：压栈（把对象压入到栈中）和弹栈（每次从栈中取出最后一个对象）。每个 Android 应用程序都由一个或多个 Activity 组成，这些 Activity 统一存放在一个栈中，这个栈被称为 Activity 栈。

Activity 作为 Android 程序最重要的成员之一，主要有以下四种状态：

- Running 状态：当一个 Activity 启动后，它将在屏幕最前端，处于栈的最顶端，此时它可以被看见并且能够获得焦点，可以接收用户输入等。如图 7-1 所示为一个 Activity 的 Running 状态。
- Paused 状态：当一个 Activity 被另外一个透明的 Activity、Toast 或 AlertDialog 等覆盖时就处于暂停状态，此时它保留了所有的状态和成员信息，并且保持和窗口管理器的连接。它是可见的，只是失去焦点，所以不可以与用户进行交互。如图 7-2 所示为一个 Activity 的 Paused 状态。
- Stopped 状态：当一个 Activity 完全被另一个 Activity 遮挡时就处于停止状态，它仍然保留着所有的状态和成员信息。只是对用户不可见，当其他地方需要内存时它往往会被系统杀掉。

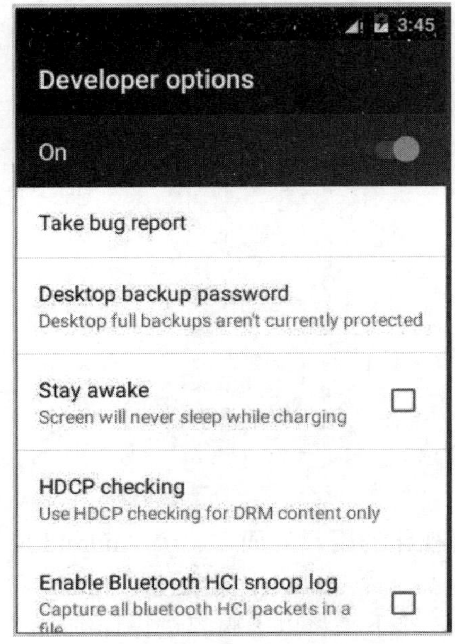

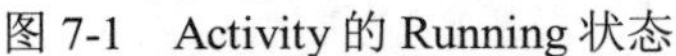

图 7-1 Activity 的 Running 状态

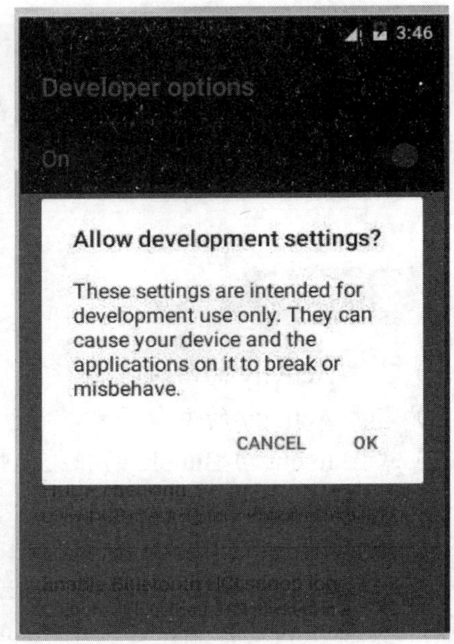


图 7-2 Activity 的 Paused 状态

- Killed 状态：Activity 尚未被启动时已经被手动终止，或已经被系统回收时处于非活动的状态，此时它被移除出 Activity 栈，需要重新启动才可以使用。

## 7.3 Activity 生命周期

在前面的学习中，我们可以发现所有继承自 Activity 的类都会重写 onCreate()方法，程序会自动加载此方法。在 Activity 类中还有 onStart()、onPause()、onStop()、onResume()、onRestart()、onDestroy()等方法，这些方法的先后执行构成 Activity 一个完整的生命周期，如图 7-3 所示为 Activity 的生命周期图。

表 7-1 对整个 Activity 生命周期中的每个方法进行了详细说明。

**【例 7.1】**在 Android 应用程序中，创建 A 与 B 两个 Activity，说明从 A 跳转到 B 以及从 B 返回到 A 过程中 Activity 生命周期各个方法调用情况。

**【说明】**

（1）运行项目后，先启动 A 这个 Activity，依次调用的方法如下：

onCreate(A)→onStart(A)→onResume(A)

（2）在 A 不关闭的情况下跳转到 B，依次调用的方法如下：

onPause(A)→onCreate(B)→onStart(B)→onResume(B)→onStop(A)

（3）按“返回”键返回到 A，依次调用的方法如下：

onPause(B)→onRestart(A)→onStart(A)→onResume(A)→onStop(B)→onDestroy(B)

（4）按“退出应用程序”按钮，依次调用的方法如下：

onPause(A)→onStop(A)→onDestroy(A)

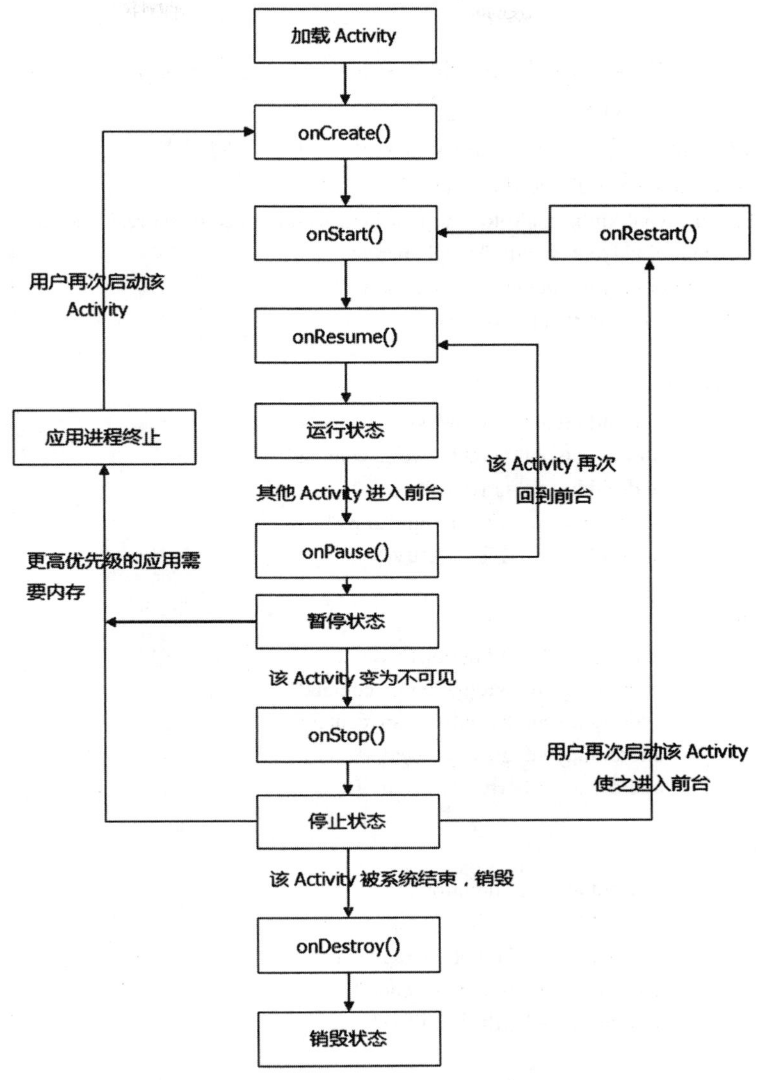


图 7-3　Activity 的生命周期

表 7-1　生命周期方法说明

| 方法 | 说明 | 下一个 |
| --- | --- | --- |
| OnCreate() | 创建 Activity 时调用，这里是做所有初始化的地方 | onStart() |
| onStart() | 当 Activity 变为在屏幕上对用户可见时调用 | onResume()或 onStop() |
| onPause() | 当系统将要启动另外一个 Activity 时调用 | onResume()或 onStop() |
| onResume() | 当 Activity 开始与用户交互时调用 | onPause() |
| onStop() | 当 Activity 被停止并转为不可见时调用 | onRestart()或 onDestroy() |
| onRestart() | 当重新启动 Activity 时调用 | onStart() |
| onDestroy() | 当 Activity 被销毁时调用 | 无 |

**【开发步骤】**

（1）创建一个名为 Activity_Lifecycle_test 的项目，包名为 com.hzu.activity_lifecycle_test，在其包下有 A 与 B 两个 Activity。

（2）在 res/layout 目录下创建 activity_a.xml 文件，其内容如下：

```
1   <?xml version="1.0" encoding="utf-8"?>
2   <LinearLayout xmlns:android="http://schemas.android.com/apk/res/android"
3           android:layout_width="match_parent"
4           android:layout_height="match_parent"
5           android:orientation="vertical">
6
7   <TextView
8                   android:layout_width="wrap_content"
9                   android:layout_height="wrap_content"
10                  android:layout_gravity="center"
11                  android:layout_marginTop="20px"
12                  android:text="当前是 Activity A" />
13
14  <Button
15                  android:id="@+id/button1"
16                  android:layout_width="wrap_content"
17                  android:layout_height="wrap_content"
18                  android:layout_gravity="center"
19                  android:text="跳转到 Activity B" />
20
21  <Button
22                  android:id="@+id/button2"
23                  android:layout_width="wrap_content"
24                  android:layout_height="wrap_content"
25                  android:layout_gravity="center"
26                  android:text="退出应用程序" />
27
28  </LinearLayout>
```

1）第 1～28 行声明了一个线性布局，方向为垂直方向。

2）第 7～12 行声明了一个 TextView，默认显示内容为“当前是 Activity A”。

3）第 14～19 行声明了一个 Button，id 值为 button1，默认显示内容为“跳转到 Activity B”。

4）第 21～26 行声明了一个 Button，id 值为 button2，默认显示内容为“退出应用程序”。

（3）在 res/layout 目录下创建 activity_b.xml 文件，其内容如下：

```
<?xml version="1.0" encoding="utf-8"?>
<LinearLayout xmlns:android="http://schemas.android.com/apk/res/android"
        android:layout_width="match_parent"
        android:layout_height="match_parent"
        android:orientation="vertical">
<TextView
            android:layout_width="wrap_content"
            android:layout_height="wrap_content"
```

```
9          android:layout_gravity="center"
10         android:layout_marginTop="20px"
11         android:text="当前是 Activity B" />
12 </LinearLayout>
```

1）第 1～12 行声明了一个线性布局，方向为垂直方向。

2）第 6～11 行声明了一个 TextView，默认显示内容为“当前是 Activity B”。

（4）在 com.hzu.activity_lifecycle_test 包下创建名为 A 的这个 Activity，其内容如下：

```
package com.hzu.activity_lifecycle_test;

import android.app.Activity;
import android.content.Intent;
import android.os.Bundle;
import android.util.Log;
import android.view.View;
import android.view.View.OnClickListener;
import android.widget.Button;

public class A extends Activity {
        Button button;
        Button exitButton;

        @Override
        protected void onCreate(Bundle savedInstanceState) {
                // TODO Auto-generated method stub
                super.onCreate(savedInstanceState);
                setContentView(R.layout.activity_a);
                button = (Button) findViewById(R.id.button1);
                exitButton = (Button) findViewById(R.id.button2);
                Log.i("Lifecycle", "A 生命周期方法 onCreate()");
                button.setOnClickListener(new OnClickListener() {
                        @Override
                        public void onClick(View v) {
                                // TODO Auto-generated method stub
                                Intent intent = new Intent();
                                intent.setClass(A.this, B.class);
                                startActivity(intent);
                        }
                });

                exitButton.setOnClickListener(new View.OnClickListener() {

                        @Override
                        public void onClick(View v) {
                                finish();

                        }
```

```
40                 });
41
42         }
43
44         @Override
45         protected void onStart() {
46                 Log.i("Lifecycle", "A 生命周期方法 onStart()");
47                 super.onStart();
48         }
49
50         @Override
51         protected void onPause() {
52                 Log.i("Lifecycle", "A 生命周期方法 onPause()");
53
54                 super.onPause();
55         }
56
57         @Override
58         protected void onRestart() {
59                 Log.i("Lifecycle", "A 生命周期方法 onRestart()");
60                 super.onRestart();
61         }
62
63         @Override
64         protected void onResume() {
65                 Log.i("Lifecycle", "A 生命周期方法 onResume()");
66                 super.onResume();
67         }
68
69         @Override
70         protected void onStop() {
71                 Log.i("Lifecycle", "A 生命周期方法 onStop()");
72                 super.onStop();
73         }
74
75         @Override
76         protected void onDestroy() {
77                 Log.i("Lifecycle", "A 生命周期方法 onDestroy()");
78                 super.onDestroy();
79         }
80
81 }
```

1）第 3～9 行引入了代码中需要使用的 Java 和 Android 相关类。

2）第 11～81 行定义了 A 这个类。

3）第 16～42 行重写了 onCreate()方法，第 20 行获取 button 对象用于跳转到 B，第 21 行获取 exitButton 对象，用于退出这个应用程序，其中 finish()方法用于关闭当前 Activity。

4）第 45～79 行分别定义了 onStart()、onPause()、onRestart()、onResume()、onStop()、onDestroy()方法。

（5）在 com.hzu.activity_lifecycle_test 包下创建名为 B 的这个 Activity，其内容如下：

```
package com.hzu.activity_lifecycle_test;

import android.app.Activity;
import android.os.Bundle;
import android.util.Log;

public class B extends Activity {
	@Override
	protected void onCreate(Bundle savedInstanceState) {
		// TODO Auto-generated method stub
		super.onCreate(savedInstanceState);
		setContentView(R.layout.activity_b);
		Log.i("Lifecycle", "B 生命周期方法 onCreate()");
	}

	@Override
	protected void onStart() {
		Log.i("Lifecycle", "B 生命周期方法 onStart()");
		super.onStart();
	}

	@Override
	protected void onPause() {
		Log.i("Lifecycle", "B 生命周期方法 onPause()");
		finish();
		super.onPause();
	}

	@Override
	protected void onRestart() {
		Log.i("Lifecycle", "B 生命周期方法 onRestart()");
		super.onRestart();
	}

	@Override
	protected void onResume() {
		Log.i("Lifecycle", "B 生命周期方法 onResume()");
		super.onResume();
	}

	@Override
	protected void onStop() {
		Log.i("Lifecycle", "B 生命周期方法 onStop()");
```

```
44              super.onStop();
45          }
46
47          @Override
48          protected void onDestroy() {
49              Log.i("Lifecycle", "B 生命周期方法 onDestroy()");
50              super.onDestroy();
51          }
52
53      }
```

1）第 3～5 行引入了代码中需要使用的 Java 与 Android 相关类。

2）第 7～53 行定义了 B 这个类。

3）第 9～14 行重写了 onCreate()方法，添加了 Log.i("Lifecycle", "B 生命周期方法 onCreate()")方法，用于展示 onCreate()方法什么时候被调用。

4）第 17～51 行分别定义了 onStart()、onPause()、onRestart()、onResume()、onStop()、onDestroy()方法。

（6）在项目下的 AndroidManifest.xml 文件中的内容如下：

```
<?xml version="1.0" encoding="utf-8"?>
<manifest xmlns:android="http://schemas.android.com/apk/res/android"
    package="com.hzu.activity_lifecycle_test"
    android:versionCode="1"
    android:versionName="1.0">

<uses-sdk
        android:minSdkVersion="14"
        android:targetSdkVersion="21" />

<application
        android:allowBackup="true"
        android:icon="@drawable/ic_launcher"
        android:label="@string/app_name"
        android:theme="@style/AppTheme">
<activity
            android:name=".A"
            android:label="@string/app_name">
<intent-filter>
<action android:name="android.intent.action.MAIN" />
<category android:name="android.intent.category.LAUNCHER" />
</intent-filter>
</activity>
<activity android:name="B">
</activity>
</application>

</manifest>
```

1）第 16～23 行设置 A 为程序启动后显示的首页。

2）第 24～25 行添加 B 到 application 结点中。

**【运行结果】**在 Eclipse 中启动 Android 模拟器，接着运行 Activity_Lifecycle_test 项目，显示效果如图 7-4 所示，点击“跳转到 Activity B”按钮后显示效果如图 7-5 所示。

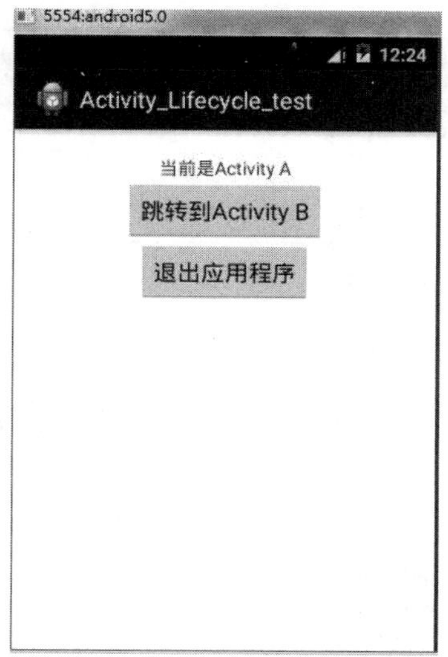

图 7-4　运行项目的效果

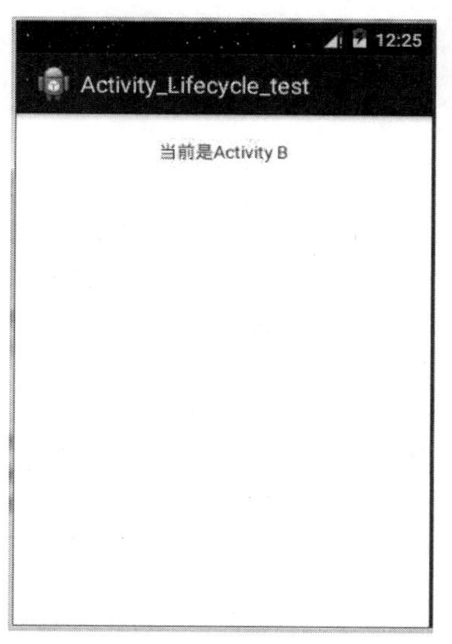


图 7-5　点击“跳转到 Activity B”后的效果

**注明：**

1）项目运行后，在 Lifecycle 过滤器中显示的效果如图 7-6 所示。

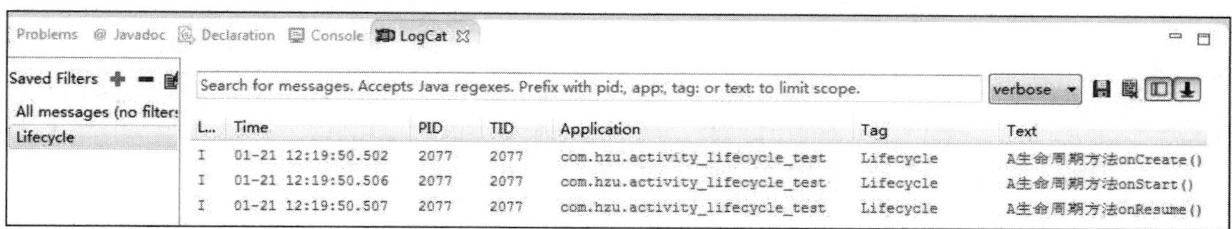

| L... | Time | PID | TID | Application | Tag | Text |
|---|---|---|---|---|---|---|
| I | 01-21 12:19:50.502 | 2077 | 2077 | com.hzu.activity_lifecycle_test | Lifecycle | A生命周期方法onCreate() |
| I | 01-21 12:19:50.506 | 2077 | 2077 | com.hzu.activity_lifecycle_test | Lifecycle | A生命周期方法onStart() |
| I | 01-21 12:19:50.507 | 2077 | 2077 | com.hzu.activity_lifecycle_test | Lifecycle | A生命周期方法onResume() |

图 7-6　启动应用程序时调用的方法

2）在点击“跳转到 Activity B”按钮后，Lifecycle 过滤器中显示的效果如图 7-7 所示。

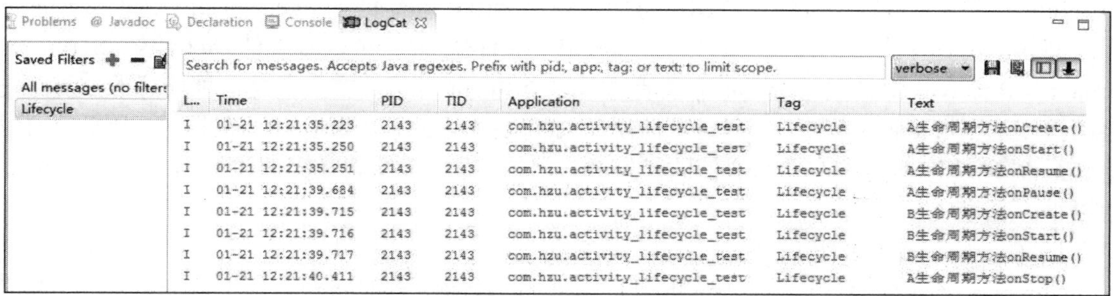

| L... | Time | PID | TID | Application | Tag | Text |
|---|---|---|---|---|---|---|
| I | 01-21 12:21:35.223 | 2143 | 2143 | com.hzu.activity_lifecycle_test | Lifecycle | A生命周期方法onCreate() |
| I | 01-21 12:21:35.250 | 2143 | 2143 | com.hzu.activity_lifecycle_test | Lifecycle | A生命周期方法onStart() |
| I | 01-21 12:21:35.251 | 2143 | 2143 | com.hzu.activity_lifecycle_test | Lifecycle | A生命周期方法onResume() |
| I | 01-21 12:21:39.684 | 2143 | 2143 | com.hzu.activity_lifecycle_test | Lifecycle | A生命周期方法onPause() |
| I | 01-21 12:21:39.715 | 2143 | 2143 | com.hzu.activity_lifecycle_test | Lifecycle | B生命周期方法onCreate() |
| I | 01-21 12:21:39.716 | 2143 | 2143 | com.hzu.activity_lifecycle_test | Lifecycle | B生命周期方法onStart() |
| I | 01-21 12:21:39.717 | 2143 | 2143 | com.hzu.activity_lifecycle_test | Lifecycle | B生命周期方法onResume() |
| I | 01-21 12:21:40.411 | 2143 | 2143 | com.hzu.activity_lifecycle_test | Lifecycle | A生命周期方法onStop() |

图 7-7　跳转 Activity 时调用的方法

3）在按“返回”物理按键后，显示效果如图 7-8 所示。

Problems @ Javadoc Declaration Console LogCat

Saved Filters

All messages (no filters)

Lifecycle

Search for messages. Accepts Java regexes. Prefix with pid:, app:, tag: or text: to limit scope. verbose

| L... | Time | PID | TID | Application | Tag | Text |
|---|---|---|---|---|---|---|
| I | 01-21 12:21:35.223 | 2143 | 2143 | com.hzu.activity_lifecycle_test | Lifecycle | A生命周期方法onCreate() |
| I | 01-21 12:21:35.250 | 2143 | 2143 | com.hzu.activity_lifecycle_test | Lifecycle | A生命周期方法onStart() |
| I | 01-21 12:21:35.251 | 2143 | 2143 | com.hzu.activity_lifecycle_test | Lifecycle | A生命周期方法onResume() |
| I | 01-21 12:21:39.684 | 2143 | 2143 | com.hzu.activity_lifecycle_test | Lifecycle | A生命周期方法onPause() |
| I | 01-21 12:21:39.715 | 2143 | 2143 | com.hzu.activity_lifecycle_test | Lifecycle | B生命周期方法onCreate() |
| I | 01-21 12:21:39.716 | 2143 | 2143 | com.hzu.activity_lifecycle_test | Lifecycle | B生命周期方法onStart() |
| I | 01-21 12:21:39.717 | 2143 | 2143 | com.hzu.activity_lifecycle_test | Lifecycle | B生命周期方法onResume() |
| I | 01-21 12:21:40.411 | 2143 | 2143 | com.hzu.activity_lifecycle_test | Lifecycle | A生命周期方法onStop() |
| I | 01-21 12:22:17.358 | 2143 | 2143 | com.hzu.activity_lifecycle_test | Lifecycle | B生命周期方法onPause() |
| I | 01-21 12:22:17.367 | 2143 | 2143 | com.hzu.activity_lifecycle_test | Lifecycle | A生命周期方法onRestart() |
| I | 01-21 12:22:17.371 | 2143 | 2143 | com.hzu.activity_lifecycle_test | Lifecycle | A生命周期方法onStart() |
| I | 01-21 12:22:17.371 | 2143 | 2143 | com.hzu.activity_lifecycle_test | Lifecycle | A生命周期方法onResume() |
| I | 01-21 12:22:17.958 | 2143 | 2143 | com.hzu.activity_lifecycle_test | Lifecycle | B生命周期方法onStop() |
| I | 01-21 12:22:17.958 | 2143 | 2143 | com.hzu.activity_lifecycle_test | Lifecycle | B生命周期方法onDestroy() |

图 7-8　按“返回”键后调用的方法

4）在图 7-4 运行项目的效果中点击“退出应用程序”按钮，显示效果如图 7-9 所示。

Problems @ Javadoc Declaration Console LogCat

Saved Filters

All messages (no filters)

Lifecycle

Search for messages. Accepts Java regexes. Prefix with pid:, app:, tag: or text: to limit scope. verbose

| L... | Time | PID | TID | Application | Tag | Text |
|---|---|---|---|---|---|---|
| I | 01-21 12:24:20.806 | 2210 | 2210 | com.hzu.activity_lifecycle_test | Lifecycle | A生命周期方法onCreate() |
| I | 01-21 12:24:20.807 | 2210 | 2210 | com.hzu.activity_lifecycle_test | Lifecycle | A生命周期方法onStart() |
| I | 01-21 12:24:20.807 | 2210 | 2210 | com.hzu.activity_lifecycle_test | Lifecycle | A生命周期方法onResume() |
| I | 01-21 12:24:23.693 | 2210 | 2210 | com.hzu.activity_lifecycle_test | Lifecycle | A生命周期方法onPause() |
| I | 01-21 12:24:23.778 | 2210 | 2210 | com.hzu.activity_lifecycle_test | Lifecycle | B生命周期方法onCreate() |
| I | 01-21 12:24:23.779 | 2210 | 2210 | com.hzu.activity_lifecycle_test | Lifecycle | B生命周期方法onStart() |
| I | 01-21 12:24:23.779 | 2210 | 2210 | com.hzu.activity_lifecycle_test | Lifecycle | B生命周期方法onResume() |
| I | 01-21 12:24:24.564 | 2210 | 2210 | com.hzu.activity_lifecycle_test | Lifecycle | A生命周期方法onStop() |
| I | 01-21 12:24:25.633 | 2210 | 2210 | com.hzu.activity_lifecycle_test | Lifecycle | B生命周期方法onPause() |
| I | 01-21 12:24:25.656 | 2210 | 2210 | com.hzu.activity_lifecycle_test | Lifecycle | A生命周期方法onRestart() |
| I | 01-21 12:24:25.657 | 2210 | 2210 | com.hzu.activity_lifecycle_test | Lifecycle | A生命周期方法onStart() |
| I | 01-21 12:24:25.658 | 2210 | 2210 | com.hzu.activity_lifecycle_test | Lifecycle | A生命周期方法onResume() |
| I | 01-21 12:24:26.810 | 2210 | 2210 | com.hzu.activity_lifecycle_test | Lifecycle | B生命周期方法onStop() |
| I | 01-21 12:24:26.810 | 2210 | 2210 | com.hzu.activity_lifecycle_test | Lifecycle | B生命周期方法onDestroy() |
| I | 01-21 12:24:27.920 | 2210 | 2210 | com.hzu.activity_lifecycle_test | Lifecycle | A生命周期方法onPause() |
| I | 01-21 12:24:28.745 | 2210 | 2210 | com.hzu.activity_lifecycle_test | Lifecycle | A生命周期方法onStop() |
| I | 01-21 12:24:28.745 | 2210 | 2210 | com.hzu.activity_lifecycle_test | Lifecycle | A生命周期方法onDestroy() |

图 7-9　退出应用程序调用的方法

## 7.4　Intent

Intent 的中文意思是目的、意图、意向，在 Android 中提供了 Intent 机制来协助处理应用程序间和应用程序内部的交互与通信。Android 应用程序的三个核心组件 Activity、Service、BroadcastReceiver 都是通过 Intent 来激活的。一个 Intent 对象其实质就是一堆信息的捆绑。它包括接收这个 Intent 对象组件感兴趣的信息，如将要执行的动作和操作的数据等。因此，可以将 Intent 理解为不同组件之间通信的“媒介”，专门提供组件间互相调用的相关信息。Intent 由以下各个组成部分：

1. Component（组件）

Component 属性明确指定 Intent 的目标组件的类名称，若 Component 有指定的属性，将直接使用它指定的组件。若没有指定，将要在 AndroidManifest.xml 文件中，使用 intent-filter 查找与该 Intent 最合适的组件。Component 属性的案例如下：

```
Intent intent = new Intent();   //创建一个意图对象
        //创建组件，通过组件来响应
  ComponentName component = new ComponentName(FirstActivity.this, SecondActivity.class);
  intent.setComponent(component);
  startActivity(intent);
```

简洁写法如下：

```
Intent intent = new Intent();
intent.setClass(FirstActivity.this, SecondActivity.class);
startActivity(intent);
```

最简结写法如下：

```
Intent intent = new Intent(FirstActivity.this, SecondActivity.class);
startActivity(intent);
```

2. Action（动作）

在日常生活中，当我们有意愿时，会有一个动词在其中。例如我想“做”作业、我要“听”音乐等。在 Intent 中 Action 就是描述做、听等动作的，在指明一个 Action 时，执行者将会按照此动作的指示，接收相关输入，表现出对应的行为，产生相应的输出。Action 是一个被执行的动作字符串，用于描述一个 Android 应用程序组件，一个 Intent 可以包含多个 Action。在 AndroidManifest.xml 的 Activity 定义中，有一个<intent-filter>结点指定一个 Action 列表用于标识 Activity 所能接收的“动作”。在 Intent 类中，定义了一批动作，如表 7-2 所示。

**表 7-2　常用的动作常量说明**

| 常量 | 目标组件 | 动作说明 |
|---|---|---|
| ACTION_BATTERY_LOW | BroadcastReceiver | 提示电池电量过低 |
| ACTION_HEADSET_PLUG | BroadcastReceiver | 耳机拔出 |
| ACTION_LOCALE_CHANGED | BroadcastReceiver | 位置改变 |
| ACTION_SCREEN_OFF | BroadcastReceiver | 屏幕已关闭 |
| ACTION_CREATE_SHORTCUT | Activity | 创建快捷方式 |
| ACTION_MAIN | Activity | 指定当前应用程序的入口 |
| ACTION.CALL | Activity | 呼叫指定的电话号码 |
| ACTION_EDIT | Activity | 指定数据给用户进行编辑 |

3. Category（类别）

Category 用来表现动作的类别，它是 Action 中要执行的动作的附加描述，可以将多个类别描述放到一个 Intent 对象中。在 Intent 类中，定义了一批类别常量，如表 7-3 所示。

**表 7-3　常用的类别常量说明**

| 常量 | 说明 |
|---|---|
| CATEGORY_DEFAULT | 设置这个种类让组件成为 Intent Filter 中定义的 Data 的默认动作。这对使用显式 Intent 启动的 Activity 是必要的 |
| CATEGORY_BROWSABLE | 指定该 Activity 能被浏览器安全调用 |

续表

| 常量 | 说明 |
| --- | --- |
| CATEGORY_HOME | 设置该 Activity 随系统启动而运行 |
| CATEGORY_LAUNCHER | Activity 显示在顶级程序列表中 |
| CATEGORY_PREFERENCE | 该 Activity 是参数面板 |

4. Data（数据）

Data 表示动作要操纵的数据，Data 属性是 Android 要访问的数据，和 Action、Category 声明方式相同，也是在<intent-filter>中。Data 是用一个 URI 对象来表示的，URI 代表数据的地址，属于一种标识符。通常情况下，我们使用 Action+Data 属性的组合来描述一个意图。

例如打开指定网页：

```
Intent intent = new Intent();
intent.setAction(Intent.ACTION_VIEW);
Uri data = Uri.parse("http://www.163.com");
intent.setData(data);
startActivity(intent);
```

5. Extras（扩展信息）

Extras 属性主要用于传递目标组件所需要的额外数据，通过 putExtras()方法设置。

6. Flags（标志）

Flags 是各种类型的标志，许多标志就是用来指定 Android 系统以哪个方式去加载一个 Activity。

Intent 可以分成显式和隐式两类。显式 Intent 通过组件名称来指定目标组件。由于其他应用程序的组件名称对于开发人员来说通常是未知的，所以显式 Intent 通常用于应用程序内部消息处理，例如一个 Activity 启动 Service 或其他 Activity。隐式 Intent 不指定组件名称，通常用于激活其他应用程序中的组件。隐式 Intent 处理机制是将 Android 中 Intent 请求内容与应用程序 AndroidManifest.xml 文件中的 IntentFilter 的过滤器比较，IntentFilter 包含系统中所有可能的供选择组件。如果 IntentFilter 中某一个组件匹配隐式 Intent 请求时，那么 Android 就选择该组件作为隐式 Intent 的目标组件。

**【例 7.2】**在 Android 应用程序中，创建 Activity A、Activity B 和 Activity C，分别完成隐式意图跳转到其他应用程序，以及在应用程序内部完成显式意图跳转与隐式意图跳转。

**【说明】**通过使用 Intent 的 setAction()方法来完成隐式意图从一个应用程序跳转到另一个应用程序；通过使用 Intent 的 setClass()方法来完成显式意图跳转。

**【开发步骤】**

（1）创建一个名为 Intent_test 的项目，包名为 com.hzu.intent_test，在其包下有 Activity A、Activity B 与 Activity C。

（2）在 res/layout 目录下创建 activity_a.xml 文件，其内容如下：

```
<LinearLayout xmlns:android="http://schemas.android.com/apk/res/android"
    xmlns:tools="http://schemas.android.com/tools"
    android:layout_width="match_parent"
    android:layout_height="match_parent"
```

```
5        android:orientation="vertical"
6    >
7
8    <TextView
9            android:id="@+id/textView1"
10           android:layout_width="wrap_content"
11           android:layout_height="wrap_content"
12           android:text="Activity A"
13           android:layout_gravity="center"
14           />
15
16   <Button
17           android:id="@+id/button1"
18           android:layout_width="fill_parent"
19           android:layout_height="wrap_content"
20           android:text="启动设置界面" />
21
22   <Button
23           android:id="@+id/button2"
24           android:layout_width="fill_parent"
25           android:layout_height="wrap_content"
26           android:text="跳转到 Activity B" />
27
28   <Button
29           android:id="@+id/button3"
30           android:layout_width="fill_parent"
31           android:layout_height="wrap_content"
32           android:text="跳转到 Activity C" />
33
34   </LinearLayout>
```

1）第 1～34 行声明了一个线性布局，方向为垂直方向。

2）第 8～14 行声明了一个 TextView，居中显示，默认显示内容为“Activity A”。

3）第 16～32 行分别声明三个 Button，其 id 分别为 button1、button2 和 button3。

（3）在 res/layout 目录下创建 activity_b.xml 文件，其内容如下：

```
<?xml version="1.0" encoding="utf-8"?>
<LinearLayout xmlns:android="http://schemas.android.com/apk/res/android"
    android:layout_width="match_parent"
    android:layout_height="match_parent"
    android:orientation="vertical">

<TextView
        android:id="@+id/textView1"
        android:layout_width="wrap_content"
        android:layout_height="wrap_content"
        android:layout_gravity="center"
        android:text="Activity B" />
```

```
13
14    </LinearLayout>
```

1）第 1～14 行声明了一个线性布局，方向为垂直方向。

2）第 7～12 行声明了一个 TextView，居中显示，默认显示内容为“Activity B”。

（4）在 res/layout 目录下创建 activity_c.xml 文件，其内容如下：

```
1     <?xml version="1.0" encoding="utf-8"?>
2     <LinearLayout xmlns:android="http://schemas.android.com/apk/res/android"
3         android:layout_width="match_parent"
4         android:layout_height="match_parent"
5         android:orientation="vertical">
6
7     <TextView
8             android:id="@+id/textView1"
9             android:layout_width="wrap_content"
10            android:layout_height="wrap_content"
11            android:layout_gravity="center"
12            android:text="Activity C" />
13
14    </LinearLayout>
```

1）第 1～14 行声明了一个线性布局，方向为垂直方向。

2）第 7～12 行声明了一个 TextView，居中显示，默认显示内容为“Activity C”。

（5）在 res/values/strings.xml 文件中内容如下：

```
<?xml version="1.0" encoding="utf-8"?>
<resources>

<string name="title_name1">ActivityA</string>
<string name="title_name2">ActivityB</string>
<string name="title_name3">ActivityC</string>

</resources>
```

（6）在 src 下包为 com.hzu.intent_test 的 ActivityA.java 文件中，编写如下内容：

```
package com.hzu.intent_test;

import android.app.Activity;
import android.content.Intent;
import android.os.Bundle;
import android.provider.Settings;
import android.view.View;
import android.widget.Button;

public class ActivityA extends Activity {
        private Button bt1, bt2, bt3;

        @Override
        protected void onCreate(Bundle savedInstanceState) {
```

```
15            super.onCreate(savedInstanceState);
16            setContentView(R.layout.activity_a);
17            bt1 = (Button) findViewById(R.id.button1);
18            bt2 = (Button) findViewById(R.id.button2);
19            bt3 = (Button) findViewById(R.id.button3);
20
21            bt1.setOnClickListener(new View.OnClickListener() {
22                @Override
23                public void onClick(View v) {
24                    Intent intent1 = new Intent();
25                    intent1.setAction(Settings.ACTION_SETTINGS);
26                    startActivity(intent1);
27                }
28            });
29
30            bt2.setOnClickListener(new View.OnClickListener() {
31                @Override
32                public void onClick(View v) {
33                    Intent intent2 = new Intent();
34                    intent2.setClass(ActivityA.this, ActivityB.class);
35                    startActivity(intent2);
36                }
37            });
38
39            bt3.setOnClickListener(new View.OnClickListener() {
40                @Override
41                public void onClick(View v) {
42                    Intent intent3 = new Intent();
43                    intent3.setAction("com.hzu.ActivityC");
44                    startActivity(intent3);
45                }
46            });
47        }
48
49    }
```

1）第 3～8 行引入代码中需要使用的 Java 与 Android 相关类。

2）第 10～49 行定义了 Activity A 类。

3）第 14～47 行重写了 onCreate()方法。

4）第 21～28 行为 bt1 添加监听方法，使用"intent1.setAction(Settings.ACTION_SETTINGS);"定义跳转到系统 Setting 菜单。

5）第 30～37 行为 bt2 添加监听方法，使用"intent2.setClass(ActivityA.this, ActivityB.class);"定义跳转到 Activity B。

6）第 39～46 行为 bt3 添加监听方法，使用"intent3.setAction("com.hzu.ActivityC");"定义隐式跳转到 Activity C。

（7）在 src 的包为 com.hzu.intent_test 的 ActivityB.java 文件中编写内容如下：

```
package com.hzu.intent_test;

import android.app.Activity;
import android.os.Bundle;

public class ActivityB extends Activity {
    @Override
    protected void onCreate(Bundle savedInstanceState) {
        super.onCreate(savedInstanceState);
        setContentView(R.layout.activity_b);
    }

}
```

（8）在 src 的包为 com.hzu.intent_test 的 ActivityC.java 文件中编写内容如下：

```
package com.hzu.intent_test;

import android.app.Activity;
import android.os.Bundle;

public class ActivityC extends Activity {
    @Override
    protected void onCreate(Bundle savedInstanceState) {
        super.onCreate(savedInstanceState);
        setContentView(R.layout.activity_c);
    }

}
```

（9）在 res/values/strings.xml 文件中的内容如下：

```
<?xml version="1.0" encoding="utf-8"?>
<resources>

<string name="title_name1">ActivityA</string>
<string name="title_name2">ActivityB</string>
<string name="title_name3">ActivityC</string>

</resources>
```

（10）在项目下 AndroidManifest.xml 文件中编写内容如下：

```
<?xml version="1.0" encoding="utf-8"?>
<manifest xmlns:android="http://schemas.android.com/apk/res/android"
    package="com.hzu.intent_test"
    android:versionCode="1"
    android:versionName="1.0">

<uses-sdk
        android:minSdkVersion="14"
```

```
9            android:targetSdkVersion="21" />
10
11    <application
12            android:allowBackup="true"
13            android:icon="@drawable/ic_launcher"
14            android:label="@string/title_name1"
15            android:theme="@style/AppTheme">
16    <activity
17                android:name=".ActivityA"
18                android:label="@string/title_name1">
19    <intent-filter>
20    <action android:name="android.intent.action.MAIN" />
21    <category android:name="android.intent.category.LAUNCHER" />
22    </intent-filter>
23    </activity>
24    <activity
25                android:name="ActivityB"
26                android:label="@string/title_name2">
27    </activity>
28    <activity
29                android:name=".ActivityC"
30                android:label="@string/title_name3">
31    <intent-filter>
32    <action android:name="com.hzu.ActivityC" />
33    <category android:name="android.intent.category.DEFAULT" />
34    </intent-filter>
35    </activity>
36    </application>
37
38    </manifest>
```

第 11～36 行是在 application 下的 activity 结点上分别添加标题名为 Activity A、Activity B、Activity C 等组件。

**【运行结果】**在 Eclipse 中启动 Android 模拟器，接着运行 Intent_test 项目，均能达到预期效果。

## 7.5 Bundle

在 Activity 生命周期中，很多地方都使用 Bundle 对象来保存数据。Bundle 类所携带的数据类似于 Map，用于存放 Key-Value（键-值对）形式的值。通过 Bundle 类强大的数据封装能力，把将要传递的数据使用 Intent 对象来传递到不同的 Activity 中，然后再进行数据拆解。Bundle 类来自于 android.os.Bundle 类，每个 Android 应用程序中都要用到 Bundle 类，所以在代码开始的地方都使用“import android.os.Bundle;”这样的语句。Bundle 类的常用方法如表 7-4 所示。

表 7-4 Bundle 类的常用方法说明

| 方法 | 说明 |
| --- | --- |
| void clear () | 从捆绑的映射中移除所有元素 |
| String getString (String key) | 返回指定键所映射的值 |
| boolean isEmpty () | 判断 Bundle 对象中是否有任何键值映射关系 |
| void putString (String key, String value) | 给指定的键关联一个指定的 String 类型值 |
| void remove (String key) | 删除 Bundle 对象中存在的该键的映射关系 |

## 7.6 Activity 传值与返回

在 Android 应用程序中，如果想通过一个 Activity A 启动 Activity B，并且从 A 传递数据到 B，通过 Intent 对象上绑定相关数据即可，这样 Intent 对象可以精准地指定到需要跳转的 Activity 上。另外一种情况是如果想通过一个 Activity A 启动 Activity B，并希望 Activity A 能在 Activity B 出栈时得到一些返回值，就需要以下三个步骤：

（1）在 Activity A 中使用 startActivityForResult()方法代替 startActivity()方法，告诉系统此 Activity A 需要返回结果。

（2）在 Activity B 中使用 setResult()方法设置返回时要传递的数据。

（3）在 Activity A 中使用 onActivityResult()方法接收 Activity B 返回的数据。

**【例 7.3】**在 Android 应用程序中，创建 MainActivity 与 SubActivity，通过“保存数据”按钮从 MainActivity 传一个数据到 SubActivity，通过“带回数据”按钮从 MainActivity 中启动 SubActivity，在 SubActivity 选择相关颜色改变 MainActivity 中“带回数据”按钮颜色。

**【说明】**使用 Intent 的 putExtra()绑定基本数据类型或使用 Intent 的 putExtras()绑定 Bundle 对象。

**【开发步骤】**

（1）创建一个名为 TransferData_test 的项目，包名为 com.hzu.transferdata_test，在其包下有 MainActivity.java 与 SubActivity.java 两个文件。

（2）在 res/layout 目录下创建 activity_main.xml 文件，其内容如下：

```
<LinearLayout xmlns:android="http://schemas.android.com/apk/res/android"
     xmlns:tools="http://schemas.android.com/tools"
     android:layout_width="match_parent"
     android:layout_height="match_parent"
     android:orientation="vertical">

<LinearLayout
          android:layout_width="match_parent"
          android:layout_height="wrap_content"
          android:orientation="horizontal">

<TextView
```

```
13                      android:layout_width="wrap_content"
14                      android:layout_height="wrap_content"
15                      android:text="请输入内容" />
16
17      <EditText
18                      android:id="@+id/et"
19                      android:layout_width="200dp"
20                      android:layout_height="wrap_content" />
21      </LinearLayout>
22
23      <Button
24                  android:id="@+id/button1"
25                  android:layout_width="match_parent"
26                  android:layout_height="wrap_content"
27                  android:text="保存数据" />
28
29      <Button
30                  android:id="@+id/button2"
31                  android:layout_width="match_parent"
32                  android:layout_height="wrap_content"
33                  android:text="带回数据" />
34
35      </LinearLayout>
```

1）第 1～35 行声明了一个线性布局，方向为垂直方向。

2）第 7～21 行添加一个子线性布局，方向为水平方向，在其线性布局中分别为 TextView 和 EditText。

3）第 23～27 行添加一个 Button，其 id 为 button1。

4）第 29～33 行添加另一个 Button，其 id 为 button2。

（3）在 res/layout 目录下创建 activity_sub.xml 文件，其内容如下：

```
<?xml version="1.0" encoding="utf-8"?>
<LinearLayout xmlns:android="http://schemas.android.com/apk/res/android"
    android:layout_width="match_parent"
    android:layout_height="match_parent"
    android:orientation="vertical">

<LinearLayout
        android:layout_width="match_parent"
        android:layout_height="wrap_content"
        android:orientation="horizontal">

<TextView
            android:layout_width="wrap_content"
            android:layout_height="wrap_content"
            android:text="您提交的内容为：" />

```

```
17  <TextView
18          android:id="@+id/tv1"
19          android:layout_width="wrap_content"
20          android:layout_height="wrap_content" />
21  </LinearLayout>
22
23  <RadioGroup
24      android:id="@+id/rg_color"
25      android:layout_width="wrap_content"
26      android:layout_height="wrap_content"
27      android:orientation="vertical">
28
29  <RadioButton
30          android:id="@+id/rb_Red"
31          android:layout_width="wrap_content"
32          android:layout_height="wrap_content"
33          android:checked="false"
34          android:text="红色"
35          android:textColor="#ff0000"
36          android:textStyle="bold" />
37
38  <RadioButton
39          android:id="@+id/rb_Green"
40          android:layout_width="wrap_content"
41          android:layout_height="wrap_content"
42          android:checked="false"
43          android:text="绿色"
44          android:textColor="#00ff00"
45          android:textStyle="bold" />
46
47  <RadioButton
48          android:id="@+id/rb_Blue"
49          android:layout_width="wrap_content"
50          android:layout_height="wrap_content"
51          android:checked="false"
52          android:text="蓝色"
53          android:textColor="#0000ff"
54          android:textStyle="bold" />
55  </RadioGroup>
56
57  </LinearLayout>
```

1）第 1～57 行声明了了一个线性布局，方向为垂直方向。

2）第 7～21 行添加了一个子线性布局，方向为水平方向，在其线性布局中分别有两个 TextView，第二个 TextView 为从 MainActivity 接收数据做准备。

3）第 23～55 行在 RadioGroup 中添加了三个 RadioButton，其 id 分别为 rb_Red、rb_Green 和 rb_Blue，“android:checked="false"”语句表示当前 RadioButton 未选中。

（4）在 src 下包为 com.hzu.transferdata_test 的 MainActivity.java 文件中编写内容如下：

```
package com.hzu.transferdata_test;

import android.app.Activity;
import android.content.Intent;
import android.os.Bundle;
import android.view.View;
import android.widget.Button;
import android.widget.EditText;

public class MainActivity extends Activity {
    private Button button1, button2;
    private EditText et;

    @Override
    protected void onCreate(Bundle savedInstanceState) {
        super.onCreate(savedInstanceState);
        setContentView(R.layout.activity_main);
        button1 = (Button) findViewById(R.id.button1);
        button2 = (Button) findViewById(R.id.button2);
        et = (EditText) findViewById(R.id.et);
        button1.setOnClickListener(new View.OnClickListener() {

            @Override
            public void onClick(View v) {
                Intent intent1 = new Intent();
                intent1.setClass(MainActivity.this, SubActivity.class);
                String msg = et.getText().toString();
                Bundle bundle = new Bundle();
                bundle.putString("msg", msg);// 使用 Bundle 封装数据
                intent1.putExtras(bundle);
                startActivity(intent1);
            }
        });

        button2.setOnClickListener(new View.OnClickListener() {

            @Override
            public void onClick(View v) {
                Intent intent2 = new Intent();
                intent2.setClass(MainActivity.this, SubActivity.class);
                startActivityForResult(intent2, 0xee);
            }
        });

    }
```

```
46
47      @Override
48      protected void onActivityResult(int requestCode, int resultCode, Intent data) {
49          if (requestCode == 0xee && resultCode == 0xff) {
50              int color = data.getIntExtra("color", 0xff000000);
51              button2.setTextColor(color);
52          }
53          super.onActivityResult(requestCode, resultCode, data);
54      }
55  }
```

1）第 3～8 行引入代码中需要使用的 Java 与 Android 相关类。

2）第 10～55 行定义了 MainActivity 类。

3）第 15～45 行重写了 onCreate()方法。

4）第 21～33 行为 bt1 添加监听方法，将用户输入的数据通过“startActivity(intent1)”传到 SubActivity，其中通过 Bundle 对象封装 msg 数据。

5）第 35～43 行为 bt2 添加监听方法，其中“startActivityForResult(intent2, 0xee);”语句表示要求带回数据到当前 Activity。

6）第 48～54 行通过 onActivityResult()方法接收从 SubActivity 传回的数据，通过 requestCode 与 resultCode 来判定是否为要接收数据的条件。然后使用 Intent 的 getIntExtra()取得数据，改变 button2 上字体的颜色。

（5）在 src 下包为 com.hzu.transferdata_test 的 SubActivity.java 文件中编写内容如下：

```
package com.hzu.transferdata_test;

import android.app.Activity;
import android.content.Intent;
import android.graphics.Color;
import android.os.Bundle;
import android.widget.RadioGroup;
import android.widget.TextView;

public class SubActivity extends Activity {
    private TextView tv;
    private RadioGroup rg;
    private Intent intent1;
    private Bundle bundle;

    @Override
    protected void onCreate(Bundle savedInstanceState) {
        // TODO Auto-generated method stub
        super.onCreate(savedInstanceState);
        setContentView(R.layout.activity_sub);
        tv = (TextView) findViewById(R.id.tv1);
        rg = (RadioGroup) findViewById(R.id.rg_color);
        bundle = getIntent().getExtras();
        String msg = null;
```

```
25              if (bundle != null) {
26                  msg = bundle.getString("msg"); // 拆分 Bundle 封装的数据
27                  tv.setText(msg);
28              }
29
30              intent1 = new Intent();
31              rg.setOnCheckedChangeListener(new RadioGroup.OnCheckedChangeListener() {
32
33                  @Override
34                  public void onCheckedChanged(RadioGroup group, int checkedId) {
35                      switch (checkedId) {
36                      case R.id.rb_Blue:
37                          intent1.putExtra("color", Color.BLUE);
38                          break;
39                      case R.id.rb_Green:
40                          intent1.putExtra("color", Color.GREEN);
41                          break;
42                      case R.id.rb_Red:
43                          intent1.putExtra("color", Color.RED);
44                          break;
45                      default:
46                          break;
47                      }
48                      setResult(0xff, intent1); // 设置返回值
49                      finish();
50                  }
51              });
52          }
53  }
```

1）第 3～8 行引入代码中需要使用的 Java 与 Android 相关类。

2）第 10～53 行定义了 SubActivity 类。

3）第 17～52 行重写了 onCreate()方法。

4）第 23～28 行接收 MainActivity 中用户提交的数据，如果不为空，显示在 tv 控件上。

5）第 31～51 行为 rg 控件添加监听方法，根据用户选择的颜色，通过“setResult(0xff, intent1);”语句带回到 MainActivity，其中 finish()方法为用户选择好颜色后，关闭当前的 SubActivity。

（6）在项目下的 AndroidManifest.xml 文件中的内容如下：

```
<?xml version="1.0" encoding="utf-8"?>
<manifest xmlns:android="http://schemas.android.com/apk/res/android"
    package="com.hzu.transferdata_test"
    android:versionCode="1"
    android:versionName="1.0">

<uses-sdk
        android:minSdkVersion="14"
```

```
            android:targetSdkVersion="21" />

<application
            android:allowBackup="true"
            android:icon="@drawable/ic_launcher"
            android:label="@string/app_name"
            android:theme="@style/AppTheme">
<activity
                android:name=".MainActivity"
                android:label="@string/app_name">
<intent-filter>
<action android:name="android.intent.action.MAIN" />
<category android:name="android.intent.category.LAUNCHER" />
</intent-filter>
</activity>
<activity android:name="SubActivity"></activity>
</application>
</manifest>
```

在 AndroidManifest.xml 文件中声明 MainActivity 与 SubActivity 两个类。

**【运行结果】**在 Eclipse 中启动 Android 模拟器，接着运行 TransferData_test 项目，显示效果如图 7-10 所示，输入“hello world!”后点击“保存数据”按钮显示如图 7-11 所示。在图 7-10 中点击“带回数据”按钮，在图 7-11 选择红色，将会把图 7-10 中“带回数据”这四个字显示为红色。

图 7-10 运行项目的效果

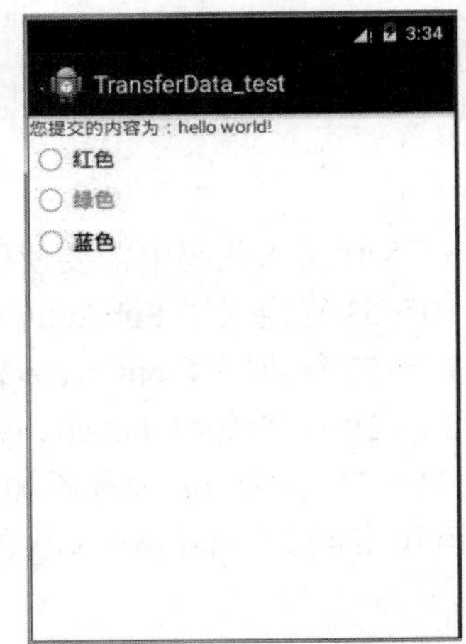


图 7-11 “保存数据”后的效果

## 7.7 本章小结

本章首先对 Activity 的用途、四种状态、生命周期进行了详细介绍，接着通过示例对 Activity 的生命周期运行机制做了讲解，最后对 Intent 与 Bundle 进行了介绍，并说明了 Bundle 在两个 Activity 之间传递数据的基本步骤。

# 第 8 章　Fragment

（1）了解 Fragment 的用途。
（2）掌握 Fragment 的生命周期、管理、通信。
（2）掌握 Fragment 的基本用法。

Fragment 的出现，解决了布局的精细化管理问题，使得 Activity 的显示更加灵活，当前很多流行的移动互联网产品都使用 Fragment 控制显示，下面将对 Fragment 进行详细介绍。

## 8.1　Fragment 概述

Fragment（碎片）是 Android 3.0 新增的功能，它与 Activity 非常相似，主要用于在 Activity 中描述一些行为或一部分用户界面，使程序更加合理和充分利用手机屏幕的空间。Fragment 使 Activity 的设计模块化，它必须被嵌入到一个 Activity 中。一个 Activity 中可以包含多个 Fragment，Fragment 同时也可被多个 Activity 使用。

Fragment 的优点如下：

- 能够将 Activity 分离成多个可重用的组件，每个都有它自己的生命周期和 UI。
- 轻松地创建动态灵活的 UI 设计，适应不同的屏幕尺寸。
- 与 Activity 绑定在一起，可以在运行中动态地移除、加入、交换等。
- 解决 Activity 间的切换不流畅问题，轻量切换。

## 8.2　创建 Fragment

创建一个 Fragment 和创建一个 Activity 类似，继承 Fragment 类或 Fragment 的子类，重写生命周期方法，主要不同之处是需要重写一个 onCreateView()方法来返回这个 Fragment 的布局。

```
public class MyFragment extends Fragment {

    @Override
    public View onCreateView(LayoutInflater inflater, ViewGroup container,
            Bundle savedInstanceState) {
        View v = inflater.inflate(R.layout.news, null);
        return v;
    }
}
```

其中 container 是 Fragment 将会被添加的组，inflate()方法的第二个参数为 null，因系统已将生成的布局加入到 container 中。

Fragment 不能单独显示在手机屏幕上，要与 Activity 绑定才能使用，其中向 Activity 中添加 Fragment 有以下两种方法：一种是静态加载，是直接在布局文件中添加，将 Fragment 作为布局中的一部分；另一种是动态加载，当 Activity 运行时，将 Fragment 放入到 Activity 中。

静态加载示例如下：

```
<fragment
        android:id="@+id/fragment1"
        android:name="com.hzu.MyFragment"
        android:layout_width="wrap_content"
        android:layout_height="wrap_content" />
```

动态加载示例如下：

```
FragmentManager fragmentManager=getFragmentManager();
FragmentTransaction transaction=fragmentManager.beginTransaction();
transaction.add(R.id.fragment1, new BlankFragment());
transaction.commit();
```

fragmentManager.beginTransaction()用于获取事件处理集，然后通过 add()方法添加 Fragment，再通过 commit()提交事务。

**【例 8.1】**设计一个静态添加 Fragment 的案例。

**【说明】**在 xml 中添加 Fragment。

**【开发步骤】**

（1）创建一个名为 Fragment_test 的项目，Activity 组件名为 MainActivity。在 res/layout 的 activity_main.xml 文件中的内容如下：

```
<LinearLayout xmlns:android="http://schemas.android.com/apk/res/android"
    xmlns:tools="http://schemas.android.com/tools"
    android:layout_width="match_parent"
    android:layout_height="match_parent"
    android:orientation="vertical">

<TextView
        android:layout_width="wrap_content"
        android:layout_height="wrap_content"
        android:text="当前是 MainActivity 中" />

<fragment
        android:id="@+id/fragment1"
        android:name="com.hzu.fragment_test.MyFragment"
        android:layout_width="wrap_content"
        android:layout_height="wrap_content"
        tools:layout="@layout/myfragment" />

</LinearLayout>
```

（2）在 res/layout 下创建 myfragment.xml 文件，其内容如下：

```
<?xml version="1.0" encoding="utf-8"?>
<LinearLayout xmlns:android="http://schemas.android.com/apk/res/android"
    android:layout_width="match_parent"
    android:layout_height="match_parent"
    android:orientation="vertical">

<TextView
        android:id="@+id/tv1"
        android:layout_width="wrap_content"
        android:layout_height="wrap_content"
        android:text="这是在 Fragment 中" />

<Button
        android:id="@+id/button1"
        android:layout_width="wrap_content"
        android:layout_height="wrap_content"
        android:text="Button" />

</LinearLayout>
```

（3）在 src 下包为 com.hzu.fragment_test 的 MainActivity.java 文件中编写内容如下：

```
package com.hzu.fragment_test;

import android.app.Activity;
import android.os.Bundle;

public class MainActivity extends Activity {

    @Override
    protected void onCreate(Bundle savedInstanceState) {
            super.onCreate(savedInstanceState);
            setContentView(R.layout.activity_main);
    }

}
```

（4）在 src 的 com.hzu.fragment_test 包下创建 MyFragment.java 文件，编写内容如下：

```
package com.hzu.fragment_test;

import android.app.Fragment;
import android.os.Bundle;
import android.view.LayoutInflater;
import android.view.View;
import android.view.ViewGroup;
import android.widget.Button;
import android.widget.Toast;

public class MyFragment extends Fragment {
```

```

    public MyFragment() {
        super();
    }

    @Override
    public View onCreateView(LayoutInflater inflater, ViewGroup container,
            Bundle savedInstanceState) {

        return inflater.inflate(R.layout.myfragment, container, false);
    }

    @Override
    public void onViewCreated(View view, Bundle savedInstanceState) {
        Button button = (Button) view.findViewById(R.id.button1);
        button.setOnClickListener(new View.OnClickListener() {

            @Override
            public void onClick(View v) {
                Toast.makeText(getActivity(), "在 MyFragment 中 button 被点击了！",
                        Toast.LENGTH_LONG).show();

            }
        });
    }

}
```

**【运行结果】**在 Eclipse 中启动 Android 模拟器，接着运行 Fragment_test 项目，显示效果如图 8-1 所示。点击图 8-1 中的 Button 按钮，会出现如图 8-2 所示的效果。

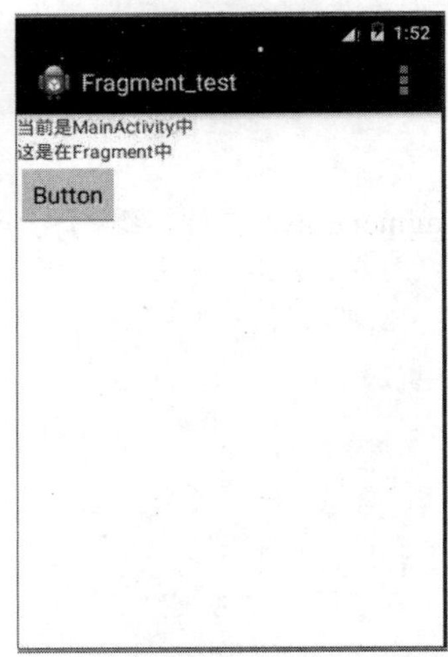


图 8-1　项目运行效果

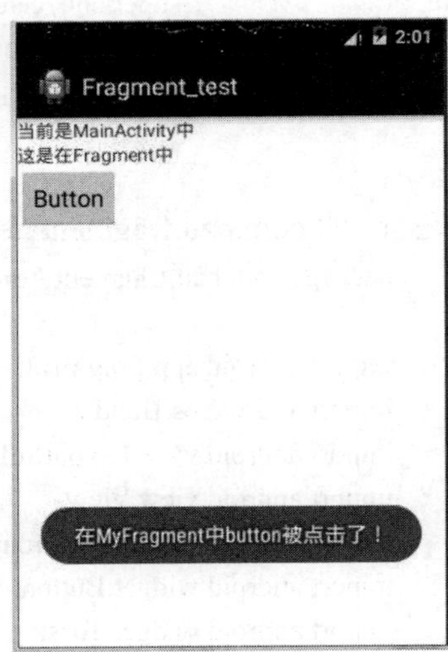


图 8-2　点击 Button 按钮的效果

**【例 8.2】** 设计一个动态添加 Fragment 的案例，效果与例 8.1 一样。

**【说明】** 在 Java 代码中添加 Fragment 对象。

**【开发步骤】**

（1）创建一个名为 Fragment2_test 的项目，Activity 组件名为 MainActivity。在 res/layout 的 activity_main.xml 文件中编写内容如下：

```
<RelativeLayout xmlns:android="http://schemas.android.com/apk/res/android"
    xmlns:tools="http://schemas.android.com/tools"
    android:layout_width="match_parent"
    android:layout_height="match_parent"
    android:paddingBottom="@dimen/activity_vertical_margin"
    android:paddingLeft="@dimen/activity_horizontal_margin"
    android:paddingRight="@dimen/activity_horizontal_margin"
    android:paddingTop="@dimen/activity_vertical_margin"
    tools:context="com.hzu.fragment2_test.MainActivity">

<TextView
        android:id="@+id/textView1"
        android:layout_width="wrap_content"
        android:layout_height="wrap_content"
        android:text="当前是 MainActivity 中" />

<LinearLayout
        android:id="@+id/linear"
        android:layout_width="match_parent"
        android:layout_height="wrap_content"
        android:layout_alignLeft="@+id/textView1"
        android:layout_below="@+id/textView1"
        android:orientation="vertical">
</LinearLayout>

</RelativeLayout>
```

（2）在 res/layout 下创建 myfragment.xml 文件，其内容如下：

```
<?xml version="1.0" encoding="utf-8"?>
<LinearLayout xmlns:android="http://schemas.android.com/apk/res/android"
    android:layout_width="match_parent"
    android:layout_height="match_parent"
    android:orientation="vertical">

<TextView
        android:id="@+id/tv1"
        android:layout_width="wrap_content"
        android:layout_height="wrap_content"
        android:text="这是在 Fragment 中" />

<Button
```

```
            android:id="@+id/button1"
            android:layout_width="wrap_content"
            android:layout_height="wrap_content"
            android:text="Button" />

</LinearLayout>
```

（3）在 src 的 com.hzu.fragment2_test 包下创建 MainActivity.java 文件，编写内容如下：

```
package com.hzu.fragment2_test;

import android.app.Activity;
import android.app.FragmentManager;
import android.app.FragmentTransaction;
import android.os.Bundle;

public class MainActivity extends Activity {

    @Override
    protected void onCreate(Bundle savedInstanceState) {
        super.onCreate(savedInstanceState);
        setContentView(R.layout.activity_main);
        FragmentManager fragmentManager = getFragmentManager();
        FragmentTransaction transaction = fragmentManager.beginTransaction();
        transaction.add(R.id.linear, new MyFragment());
        transaction.commit();
    }

}
```

（4）在 src 的 com.hzu. fragment2_test 包下创建 MyFragment.java 文件，编写内容如下：

```
package com.hzu.fragment2_test;

import android.app.Fragment;
import android.os.Bundle;
import android.view.LayoutInflater;
import android.view.View;
import android.view.ViewGroup;
import android.widget.Button;
import android.widget.Toast;

public class MyFragment extends Fragment {

    public MyFragment() {
        super();
    }

    @Override
    public View onCreateView(LayoutInflater inflater, ViewGroup container,
```

```
                Bundle savedInstanceState) {
        return inflater.inflate(R.layout.myfragment, container, false);
    }
    @Override
    public void onViewCreated(View view, Bundle savedInstanceState) {
        Button button = (Button) view.findViewById(R.id.button1);
        button.setOnClickListener(new View.OnClickListener() {
            @Override
            public void onClick(View v) {
                Toast.makeText(getActivity(), "在 MyFragment 中 button 被点击了！",
                        Toast.LENGTH_LONG).show();
            }
        });
    }
}
```

**【运行结果】**在 Eclipse 中启动 Android 模拟器，接着运行 Fragment2_test 项目，显示效果与例 8.1 相同。

## 8.3　Fragment 生命周期

Fragment 必须是依赖于 Activity 而存在，因此 Activity 的生命周期会直接影响到 Fragment 的生命周期。FragmentTransaction 提供了很多操作 Fragment 的方法，如 add()、replace()、attach()等，调用这些方法会触发 Fragment 不同的生命周期。如果程序开发人员使用以上这些方法时不知道 Fragment 当前处于何种状态是非常可怕的事情，直接影响到 Android 应用程序的质量与执行效率。

**【例 8.3】**设计一个查看 Activity 与 Fragment 各自生命周期调用方法的应用程序。

**【说明】**查看 Activity 中生命周期调用方法与 Fragment 中生命周期调用方法之间的相关关系，Activity 与 Fragment 各自生命周期如图 8-3 所示。

**【开发步骤】**

（1）创建一个名为 FragmentLifecycle_test 的项目，Activity 组件名为 MainActivity。在 res/layout 的 activity_main.xml 文件中的内容如下：

```
<RelativeLayout xmlns:android="http://schemas.android.com/apk/res/android"
    xmlns:tools="http://schemas.android.com/tools"
    android:layout_width="match_parent"
    android:layout_height="match_parent"
    android:paddingBottom="@dimen/activity_vertical_margin"
    android:paddingLeft="@dimen/activity_horizontal_margin"
    android:paddingRight="@dimen/activity_horizontal_margin"
```

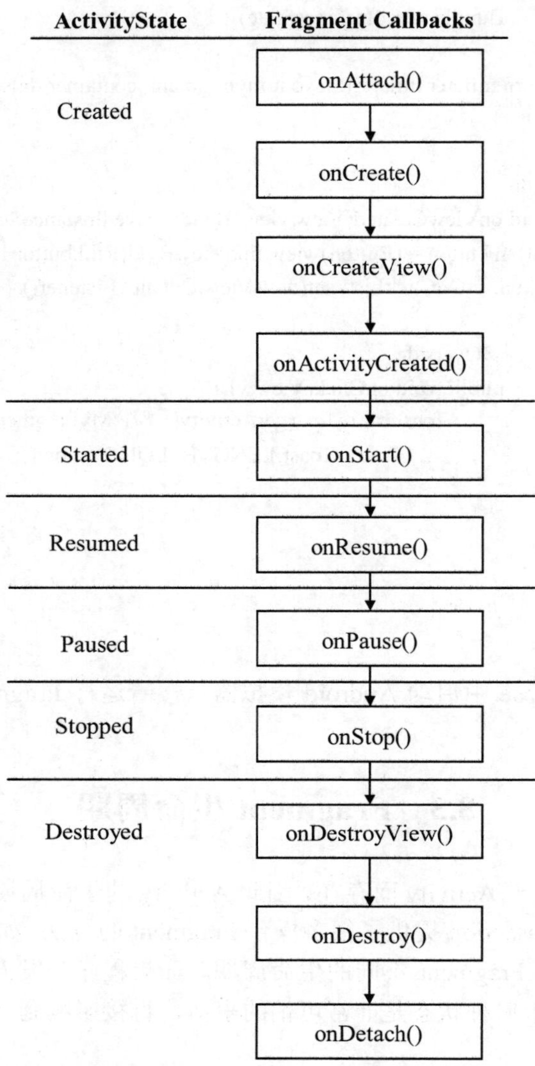


图 8-3　Activity 与 Fragment 各自的生命周期

```
        android:paddingTop="@dimen/activity_vertical_margin"
        tools:context="com.hzu.fragmentlifecycle_test.MainActivity">

    <TextView
            android:id="@+id/textView1"
            android:layout_width="wrap_content"
            android:layout_height="wrap_content"
            android:text="当前是 MainActivity 中" />

    <LinearLayout
            android:id="@+id/linear"
            android:layout_width="match_parent"
            android:layout_height="wrap_content"
```

```
            android:layout_alignLeft="@+id/textView1"
            android:layout_below="@+id/textView1"
            android:orientation="vertical">
</LinearLayout>

</RelativeLayout>
```

（2）在 res/layout 下创建 myfragment.xml 文件，其内容如下：

```
<?xml version="1.0" encoding="utf-8"?>
<LinearLayout xmlns:android="http://schemas.android.com/apk/res/android"
        android:layout_width="match_parent"
        android:layout_height="match_parent"
        android:orientation="vertical">

<TextView
            android:id="@+id/textView1"
            android:layout_width="wrap_content"
            android:layout_height="wrap_content"
            android:text="这是在 Fragment 中" />

</LinearLayout>
```

（3）在 src 的 com.hzu.fragmentlifecycle_test 包下创建 MainActivity.java 文件，编写内容如下：

```
package com.hzu.fragmentlifecycle_test;

import android.app.Activity;
import android.app.FragmentManager;
import android.app.FragmentTransaction;
import android.os.Bundle;
import android.util.Log;

public class MainActivity extends Activity {

        private static final String TAG = "fragmentlifecycle";

        @Override
        protected void onCreate(Bundle savedInstanceState) {
                super.onCreate(savedInstanceState);
                setContentView(R.layout.activity_main);
                Log.i(TAG, "MainActivity 中 onCreate()方法");
                FragmentManager fragmentManager = getFragmentManager();
                FragmentTransaction transaction = fragmentManager.beginTransaction();
                transaction.add(R.id.linear, new MyFragment());
                transaction.commit();
        }

        @Override
```

```
    protected void onStart() {
        super.onStart();
        Log.i(TAG, "MainActivity 中 onStart()方法");
    }

    @Override
    protected void onResume() {
        super.onResume();
        Log.i(TAG, "MainActivity 中 onResume()方法");
    }

    @Override
    protected void onPause() {
        super.onPause();
        Log.i(TAG, "MainActivity 中 onPause()方法");
    }

    @Override
    protected void onStop() {
        super.onStop();
        Log.i(TAG, "MainActivity 中 onStop()方法");
    }

    @Override
    protected void onDestroy() {
        super.onDestroy();
        Log.i(TAG, "MainActivity 中 onDestroy()方法");
    }
}
```

（4）在 src 的 com.hzu.fragmentlifecycle_test 包下创建 MyFragment.java 文件，编写内容如下：

```
package com.hzu.fragmentlifecycle_test;

import android.app.Activity;
import android.app.Fragment;
import android.os.Bundle;
import android.util.Log;
import android.view.LayoutInflater;
import android.view.View;
import android.view.ViewGroup;

public class MyFragment extends Fragment {
    private static final String TAG = "fragmentlifecycle";

    public MyFragment() {
        super();
```

```
    }

    @Override
    public void onAttach(Activity activity) {
        super.onAttach(activity);
        Log.d(TAG, "MyFragment 中 onAttach()方法");
    }

    @Override
    public void onCreate(Bundle savedInstanceState) {
        super.onCreate(savedInstanceState);
        Log.d(TAG, "MyFragment 中 onCreate()方法");
    }

    @Override
    public View onCreateView(LayoutInflater inflater, ViewGroup container,
            Bundle savedInstanceState) {
        Log.d(TAG, "MyFragment 中 onCreateView()方法");
        return inflater.inflate(R.layout.myfragment, container, false);
    }

    @Override
    public void onActivityCreated(Bundle savedInstanceState) {
        super.onActivityCreated(savedInstanceState);
        Log.d(TAG, "MyFragment 中 onActivityCreated()方法");
    }

    @Override
    public void onStart() {
        super.onStart();
        Log.d(TAG, "MyFragment 中 onStart()方法");
    }

    @Override
    public void onResume() {
        super.onResume();
        Log.d(TAG, "MyFragment 中 onResume()方法");
    }

    @Override
    public void onPause() {
        super.onPause();
        Log.d(TAG, "MyFragment 中 onPause()方法");
    }

    @Override
    public void onStop() {
```

```
            super.onStop();
            Log.d(TAG, "MyFragment 中 onStop()方法");
        }

        @Override
        public void onDestroyView() {
            super.onDestroyView();
            Log.d(TAG, "MyFragment 中 onDestroyView()方法");
        }

        @Override
        public void onDestroy() {
            super.onDestroy();
            Log.d(TAG, "MyFragment 中 onDestroy()方法");
        }

        @Override
        public void onDetach() {
            super.onDetach();
            Log.d(TAG, "MyFragment 中 onDetach()方法");
        }

    }
```

**【运行结果】**在 Eclipse 中启动 Android 模拟器，接着运行 FragmentLifecycle_test 项目，在过滤器 fragmentlifecycle 中显示效果如图 8-4 所示。

| Level | Time | PID | TID | Application | Tag | Text |
|---|---|---|---|---|---|---|
| I | 01-24 07:14:58.467 | 2209 | 2209 | | fragmentlifecycle | MainActivity中onCreate()方法 |
| D | 01-24 07:14:58.470 | 2209 | 2209 | com.hzu.fragmentlifecycle_test | fragmentlifecycle | MyFragment中onAttach()方法 |
| D | 01-24 07:14:58.470 | 2209 | 2209 | com.hzu.fragmentlifecycle_test | fragmentlifecycle | MyFragment中onCreate()方法 |
| D | 01-24 07:14:58.470 | 2209 | 2209 | com.hzu.fragmentlifecycle_test | fragmentlifecycle | MyFragment中onCreateView()方法 |
| D | 01-24 07:14:58.471 | 2209 | 2209 | com.hzu.fragmentlifecycle_test | fragmentlifecycle | MyFragment中onActivityCreated()方法 |
| I | 01-24 07:14:58.472 | 2209 | 2209 | com.hzu.fragmentlifecycle_test | fragmentlifecycle | MainActivity中onStart()方法 |
| D | 01-24 07:14:58.472 | 2209 | 2209 | com.hzu.fragmentlifecycle_test | fragmentlifecycle | MyFragment中onStart()方法 |
| I | 01-24 07:14:58.472 | 2209 | 2209 | com.hzu.fragmentlifecycle_test | fragmentlifecycle | MainActivity中onResume()方法 |
| D | 01-24 07:14:58.472 | 2209 | 2209 | com.hzu.fragmentlifecycle_test | fragmentlifecycle | MyFragment中onResume()方法 |

图 8-4　MyFragment 启动调用的方法

按下模拟器上 HOME 键显示效果如图 8-5 所示。

| Level | Time | PID | TID | Application | Tag | Text |
|---|---|---|---|---|---|---|
| I | 01-24 07:14:58.467 | 2209 | 2209 | | fragmentlifecycle | MainActivity中onCreate()方法 |
| D | 01-24 07:14:58.470 | 2209 | 2209 | com.hzu.fragmentlifecycle_test | fragmentlifecycle | MyFragment中onAttach()方法 |
| D | 01-24 07:14:58.470 | 2209 | 2209 | com.hzu.fragmentlifecycle_test | fragmentlifecycle | MyFragment中onCreate()方法 |
| D | 01-24 07:14:58.470 | 2209 | 2209 | com.hzu.fragmentlifecycle_test | fragmentlifecycle | MyFragment中onCreateView()方法 |
| D | 01-24 07:14:58.471 | 2209 | 2209 | com.hzu.fragmentlifecycle_test | fragmentlifecycle | MyFragment中onActivityCreated()方法 |
| I | 01-24 07:14:58.472 | 2209 | 2209 | com.hzu.fragmentlifecycle_test | fragmentlifecycle | MainActivity中onStart()方法 |
| D | 01-24 07:14:58.472 | 2209 | 2209 | com.hzu.fragmentlifecycle_test | fragmentlifecycle | MyFragment中onStart()方法 |
| I | 01-24 07:14:58.472 | 2209 | 2209 | com.hzu.fragmentlifecycle_test | fragmentlifecycle | MainActivity中onResume()方法 |
| D | 01-24 07:14:58.472 | 2209 | 2209 | com.hzu.fragmentlifecycle_test | fragmentlifecycle | MyFragment中onResume()方法 |
| D | 01-24 07:16:00.285 | 2209 | 2209 | com.hzu.fragmentlifecycle_test | fragmentlifecycle | MyFragment中onPause()方法 |
| I | 01-24 07:16:00.286 | 2209 | 2209 | com.hzu.fragmentlifecycle_test | fragmentlifecycle | MainActivity中onPause()方法 |
| D | 01-24 07:16:00.444 | 2209 | 2209 | com.hzu.fragmentlifecycle_test | fragmentlifecycle | MyFragment中onStop()方法 |
| I | 01-24 07:16:00.444 | 2209 | 2209 | com.hzu.fragmentlifecycle_test | fragmentlifecycle | MainActivity中onStop()方法 |

图 8-5　HOME 键按下调用的方法

再次长按模拟器上 HOME 键，选择 FragmentLifecycle_test 程序时的显示效果如图 8-6 所示。

| Level | Time | PID | TID | Application | Tag | Text |
|---|---|---|---|---|---|---|
| I | 01-24 07:14:58.467 | 2209 | 2209 |  | fragmentlifecycle | MainActivity中onCreate()方法 |
| D | 01-24 07:14:58.470 | 2209 | 2209 | com.hzu.fragmentlifecycle_test | fragmentlifecycle | MyFragment中onAttach()方法 |
| D | 01-24 07:14:58.470 | 2209 | 2209 | com.hzu.fragmentlifecycle_test | fragmentlifecycle | MyFragment中onCreate()方法 |
| D | 01-24 07:14:58.470 | 2209 | 2209 | com.hzu.fragmentlifecycle_test | fragmentlifecycle | MyFragment中onCreateView()方法 |
| D | 01-24 07:14:58.471 | 2209 | 2209 | com.hzu.fragmentlifecycle_test | fragmentlifecycle | MyFragment中onActivityCreated()方法 |
| I | 01-24 07:14:58.472 | 2209 | 2209 | com.hzu.fragmentlifecycle_test | fragmentlifecycle | MainActivity中onStart()方法 |
| D | 01-24 07:14:58.472 | 2209 | 2209 | com.hzu.fragmentlifecycle_test | fragmentlifecycle | MyFragment中onStart()方法 |
| I | 01-24 07:14:58.472 | 2209 | 2209 | com.hzu.fragmentlifecycle_test | fragmentlifecycle | MainActivity中onResume()方法 |
| D | 01-24 07:14:58.472 | 2209 | 2209 | com.hzu.fragmentlifecycle_test | fragmentlifecycle | MyFragment中onResume()方法 |
| D | 01-24 07:16:00.285 | 2209 | 2209 | com.hzu.fragmentlifecycle_test | fragmentlifecycle | MyFragment中onPause()方法 |
| I | 01-24 07:16:00.286 | 2209 | 2209 | com.hzu.fragmentlifecycle_test | fragmentlifecycle | MainActivity中onPause()方法 |
| D | 01-24 07:16:00.444 | 2209 | 2209 | com.hzu.fragmentlifecycle_test | fragmentlifecycle | MyFragment中onStop()方法 |
| I | 01-24 07:16:00.444 | 2209 | 2209 | com.hzu.fragmentlifecycle_test | fragmentlifecycle | MainActivity中onStop()方法 |
| I | 01-24 07:17:01.178 | 2209 | 2209 | com.hzu.fragmentlifecycle_test | fragmentlifecycle | MainActivity中onStart()方法 |
| D | 01-24 07:17:01.178 | 2209 | 2209 | com.hzu.fragmentlifecycle_test | fragmentlifecycle | MyFragment中onStart()方法 |
| I | 01-24 07:17:01.178 | 2209 | 2209 | com.hzu.fragmentlifecycle_test | fragmentlifecycle | MainActivity中onResume()方法 |
| D | 01-24 07:17:01.178 | 2209 | 2209 | com.hzu.fragmentlifecycle_test | fragmentlifecycle | MyFragment中onResume()方法 |

图 8-6　回到程序时调用的方法

按模拟器上“返回”键，退出 FragmentLifecycle_test 程序时显示效果如图 8-7 所示。

| Level | Time | PID | TID | Application | Tag | Text |
|---|---|---|---|---|---|---|
| D | 01-24 07:14:58.472 | 2209 | 2209 | com.hzu.fragmentlifecycle_test | fragmentlifecycle | MyFragment中onStart()方法 |
| I | 01-24 07:14:58.472 | 2209 | 2209 | com.hzu.fragmentlifecycle_test | fragmentlifecycle | MainActivity中onResume()方法 |
| D | 01-24 07:14:58.472 | 2209 | 2209 | com.hzu.fragmentlifecycle_test | fragmentlifecycle | MyFragment中onResume()方法 |
| D | 01-24 07:16:00.285 | 2209 | 2209 | com.hzu.fragmentlifecycle_test | fragmentlifecycle | MyFragment中onPause()方法 |
| I | 01-24 07:16:00.286 | 2209 | 2209 | com.hzu.fragmentlifecycle_test | fragmentlifecycle | MainActivity中onPause()方法 |
| D | 01-24 07:16:00.444 | 2209 | 2209 | com.hzu.fragmentlifecycle_test | fragmentlifecycle | MyFragment中onStop()方法 |
| I | 01-24 07:16:00.444 | 2209 | 2209 | com.hzu.fragmentlifecycle_test | fragmentlifecycle | MainActivity中onStop()方法 |
| I | 01-24 07:17:01.178 | 2209 | 2209 | com.hzu.fragmentlifecycle_test | fragmentlifecycle | MainActivity中onStart()方法 |
| D | 01-24 07:17:01.178 | 2209 | 2209 | com.hzu.fragmentlifecycle_test | fragmentlifecycle | MyFragment中onStart()方法 |
| I | 01-24 07:17:01.178 | 2209 | 2209 | com.hzu.fragmentlifecycle_test | fragmentlifecycle | MainActivity中onResume()方法 |
| D | 01-24 07:17:01.178 | 2209 | 2209 | com.hzu.fragmentlifecycle_test | fragmentlifecycle | MyFragment中onResume()方法 |
| D | 01-24 07:17:40.093 | 2209 | 2209 | com.hzu.fragmentlifecycle_test | fragmentlifecycle | MyFragment中onPause()方法 |
| I | 01-24 07:17:40.093 | 2209 | 2209 | com.hzu.fragmentlifecycle_test | fragmentlifecycle | MainActivity中onPause()方法 |
| D | 01-24 07:17:40.820 | 2209 | 2209 | com.hzu.fragmentlifecycle_test | fragmentlifecycle | MyFragment中onStop()方法 |
| I | 01-24 07:17:40.820 | 2209 | 2209 | com.hzu.fragmentlifecycle_test | fragmentlifecycle | MainActivity中onStop()方法 |
| D | 01-24 07:17:40.821 | 2209 | 2209 | com.hzu.fragmentlifecycle_test | fragmentlifecycle | MyFragment中onDestroyView()方法 |
| D | 01-24 07:17:40.862 | 2209 | 2209 | com.hzu.fragmentlifecycle_test | fragmentlifecycle | MyFragment中onDestroy()方法 |
| D | 01-24 07:17:40.862 | 2209 | 2209 | com.hzu.fragmentlifecycle_test | fragmentlifecycle | MyFragment中onDetach()方法 |
| I | 01-24 07:17:40.862 | 2209 | 2209 | com.hzu.fragmentlifecycle_test | fragmentlifecycle | MainActivity中onDestroy()方法 |

图 8-7　退出程序时调用的方法

## 8.4　Fragment 管理

FragmentTransaction 允许将一系列的 Fragment 的操作在一次处理事务（Transaction）中完成。每个事务都是一组想要同时执行的方法，例如在一个事物中执行 add()、remove()和 replace()方法，然后调用 commit()方法，把事务的执行结果反映到 Activity 中。注意，在调用 commit()方法之前，为了把事务添加到 Fragment 事务的回退堆栈中，可能调用 addToBackStack()方法。这个回退堆栈被 Activity 管理，并且允许用户通过按返回键返回到先前的 Fragment 状态。以下是添加、替换、隐藏和移除 Fragment 的方法：

- 添加：add(int containerViewId, Fragment fragment, String tag)
- 替换：replace(int containerViewId, Fragment fragment, String tag)

- 隐藏：hide(Fragment fragment)
- 移除：remove(Fragment fragment)

【例 8.4】通过 Fragment 设置贺州新闻、贺州旅游和旅游咨询三块内容。

【说明】点击不同按键时贺州新闻、贺州旅游和旅游咨询三块内容可自由切换。

【开发步骤】

（1）创建一个名为 FragmentManage_test 的项目，Activity 组件名为 MainActivity。

（2）在 res/layout 的 activity_main.xml 文件中的内容如下：

```
<RelativeLayout xmlns:android="http://schemas.android.com/apk/res/android"
      xmlns:tools="http://schemas.android.com/tools"
      android:layout_width="match_parent"
      android:layout_height="match_parent"
      tools:context="com.hzu.fragmentmanage_test.MainActivity">

<FrameLayout
         android:id="@+id/frame"
         android:layout_width="match_parent"
         android:layout_height="match_parent"
         android:layout_alignParentLeft="true"
         android:layout_alignParentStart="true"
         android:layout_below="@+id/btn_hzNews">
</FrameLayout>

<Button
         android:id="@+id/btn_hzNews"
         android:layout_width="wrap_content"
         android:layout_height="wrap_content"
         android:layout_alignParentLeft="true"
         android:layout_alignParentStart="true"
         android:layout_alignParentTop="true"
         android:text="贺州新闻" />

<Button
         android:id="@+id/btn_hzTour"
         android:layout_width="wrap_content"
         android:layout_height="wrap_content"
         android:layout_above="@+id/frame"
         android:layout_centerHorizontal="true"
         android:text="贺州旅游" />

<Button
         android:id="@+id/btn_trConsult"
         android:layout_width="wrap_content"
         android:layout_height="wrap_content"
         android:layout_alignParentEnd="true"
         android:layout_alignParentRight="true"
```

```
            android:layout_alignParentTop="true"
            android:text="旅游咨询" />

</RelativeLayout>
```

（3）在 res/layout 下创建 myfragment.xml 文件，其内容如下：

```
<?xml version="1.0" encoding="utf-8"?>
<LinearLayout xmlns:android="http://schemas.android.com/apk/res/android"
      android:layout_width="match_parent"
      android:layout_height="match_parent"
      android:orientation="vertical">

<TextView
            android:id="@+id/textView1"
            android:layout_width="wrap_content"
            android:layout_height="wrap_content"
            android:text="这是在 Fragment 中" />

</LinearLayout>
```

（4）在 res/layout 下创建 fragment_consult.xml 文件，其内容如下：

```
<?xml version="1.0" encoding="utf-8"?>
<LinearLayout xmlns:android="http://schemas.android.com/apk/res/android"
      android:layout_width="match_parent"
      android:layout_height="match_parent"
      android:orientation="vertical">

<TextView
            android:layout_width="match_parent"
            android:layout_height="match_parent"
            android:gravity="center"
            android:text="旅游咨询"
            android:textSize="40sp" />

</LinearLayout>
```

（5）在 res/layout 下创建 fragment_news.xml 文件，其内容如下：

```
<?xml version="1.0" encoding="utf-8"?>
<LinearLayout xmlns:android="http://schemas.android.com/apk/res/android"
      android:layout_width="match_parent"
      android:layout_height="match_parent"
      android:orientation="vertical">

<TextView
            android:layout_width="match_parent"
            android:layout_height="match_parent"
            android:gravity="center"
             android:text="贺州新闻"
            android:textSize="40sp" />
```

```

</LinearLayout>
```

（6）在 res/layout 下创建 fragment_tour.xml 文件，其内容如下：

```
<?xml version="1.0" encoding="utf-8"?>
<LinearLayout xmlns:android="http://schemas.android.com/apk/res/android"
        android:layout_width="match_parent"
        android:layout_height="match_parent"
        android:orientation="vertical">

<TextView
            android:layout_width="match_parent"
            android:layout_height="match_parent"
            android:gravity="center"
            android:text="贺州旅游"
            android:textSize="40sp" />

</LinearLayout>
```

（7）在 src 的 com.hzu.fragmentmanage_test 包下创建 MainActivity.java 文件，编写内容如下：

```
package com.hzu.fragmentmanage_test;

import android.app.Activity;
import android.app.FragmentManager;
import android.app.FragmentTransaction;
import android.os.Bundle;
import android.util.Log;
import android.view.View;
import android.view.View.OnClickListener;
import android.widget.Button;

public class MainActivity extends Activity implements OnClickListener {
        private Button btn_news;
        private Button btn_tour;
        private Button btn_consult;

        @Override
        protected void onCreate(Bundle savedInstanceState) {
                super.onCreate(savedInstanceState);
                setContentView(R.layout.activity_main);
                btn_news = (Button) findViewById(R.id.btn_hzNews);
                btn_tour = (Button) findViewById(R.id.btn_hzTour);
                btn_consult = (Button) findViewById(R.id.btn_trConsult);

                btn_news.setOnClickListener(this);
                btn_tour.setOnClickListener(this);
                btn_consult.setOnClickListener(this);
```

```
        // 获取管理器
        FragmentManager manager = getFragmentManager();
        // 拿到事件处理集
        FragmentTransaction transaction = manager.beginTransaction();
        // 开始添加操作
        transaction.add(R.id.frame, new MyFragment());
        // 最后一定要记得提交修改
        transaction.commit();

        manager.addOnBackStackChangedListener
                    (new FragmentManager.OnBackStackChangedListener() {
            @Override
            public void onBackStackChanged() {
                // 将回退栈中的 Fragment 数量打印出来
                int count = getFragmentManager().getBackStackEntryCount();
                Log.i("MainActivity", "回退栈中的数量是" + count);
            }
        });
    }

    @Override
    public void onClick(View v) {
        FragmentManager manager = getFragmentManager();
        FragmentTransaction transaction = manager.beginTransaction();
        switch (v.getId()) {// 替换 Fragment
        case R.id.btn_hzNews:
            transaction.replace(R.id.frame, new NewsFragment());
            break;
        case R.id.btn_hzTour:
            transaction.replace(R.id.frame, new TourFragment());
            break;
        case R.id.btn_trConsult:
            transaction.replace(R.id.frame, new ConsultFragment());
            break;
        default:
            break;
        }
        // 将事件添加到回退栈，在用户点击返回键时，会返回到上一个出现的 Fragment
        transaction.addToBackStack(null);
        transaction.commit();

    }

}
```

（8）在 src 的 com.hzu.fragmentmanage_test 包下创建 MyFragment.java 文件，编写内容如下：

```
package com.hzu.fragmentmanage_test;

import android.app.Fragment;
import android.os.Bundle;
import android.view.LayoutInflater;
import android.view.View;
import android.view.ViewGroup;

public class MyFragment extends Fragment {

    public MyFragment() {
        super();
    }

    @Override
    public View onCreateView(LayoutInflater inflater, ViewGroup container,
            Bundle savedInstanceState) {
        return inflater.inflate(R.layout.myfragment, container, false);
    }

}
```

（9）在 src 的 com.hzu.fragmentmanage_test 包下创建 NewsFragment.java 文件，编写内容如下：

```
package com.hzu.fragmentmanage_test;

import android.app.Fragment;
import android.os.Bundle;
import android.view.LayoutInflater;
import android.view.View;
import android.view.ViewGroup;

public class NewsFragment extends Fragment {

    public NewsFragment() {
        // TODO Auto-generated constructor stub
    }

    @Override
    public View onCreateView(LayoutInflater inflater, ViewGroup container,
            Bundle savedInstanceState) {
        // Inflate the layout for this fragment
        return inflater.inflate(R.layout.fragment_news, container, false);
    }
}
```

（10）在 src 的 com.hzu.fragmentmanage_test 包下创建 TourFragment.java 文件，编写内容如下：

```
package com.hzu.fragmentmanage_test;

import android.app.Fragment;
import android.os.Bundle;
import android.view.LayoutInflater;
import android.view.View;
import android.view.ViewGroup;

public class TourFragment extends Fragment {

        public TourFragment() {
                // TODO Auto-generated constructor stub
        }

        @Override
        public View onCreateView(LayoutInflater inflater, ViewGroup container,
                        Bundle savedInstanceState) {
                // Inflate the layout for this fragment
                return inflater.inflate(R.layout.fragment_tour, container, false);
        }
}
```

（11）在 src 的 com.hzu.fragmentmanage_test 包下创建 ConsultFragment.java 文件，编写内容如下：

```
package com.hzu.fragmentmanage_test;

import android.app.Fragment;
import android.os.Bundle;
import android.view.LayoutInflater;
import android.view.View;
import android.view.ViewGroup;

public class ConsultFragment extends Fragment {

        public ConsultFragment() {
                // TODO Auto-generated constructor stub
        }

        @Override
        public View onCreateView(LayoutInflater inflater, ViewGroup container,
                        Bundle savedInstanceState) {
                // Inflate the layout for this fragment
                return inflater.inflate(R.layout.fragment_consult, container, false);
        }
}
```

**【运行结果】**在 Eclipse 中启动 Android 模拟器，接着运行 FragmentManage_test 项目，项目运行显示的效果如图 8-8 所示，点击图 8-8 中的“贺州新闻”按钮，显示效果如图 8-9 所示。

图 8-8 项目运行的效果

图 8-9 点击"贺州新闻"按钮效果

在过滤器 MainActivity 中显示的效果如图 8-10 所示。

| Level | Time | PID | TID | Application | Tag | Text |
|---|---|---|---|---|---|---|
| I | 01-24 11:27:59.594 | 2741 | 2741 | com.hzu.fragmentmanage_test | MainActivity | 回退栈中的数量是1 |

图 8-10 过滤器中显示的效果

## 8.5 Fragment 之间通信

Fragment 经常需要在初始化或运行过程中与其他的 Fragment 进行数据传递。为了提高代码的重用性，降低耦合度，一般在 Fragment 间不直接进行数据传递，而是使用 Activity 作为中介来进行数据传递。第一种情况：在初始化 Fragment 时传值，在 Activity 中使用 Fragment 的 setArguments()方法将需要使用的数据通过绑定到 Bundle 对象上进行传递，而在 Fragment 中则使用 getArguments()方法接收 Bundle 对象中的数据。第二种情况：在程序运行过程中传值到 Fragment，在 Fragment 定义一个用于接收数据的方法。在 Activity 中调用 Fragment 中已经定义好的方法将数据传递到 Fragment 中。

**【例 8.5】**设计程序：分别在初始化 Fragment 时传值到 Fragment 中与在程序运行过程中传值到 Fragment 中的程序。

**【说明】**主要使用 setArguments()、getArguments()与自定义的方法进行数据处理。

**【开发步骤】**

（1）创建一个名为 FragmentTransferData_test 的项目，Activity 组件名为 MainActivity。

（2）在 res/layout 的 activity_main.xml 文件中的内容如下：

```
<LinearLayout xmlns:android="http://schemas.android.com/apk/res/android"
    xmlns:tools="http://schemas.android.com/tools"
    android:layout_width="match_parent"
```

```
        android:layout_height="match_parent"
        android:orientation="vertical">

<LinearLayout
            android:layout_width="match_parent"
            android:layout_height="wrap_content"
            android:orientation="horizontal">

<TextView
                android:layout_width="wrap_content"
                android:layout_height="wrap_content"
                android:text="请输入内容：" />

<EditText
                android:id="@+id/et"
                android:layout_width="200dp"
                android:layout_height="wrap_content"
                android:text="默认内容" />
</LinearLayout>

<Button
            android:id="@+id/button1"
            android:layout_width="match_parent"
            android:layout_height="wrap_content"
            android:text="添加内容" />

<Button
            android:id="@+id/button2"
            android:layout_width="match_parent"
            android:layout_height="wrap_content"
            android:text="改变内容" />

<FrameLayout
            android:id="@+id/frame"
            android:layout_width="match_parent"
            android:layout_height="match_parent"
            android:layout_below="@+id/btn_change">
</FrameLayout>

</LinearLayout>
```

（3）在 res/layout 下创建 myfragment.xml 文件，其内容如下：

```
<?xml version="1.0" encoding="utf-8"?>
<LinearLayout xmlns:android="http://schemas.android.com/apk/res/android"
        android:layout_width="match_parent"
        android:layout_height="match_parent"
        android:orientation="vertical">
```

```
<TextView
            android:id="@+id/tv1"
            android:layout_width="wrap_content"
            android:layout_height="wrap_content"
            android:text="这是在 Fragment 中" />

</LinearLayout>
```

（4）在 src 的 com.hzu.fragmenttransferdata_test 包下创建 MainActivity.java 文件，编写内容如下：

```
package com.hzu.fragmenttransferdata_test;

import android.app.Activity;
import android.os.Bundle;
import android.view.View;
import android.widget.Button;
import android.widget.EditText;

public class MainActivity extends Activity {
    private Button bt1;
    private Button bt2;
    private EditText et1;

    @Override
    protected void onCreate(Bundle savedInstanceState) {
        super.onCreate(savedInstanceState);
        setContentView(R.layout.activity_main);
        bt1 = (Button) findViewById(R.id.button1);
        bt2 = (Button) findViewById(R.id.button2);
        et1 = (EditText) findViewById(R.id.et);

        bt1.setOnClickListener(new View.OnClickListener() {

            @Override
            public void onClick(View v) {
                // 获取用户输入的内容
                String content = et1.getText().toString();
                Bundle bundle = new Bundle();
                bundle.putString("content", content);
                MyFragment fragment = new MyFragment();
                // 将需要传递的数据添加到 Fragment 中
                fragment.setArguments(bundle);
                getFragmentManager().beginTransaction()
                        .replace(R.id.frame, fragment).addToBackStack(null)
                        .commit();
            }
```

```
            });

            bt2.setOnClickListener(new View.OnClickListener() {

                @Override
                public void onClick(View v) {
                    String content = et1.getText().toString();
                    // 找到对应的 Fragment
                    MyFragment fragment = (MyFragment) getFragmentManager()
                            .findFragmentById(R.id.frame);
                    // 调用 Fragment 中对应的方法将数据传递到 Fragment 中
                    fragment.changeContent(content);

                }
            });

        }

}
```

（5）在 src 的 com.hzu.fragmenttransferdata_test 包下创建 MyFragment.java 文件，编写内容如下：

```
package com.hzu.fragmenttransferdata_test;

import android.app.Fragment;
import android.os.Bundle;
import android.view.LayoutInflater;
import android.view.View;
import android.view.ViewGroup;
import android.widget.TextView;

public class MyFragment extends Fragment {

        private TextView tv;

        public MyFragment() {
                super();
        }

        @Override
        public View onCreateView(LayoutInflater inflater, ViewGroup container,
                        Bundle savedInstanceState) {
                return inflater.inflate(R.layout.myfragment, container, false);
        }

        @Override
        public void onViewCreated(View view, Bundle savedInstanceState) {
```

```

        tv = (TextView) view.findViewById(R.id.tv1);
        // 获取初始化的数据
        Bundle bundle = getArguments();
        if (bundle != null) {
            String content = bundle.getString("content");
            tv.setText(content);
        }
    }

    // 定义一个方法用于接收 Activity 传递过来的数据
    public void changeContent(String content) {
        tv.setText(content);
    }

}
```

**【运行结果】**在 Eclipse 中启动 Android 模拟器，接着运行 FragmentManage_test 项目，项目运行显示的效果如图 8-11 所示，改变内容后显示的效果如图 8-12 所示。

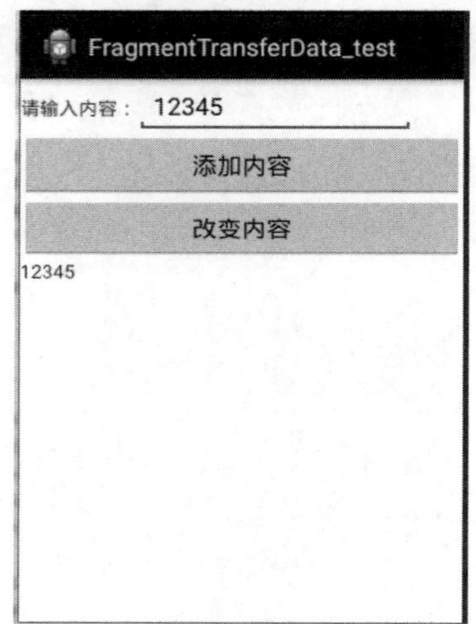


图 8-11　运行时的效果

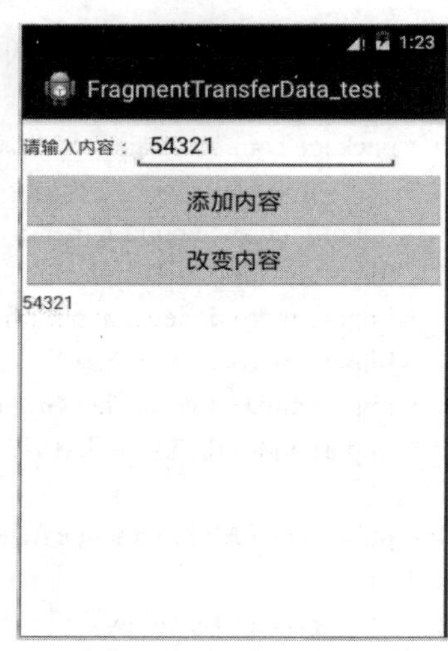


图 8-12　运行过程中传值

## 8.6　本章小结

本章主要讲解了 Fragment 的用途、Fragment 的创建、Fragment 的生命周期、Fragment 的管理和 Fragment 之间的通信，在现代移动互联网产品中应该大量使用 Fragment，使用户体验更好，界面结构设计更加合理。

# 第 9 章　Android 后台处理

学习目标

（1）了解 Service 与 Notification 的用途。

（2）掌握 BroadcastReceiver 的使用。

在 Android 应用程序中，有一些应用程序并不需要有界面显示，并且可以在其运行的同时让手机运行其他程序，例如，播放音乐、接收与发送广播消息等，这些都是直接在后台运行代码的，不需要设计布局文件。

## 9.1　Service

Service（服务）是一种 Android 组件，可以一直在后台长时间运行，但不提供界面交互功能。Service 可以由其他应用组件启动，当用户切换到其他应用时，服务仍将在后台继续运行。此外组件可以绑定到服务，也可以与之进行交互，甚至可以执行进程间的通信。Service 分为本地服务与远程服务两种类型：本地服务是指在同一个应用程序中调用，远程服务是指在不同包之间的服务调用。只要通过正确的方式，不同应用程序间就能够通过远程服务来传递数据。

**注意**：Service 是一个组件，它不同于一般服务概念，不会创建线程和进程，在默认的情况下，Service 总是运行在主线程中。启动 Service 有两种方式，一种是通过调用 startService()方法启动一个 Service；另外一种是通过使用 bindService()方法来绑定一个存在的 Service。

1. 通过 startService 启动

在一个应用程序中调用了 startService()方法就表示启动 Service，然后系统会调用 onCreate()方法以及 onStart()方法（或 onStartCommand()方法）。这样启动的 Service 会一直在后台运行，直到调用 onStop()方法才会停止，服务结束时会调用 onDestroy()方法。如果一个 Service 已经启动，其他代码再试图调用 startService()方法，此时虽然不会执行 onCreate()方法，但会重新执行一次 onStart()方法。

2. 通过 bindService 启动

通过 bindService 来启动 Service，在启动时只调用 onCreate()方法，此时，该 Service 和调用 Service 的客户类捆绑起来。如果调用 bindService()方法前服务已经被绑定，则多次调用 bindService()方法并不会导致多次创建服务及绑定操作。如果客户类被销毁，则 Service 也会调用 onUnbind()方法以及 onDestroy()方法。

Service 的常用方法如下：

- onStartCommand()方法：系统中其他组件如 Activity 通过调用 startService()方法就会调用此方法，一旦此方法执行，Service 就会启动且长期在后台运行。
- OnBind()方法：当组件调用 bindService()想要绑定到 Service 时，系统将调用此方法。
- OnCreate()方法：系统在 Service 第一次创建时执行此方法，来执行只运行一次的初始化工作。如果 Service 已经运行，这个方法将不会被调用。
- OnDestroy()方法：系统在 Service 不再被使用并要销毁时调用此方法。此方法主要用于释放资源，比如线程、已注册的监听器、接收器等。

客户类与绑定的服务进行交互的步骤如下：

（1）服务绑定后，会获得一个接口 IBinder，可以认为是客户类通过 Binder 来调用 Service 提供的功能。

（2）实现客户类的 ServiceConnection 接口。

（3）实现服务类的 OnBind()方法，返回一个接口的实例。

（4）当客户类调用服务接口提供的方法时，即服务在执行响应的代码。

使用这两种方式启动 Service，其生命周期如图 9-1 所示。

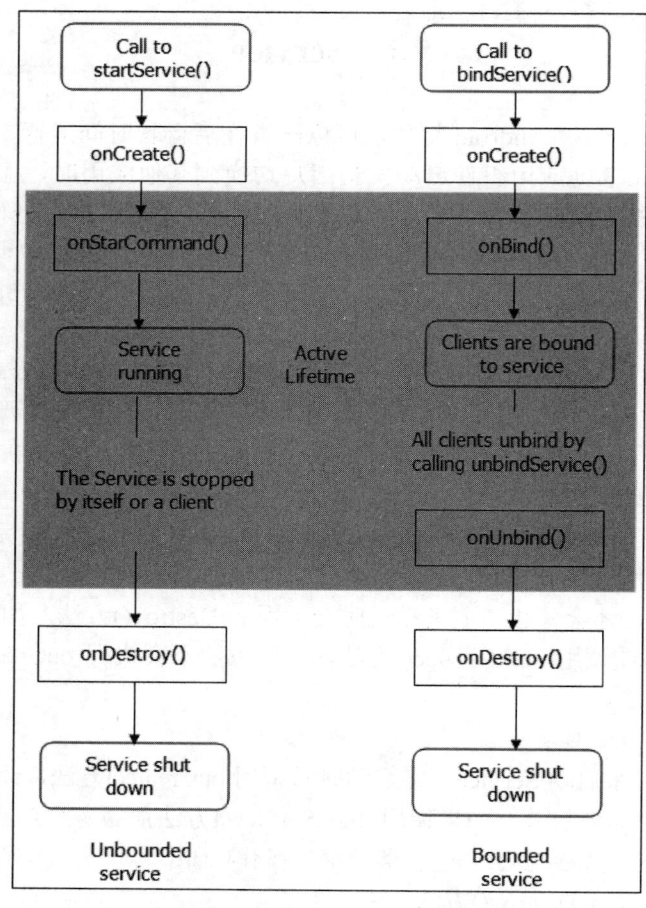


图 9-1　两种启动 Service 的生命周期方式

【例 9.1】设计一个显示两种启动 Service 的生命周期中各种方法调用情况的程序。

【说明】通过 startService 启动与通过 bindService 启动所使用的方法不相同。

【开发步骤】

（1）创建一个名为 Service_test 的项目，Activity 组件名为 MainActivity。在 res/layout 下创建一个 activity_main.xml 文件，内容如下：

```
1    <LinearLayout xmlns:android="http://schemas.android.com/apk/res/android"
2        xmlns:tools="http://schemas.android.com/tools"
3        android:layout_width="match_parent"
4        android:layout_height="match_parent"
5        android:orientation="vertical">
6
7    <Button
8            android:id="@+id/Btn_StartService"
9            android:layout_width="match_parent"
10           android:layout_height="wrap_content"
11           android:layout_alignParentLeft="true"
12           android:layout_alignParentTop="true"
13           android:text="StartService" />
14
15   <Button
16           android:id="@+id/Btn_StopService"
17           android:layout_width="match_parent"
18           android:layout_height="wrap_content"
19           android:layout_alignParentLeft="true"
20           android:layout_below="@+id/button1"
21           android:text="StopService" />
22
23   <Button
24           android:id="@+id/Btn_BindService"
25           android:layout_width="match_parent"
26           android:layout_height="wrap_content"
27           android:layout_alignLeft="@+id/button2"
28           android:layout_below="@+id/button2"
29           android:text="BindService" />
30
31   <Button
32           android:id="@+id/Btn_UnbindService"
33           android:layout_width="match_parent"
34           android:layout_height="wrap_content"
35           android:layout_alignLeft="@+id/button3"
36           android:layout_below="@+id/button3"
37           android:text="UnbindService" />
38
39   </LinearLayout>
```

1）第 1～39 行声明了一个线性布局，方向为垂直方向。

2）第 7～37 行分别声明了四个 Button。

（2）在 src 的 com.hzu.service_test 包下创建 MainActivity.java 文件，编写内容如下：

```
package com.hzu.service_test;

import android.app.Activity;
import android.content.ComponentName;
import android.content.Intent;
import android.content.ServiceConnection;
import android.os.Bundle;
import android.os.IBinder;
import android.view.View;
import android.view.View.OnClickListener;
import android.widget.Button;
import android.widget.Toast;

public class MainActivity extends Activity implements OnClickListener {

    private Button bt_startService, bt_stopService, bt_bindService,
            bt_unbindService;
    private ServiceConnection conn = new ServiceConnection() {

        @Override
        public void onServiceConnected(ComponentName name, IBinder service) {
            // TODO Auto-generated method stub

        }

        @Override
        public void onServiceDisconnected(ComponentName name) {
            // TODO Auto-generated method stub

        }
    };

    @Override
    protected void onCreate(Bundle savedInstanceState) {
        super.onCreate(savedInstanceState);
        setContentView(R.layout.activity_main);
        bt_startService = (Button) findViewById(R.id.Btn_StartService);
        bt_stopService = (Button) findViewById(R.id.Btn_StopService);
        bt_bindService = (Button) findViewById(R.id.Btn_BindService);
        bt_unbindService = (Button) findViewById(R.id.Btn_UnbindService);
        bt_startService.setOnClickListener(this);
        bt_stopService.setOnClickListener(this);
        bt_bindService.setOnClickListener(this);
        bt_unbindService.setOnClickListener(this);
```

```
45          }
46
47          @Override
48          public void onClick(View v) {
49                Intent intent = new Intent(MainActivity.this, MyService.class);
50                switch (v.getId()) {
51                case R.id.Btn_StartService:
52                      startService(intent);
53                      Toast.makeText(MainActivity.this, "启动服务成功", Toast.LENGTH_SHORT)
54                                  .show();
55                      break;
56                case R.id.Btn_StopService:
57                      stopService(intent);
58                      Toast.makeText(MainActivity.this, "停止服务成功", Toast.LENGTH_SHORT)
59                                  .show();
60                      break;
61                case R.id.Btn_BindService:
62                      bindService(intent, conn, BIND_AUTO_CREATE);
63                      Toast.makeText(MainActivity.this, "绑定服务成功", Toast.LENGTH_SHORT)
64                                  .show();
65                      break;
66                case R.id.Btn_UnbindService:
67                      unbindService(conn);
68                      Toast.makeText(MainActivity.this, "解除绑定服务成功", Toast.LENGTH_SHORT)
69                                  .show();
70                      break;
71                }
72          }
73    }
```

1）第 3～12 行引入了代码中需要使用的 Android 相关类。

2）第 18～31 行实例化 ServiceConnection 类，用于第 62 行的 BindService 参数。

3）第 48～71 行分别用于启动服务、停止服务、绑定服务与解绑服务。

（3）在 src 的 com.hzu.service_test 包下创建 MyService.java 文件，编写内容如下：

```
package com.hzu.service_test;

import android.app.Service;
import android.content.Intent;
import android.content.ServiceConnection;
import android.os.IBinder;
import android.util.Log;

public class MyService extends Service {
      @Override
      public void onCreate() {
            Log.i("MyService", "onCreate()方法已启动");
            super.onCreate();
```

```
14          }
15
16          @Override
17          public int onStartCommand(Intent intent, int flags, int startId) {
18                  Log.i("MyService", "onStartCommand()方法已启动");
19                  return super.onStartCommand(intent, flags, startId);
20          }
21
22          @Override
23          public void onDestroy() {
24                  Log.i("MyService", "onDestroy()方法已启动");
25                  super.onDestroy();
26          }
27
28          @Override
29          public IBinder onBind(Intent intent) {
30                  Log.i("MyService", "onBind()方法已启动");
31                  return null;
32          }
33
34          @Override
35          public boolean onUnbind(Intent intent) {
36                  Log.i("MyService", "onUnbind()方法已启动");
37                  return super.onUnbind(intent);
38          }
39
40  }
```

1）第 3～7 行引入了代码中需要使用的 Android 相关类。

2）第 11～14 行重写了 Service 类的 onCreate()方法。

3）第 17～38 行依次重写 onStartCommand()方法、onDestroy()方法、onBind()方法和 onUnbind 方法。

（4）在 Service_test 的项目下创建 AndroidManifest.xml 文件，编写内容如下：

```
<?xml version="1.0" encoding="utf-8"?>
<manifest xmlns:android="http://schemas.android.com/apk/res/android"
        package="com.hzu.service_test"
        android:versionCode="1"
        android:versionName="1.0">

<uses-sdk
                android:minSdkVersion="14"
                android:targetSdkVersion="21" />

<application
                android:allowBackup="true"
                android:icon="@drawable/ic_launcher"
```

```
14                  android:isGame="true"
15                  android:label="@string/app_name"
16                  android:multiArch="true"
17                  android:theme="@style/AppTheme">
18      <activity
19                      android:name=".MainActivity"
20                      android:label="@string/app_name">
21      <intent-filter>
22      <action android:name="android.intent.action.MAIN" />
23
24      <category android:name="android.intent.category.LAUNCHER" />
25      </intent-filter>
26      </activity>
27
28      <service android:name="MyService">
29      </service>
30      </application>
31
32      </manifest>
```

第 28～29 行添加<service>标签，然后把 MyService 类加入到此标签中。

**【运行结果】**在 Eclipse 中启动 Android 模拟器，接着运行 Service_test 项目，显示效果如图 9-2 所示，点击 startService 按钮显示的效果如图 9-3 所示。

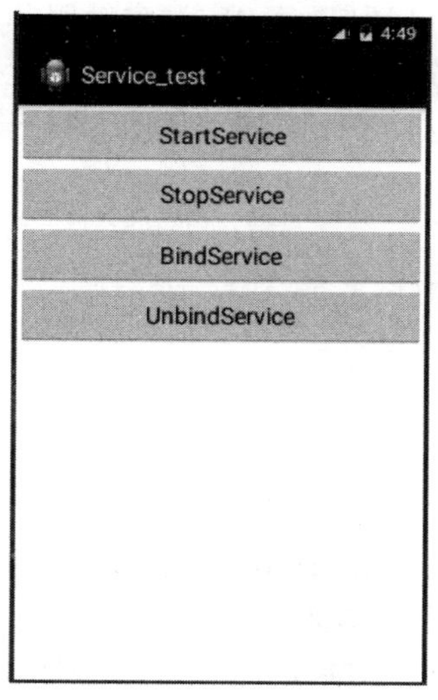


图 9-2　项目运行效果

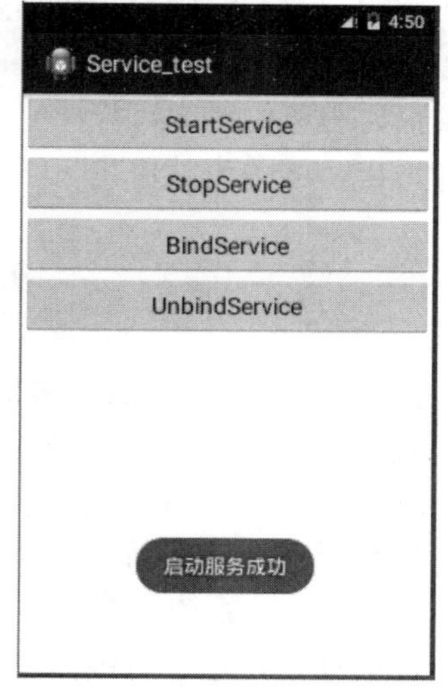


图 9-3　启动服务效果

在点击 StartService 按钮后，使用 MyService 作为过滤器，显示在 LogCat 的效果如图 9-4 所示。

| Level | Time | PID | TID | Application | Tag | Text |
|---|---|---|---|---|---|---|
| I | 02-08 04:51:26.736 | 2312 | 2312 | com.hzu.service_test | MyService | onCreate()方法已启动 |
| I | 02-08 04:51:26.736 | 2312 | 2312 | com.hzu.service_test | MyService | onStartCommand()方法已启动 |

图 9-4　StartService 启动服务后调用方法

在点击 StopService 按钮后，使用 MyService 作为过滤器，显示在 LogCat 的效果如图 9-5 所示。

| Level | Time | PID | TID | Application | Tag | Text |
|---|---|---|---|---|---|---|
| I | 02-08 04:51:26.736 | 2312 | 2312 | com.hzu.service_test | MyService | onCreate()方法已启动 |
| I | 02-08 04:51:26.736 | 2312 | 2312 | com.hzu.service_test | MyService | onStartCommand()方法已启动 |
| I | 02-08 04:58:55.649 | 2312 | 2312 | com.hzu.service_test | MyService | onDestroy()方法已启动 |

图 9-5　StopService 后调用方法

在点击 BindService 按钮后，使用 MyService 作为过滤器，显示在 LogCat 的效果如图 9-6 所示。

| Level | Time | PID | TID | Application | Tag | Text |
|---|---|---|---|---|---|---|
| I | 02-08 04:59:20.401 | 2312 | 2312 | com.hzu.service_test | MyService | onCreate()方法已启动 |
| I | 02-08 04:59:20.401 | 2312 | 2312 | com.hzu.service_test | MyService | onBind()方法已启动 |

图 9-6　BindService 后调用方法

在点击 UnbindService 按钮后，使用 MyService 作为过滤器，显示在 LogCat 的效果如图 9-7 所示。

| Level | Time | PID | TID | Application | Tag | Text |
|---|---|---|---|---|---|---|
| I | 02-08 04:59:20.401 | 2312 | 2312 | com.hzu.service_test | MyService | onCreate()方法已启动 |
| I | 02-08 04:59:20.401 | 2312 | 2312 | com.hzu.service_test | MyService | onBind()方法已启动 |
| I | 02-08 04:59:39.081 | 2312 | 2312 | com.hzu.service_test | MyService | onUnbind()方法已启动 |
| I | 02-08 04:59:39.081 | 2312 | 2312 | com.hzu.service_test | MyService | onDestroy()方法已启动 |

图 9-7　UnbindService 后调用方法

服务不能自己运行，需要使用 startService()方法或 bindService()方法启动服务，以上两种方法都可以启动 Service。使用 startService()方法启动服务时，调用者与服务之间互不关联，即调用者退出后，服务仍然在运行；使用 bindService()方法启动服务时，调用者与服务绑定在一起，调用者退出，服务也就停止。

## 9.2　Notification

Toast 是一种消息提示方式，它只能保存一段时间，用户不能控制显示时长。Notification 是在不妨碍用户操作的前提下，在屏幕的最上方把消息显示在通知栏中。Notification 无需 Activity，只要用户按住状态栏并下拉，即可打开状态栏查看相关的消息。

通知一般通过 NotificationManager 服务来发送一个 Notification 对象来完成，Notification Manager 是一个重要的系统级服务，该对象位于应用程序的框架层中，应用程序可以通过它向系统发送全局的通知，这个时候需要创建一个 Notification 对象，用于封装通知的内容。一般情况下，不会直接构建 Notification 对象，而是使用它的一个内部类 NotificationCompat.Builder 来实例化一个对象，再设置通知的各种属性，最后通过 NotificationCompat.Builder.build()方法得到一个 Notification 对象。当得到这个对象之后，使用 NotificationManager.notify()方法发送通知。

NotificationManager（通知管理器），这个对象是由系统维护的服务，是以单例模式获得，一般不直接实例化这个对象，在 Activity 中使用 Activity.getSystemService()方法获得 Notification Manager 对象。

pendingIntent 是一种特殊的 Intent，它位于 android.app 包下。Intent 是表示立刻执行的动作，而 pendingIntent 表示即将发生的动作。

**【例 9.2】** 在通知栏生成一条通知，用户通过点击这条通知可以进入到这个通知的具体页面中。

**【说明】** 主要是通过使用 NotificationCompat 对象下的 Builder 来封装通知中的相关内容，然后使用 PendingIntent 设置要跳转的页面，最后使用 NotificationManager 来发送通知。

**【开发步骤】**

（1）创建一个名为 Notification_test 的项目，Activity 组件名为 MainActivity。

（2）将 map_large.ico 与 map_small.ico 两张图片复制到本项目的 res/drawable-hdpi 目录下。

（3）在 res/layout 下创建一个 activity_main.xml 文件，内容如下：

```
1   <RelativeLayout xmlns:android="http://schemas.android.com/apk/res/android"
2       xmlns:tools="http://schemas.android.com/tools"
3       android:layout_width="match_parent"
4       android:layout_height="match_parent">
5
6   <Button
7           android:id="@+id/button1"
8           android:layout_width="match_parent"
9           android:layout_height="wrap_content"
10          android:layout_alignParentTop="true"
11          android:layout_centerHorizontal="true"
12          android:text="SendNotification" />
13
14  </RelativeLayout>
```

1）第 1～14 行声明了一个相对布局。

2）第 6～12 行声明了一个 Button，其 id 为 button1。

（4）在 res/layout 下创建一个 show.xml 文件，内容如下：

```
1   <?xml version="1.0" encoding="utf-8"?>
2   <LinearLayout xmlns:android="http://schemas.android.com/apk/res/android"
3       android:layout_width="match_parent"
4       android:layout_height="match_parent"
5       android:orientation="vertical">
6
7   <TextView
8           android:id="@+id/tv"
9           android:layout_width="wrap_content"
10          android:layout_height="wrap_content"
11          android:layout_gravity="center"
12          android:text="显示的具体内容" />
13
14  </LinearLayout>
```

1）第 1～14 行声明了一个线性布局，其方向为垂直方向。

2）第 7～12 行声明了一个 TextView，其 id 为 tv。

（5）在 src 的 com.hzu.notification_test 包下创建 MainActivity.java 文件，内容如下：

```
package com.hzu.notification_test;

import android.app.Activity;
import android.app.NotificationManager;
import android.app.PendingIntent;
import android.content.Intent;
import android.graphics.BitmapFactory;
import android.os.Bundle;
import android.support.v4.app.NotificationCompat;
import android.view.View;
import android.widget.Button;

public class MainActivity extends Activity {
    private Button bt_notification;

    @Override
    protected void onCreate(Bundle savedInstanceState) {
        super.onCreate(savedInstanceState);
        setContentView(R.layout.activity_main);
        bt_notification = (Button) findViewById(R.id.button1);
        bt_notification.setOnClickListener(new View.OnClickListener() {

            @Override
            public void onClick(View v) {
                NotificationCompat.Builder builder = new NotificationCompat.Builder(
                        MainActivity.this);
                builder.setContentTitle("通知标题");
                builder.setContentText("文本中要显示的信息")
                // 设置小图标
                        .setSmallIcon(R.drawable.map_small)
                        // 设置大图标
                        .setLargeIcon(
                                BitmapFactory.decodeResource(getResources(),
                                        R.drawable.map_large));
                Intent intent = new Intent();
                intent.setClass(MainActivity.this, ShowActivity.class);
                // 创建新的 PendingIntent 实例
                PendingIntent pIntent = PendingIntent.getActivity(
                        MainActivity.this, 0xee, intent,
                        PendingIntent.FLAG_UPDATE_CURRENT);
                // 设置点击后的事件
                builder.setContentIntent(pIntent);
                // 设置自动取消
                builder.setAutoCancel(true);
                // 通过系统服务将通知显示在状态栏上
```

```
46                          NotificationManager manager = (NotificationManager)
47                                   getSystemService(NOTIFICATION_SERVICE);
48                          manager.notify(1, builder.build());
49
50                   }
51             });
52       }
53 }
```

1）第 3～11 行引入代码中需要使用的 Android 相关类。

2）第 17～52 行重写了 Activity 类中的 onCreate()方法。

3）第 25 行创建了 NotificationCompat.Builder 对象。

4）第 27 行设置通知中的标题。

5）第 28 行设置通知中的文本内容。

6）第 30 行设置通知中的小图标。

7）第 32～34 行设置通知中的大图标。

8）第 38～40 行创建 PendingIntent 对象，其中 getActivity()中包含有四个参数，第一个表示上下文对象，第二个参数为请求码，第三个参数 Intent 用来存储信息，第四个参数为对参数的操作标识，常用 FLAG_CANCEL_CURRENT 和 FLAG_UPDATE_CURRENT。

9）第 42 行设置点击后的事件。

10）第 44 行设置自动取消，第 46～47 行获取 NotificationManager 对象，第 48 行发出通知。

（6）在 src 的 com.hzu.notification_test 包下创建 ShowActivity.java 文件，其内容如下：

```
1   package com.hzu.notification_test;
2
3   import android.app.Activity;
4   import android.os.Bundle;
5
6   public class ShowActivity extends Activity {
7       @Override
8       protected void onCreate(Bundle savedInstanceState) {
9           // TODO Auto-generated method stub
10          super.onCreate(savedInstanceState);
11          setContentView(R.layout.show);
12      }
13
14  }
```

第 11 行设置用户点击通知后显示的内容。

（7）在 Notification_test 的项目下创建 AndroidManifest.xml 文件，编写内容如下：

```
<?xml version="1.0" encoding="utf-8"?>
<manifest xmlns:android="http://schemas.android.com/apk/res/android"
    package="com.hzu.notification_test"
    android:versionCode="1"
    android:versionName="1.0">
```

```
6
7   <uses-sdk
8           android:minSdkVersion="14"
9           android:targetSdkVersion="21" />
10
11  <application
12          android:allowBackup="true"
13          android:hardwareAccelerated="false"
14          android:icon="@drawable/ic_launcher"
15          android:label="@string/app_name"
16          android:theme="@style/AppTheme"
17          android:vmSafeMode="true">
18  <activity
19              android:name=".MainActivity"
20              android:label="@string/app_name">
21  <intent-filter>
22  <action android:name="android.intent.action.MAIN" />
23
24  <category android:name="android.intent.category.LAUNCHER" />
25  </intent-filter>
26  </activity>
27  <activity android:name="ShowActivity">
28  </activity>
29  </application>
30  </manifest>
```

第 27～28 行添加<activity>标签，然后把 ShowActivity 类加入到此标签中。

**【运行结果】**在 Eclipse 中启动 Android 模拟器，接着运行 Notification_test 项目，显示效果如图 9-8 所示，点击 SendNotification 按钮，效果如图 9-9 所示。

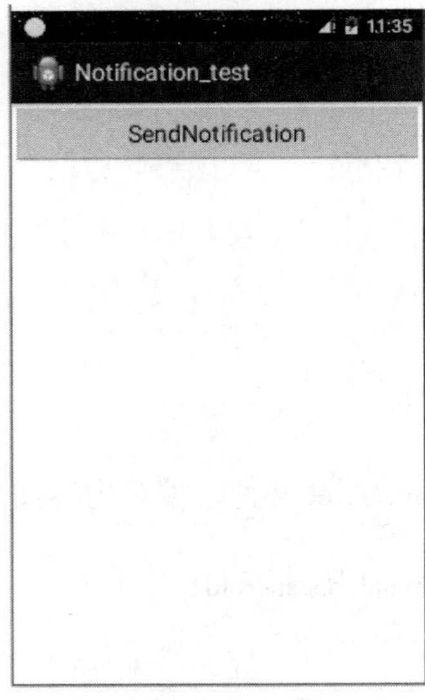


图 9-8　项目运行效果

图 9-9　创建的通知栏效果

## 9.3　BroadcastReceiver

Broadcast（广播）是指当系统发生某个事件时，就会向外发出广播，广播本身并不知道有多少个应用在关注自己。一个应用程序也可以发出广播，通过广播中的关键字段，系统可以寻找所有关注这个广播的应用，并触发它们注册的 Receiver（接收者）。

Broadcast 分为有序广播和无序广播两类。有序广播可以按照一定的顺序接收广播，同时在传递广播时可以添加数据，广播接收者可以在中途截断广播的传递。无序广播接收者的接收顺序无法指定，不能截断广播与添加数据。

BroadcastReceiver（广播接收者）是指用来接收系统或其他应用程序发送过来的广播。它是 Android 系统的四大组件之一，本质是一个系统全局的监听者，可以非常方便地实现系统中不同的组件间以及不同应用程序间的通信。BroadcastReceiver 的运行机制是：先构建一个 Intent 对象，然后调用 sendBroadcast()方法将广播发送出去。当应用程序注册了 BroadcastReceiver 之后，系统或其它应用程序发出广播时，注册接收器将检测发送过来的广播，若匹配则调用 onReceiver()方法，在此方法中响应相关事件，若不匹配则不做处理。

1. 发送广播的方式

（1）使用 sendBroadcast()方法发送的广播。

使用 sendBroadcast()方法发送广播，就是所有满足条件的接收者都会执行其 onReceive()方法来处理后续逻辑代码，在此过程中，对于以上满足条件的接收者执行 onReceive()方法的顺序是不固定的。

（2）使用 sendOrderedBroadcast()方法发送的广播。

使用 sendOrderedBroadcast()方法发送的 Intent，就会根据 BroadcastReceiver 注册时 IntentFilter 中设置的优先级（1000～-1000）来决定执行 onReceive()方法的顺序。如果设置相同优先级 BroadcastReceiver，对于满足条件的接收者执行 onReceive()方法的顺序是不固定的。

2. 接收广播的方式

BroadcastReceiver 接收过程如下：

（1）编写 BroadcastReceiver 类的子类，重写 onReceive()方法。

```
public class MyReceiver extends BroadcastReceiver {
        @Override
        public void onReceive(Context context, Intent intent) {
            // 在此编写逻辑代码
    }
}
```

**注意：**在 onReceive()方法中处理的逻辑代码最好不要超过 5s，如果超过 5s，系统会提示超时操作，可以考虑把超时的操作放到多线程中进行处理。

（2）注册 BroadcastReceiver 对象。

在系统或应用程序广播的 Intent，接收者如果想接收此广播，都需要注册一个 BroadcastReceiver，再给注册的这个 BroadcastReceiver 设置一个 IntentFilter 来决定当前的 BroadcastReceiver 对哪些 Intent 进行监听。注册 BroadcastReceiver 有以下两种方式。

1）动态注册方式。

动态注册方式是使用 Activity 中 registerReceiver()方法来注册，注册代码如下：

```
//创建注册前需要的对象
MyReceiver myReceiver=new MyReceiver();
IntentFilter filter=new IntentFilter();
filter.addAction("com.hzu.action.receiver");
//动态注册 BroadcastReceiver
registerReceiver(myReceiver, filter);
```

registerReceiverr()方法包含两个参数，第一个参数指明接收对象，第二个参数是 IntentFilter。

2）静态注册方式。

静态注册方式是在该项目下的 AndroidManifest.xml 文件中的<application>标签下添加<receiver>标签，并设置要接收的 action。注册代码如下：

```
<receiver android:name=".MyReceiver">
<intent-filter>
<action android:name="com.hzu.action.receiver"></action>
</intent-filter>
</receiver>
```

Android 系统中还提供了很多标准广播 Action，这些广播是系统自带的，可以在 AndroidManifest.xml 中<intent-filter>标签下直接使用。常用的标准广播 Action 常量如下表 9-1 所示。

**表 9-1 常用的标准广播 Action 常量**

| 常量名 | 常量值 | 说明 |
|---|---|---|
| ACTION_BOOT_COMPLETED | android.intent.action.BOOT_COMPLETED | 系统启动完成 |
| ACTION_TIME_CHANGED | android.intent.action.ACTION_TIME_CHANGED | 时间改变 |
| ACTION_DATE_CHANGED | android.intent.action.ACTION_DATE_CHANGED | 日期改变 |
| ACTION_TIMEZONE_CHANGED | android.intent.action.ACTION_TIMEZONE_CHANGED | 时区改变 |
| ACTION_BATTERY_LOW | android.intent.action. ACTION_BATTERY_LOW | 电量低 |
| ACTION_SCREEN_ON | android.intent.action. ACTION_SCREEN_ON | 屏幕亮起 |
| ACTION_REBOOT | android.intent.action. ACTION_REBOOT | 系统重启 |

**【例 9.3】**设计一个无序广播。

**【说明】**主要是使用动态注册方式，用 Activity 中的 registerReceiver()方法来完成注册，在 Activity 销毁时要解除注册。

【开发步骤】

（1）创建一个名为 BroadcastReceiver_test 的项目，Activity 组件名为 MainActivity。在 res/layout 下创建一个 activity_main.xml 文件，内容如下：

```
1   <RelativeLayout xmlns:android="http://schemas.android.com/apk/res/android"
2       xmlns:tools="http://schemas.android.com/tools"
3       android:layout_width="match_parent"
4       android:layout_height="match_parent"
5       android:paddingBottom="@dimen/activity_vertical_margin"
6       android:paddingLeft="@dimen/activity_horizontal_margin"
7       android:paddingRight="@dimen/activity_horizontal_margin"
8       android:paddingTop="@dimen/activity_vertical_margin"
9       tools:context="com.hzu.broadcastreceiver_test.MainActivity">
10
11  <Button
12          android:id="@+id/button1"
13          android:layout_width="wrap_content"
14          android:layout_height="wrap_content"
15          android:layout_alignParentTop="true"
16          android:layout_centerHorizontal="true"
17          android:text="发送广播" />
18
19  </RelativeLayout>
```

1）第 1～19 行声明了一个相对布局。

2）第 11～17 行声明了一个 Button，其 id 为 button1。

（2）在 src 的 com.hzu.broadcastreceiver_test 包下创建 MyReceiver.java 文件，编写内容如下：

```
1   package com.hzu.broadcastreceiver_test;
2
3   import android.content.BroadcastReceiver;
4   import android.content.Context;
5   import android.content.Intent;
6   import android.widget.Toast;
7
8   public class MyReceiver extends BroadcastReceiver {
9       @Override
10      public void onReceive(Context context, Intent intent) {
11          Toast.makeText(context, "接收到一条广播！ ", Toast.LENGTH_LONG).show();
12      }
13  }
```

1）第 3～6 行引入代码中需要使用的 Android 的相关类。

2）第 10～12 行重写 BroadcastReceiver 类的 onReceive ()方法，用于接收广播后相应的逻辑处理代码。

（3）在 src 的 com.hzu.broadcastreceiver_test 包下创建 MainActivity.java 文件，编写内容如下：

```
package com.hzu.broadcastreceiver_test;

import android.app.Activity;
```

```
4   import android.content.Intent;
5   import android.content.IntentFilter;
6   import android.os.Bundle;
7   import android.view.View;
8   import android.widget.Button;
9
10  public class MainActivity extends Activity {
11      private Button bt_sendBroadcast;
12      private MyReceiver myReceiver;
13
14      @Override
15      protected void onCreate(Bundle savedInstanceState) {
16          super.onCreate(savedInstanceState);
17          setContentView(R.layout.activity_main);
18
19          myReceiver = new MyReceiver(); // 创建注册前需要的对象
20          IntentFilter filter = new IntentFilter();
21          filter.addAction("com.hzu.action.receiver");
22          registerReceiver(myReceiver, filter);          // 动态注册 BroadcastReceiver
23
24          bt_sendBroadcast = (Button) findViewById(R.id.button1);
25          bt_sendBroadcast.setOnClickListener(new View.OnClickListener() {
26
27              public void onClick(View v) {
28                  Intent intent = new Intent();
29                  intent.setAction("com.hzu.action.receiver");
30                  sendBroadcast(intent);
31              }
32          });
33      }
34
35      protected void onDestroy() {
36          unregisterReceiver(myReceiver);
37          super.onDestroy();
38      }
39  }
```

1）第 3～8 行引入代码中需要使用的 Android 相关类。

2）第 15～33 行重写了 Activity 类中的 onCreate ()方法。

3）第 19～20 行创建注册前需要的对象，第 22 行动态注册 BroadcastReceiver。

4）第 27～32 行添加“发送广播”按钮的监听器，发送广播。

5）第 35～38 行在 onDestroy()方法中注销广播。

**【运行结果】**在 Eclipse 中启动 Android 模拟器，接着运行 BroadcastReceiver_test 项目，显示效果如图 9-10 所示，点击“发送广播”按钮，效果如图 9-11 所示。

图 9-10　项目运行效果

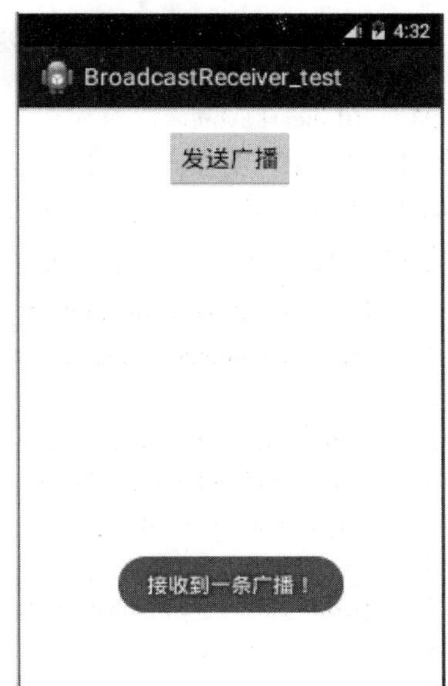


图 9-11　发送广播效果

**【例 9.4】**设计一个自动计数器，每隔 1 秒加 1。

**【说明】**在 Service 组件中创建一个线程，在此线程中产生数值，每隔 1 秒加 1，然后把更新值显示到界面上。

**【开发步骤】**

（1）创建一个名为 CountShow_test 的项目，Activity 组件名为 MainActivity。在 res/layout 下创建一个 activity_main.xml 文件，内容如下：

```
<LinearLayout xmlns:android="http://schemas.android.com/apk/res/android"
      xmlns:tools="http://schemas.android.com/tools"
      android:layout_width="match_parent"
      android:layout_height="match_parent"
      android:orientation="vertical">

<Button
          android:id="@+id/button"
          android:layout_width="match_parent"
          android:layout_height="wrap_content"
          android:layout_alignParentLeft="true"
          android:layout_alignParentTop="true"
          android:text="启动计数" />

<TextView
          android:id="@+id/tv"
          android:layout_width="wrap_content"
          android:layout_height="wrap_content"
          android:layout_gravity="center"
          android:text="0"
```

```
21              android:textSize="25px" />
22
23  </LinearLayout>
```

1）第 1～23 行声明了一个线性布局。

2）第 7～13 行声明了一个 Button，其 id 为 button。

3）第 15～21 行声明了一个 TextView，其 id 为 tv。

（2）在 src 的 com.hzu.countshow_test 包下创建 CountService.java 文件，编写内容如下：

```
package com.hzu.countshow_test;

import android.app.Service;
import android.content.Intent;
import android.os.IBinder;

public class CountService extends Service {
    private int x = 0;
    private boolean flag = false;

    @Override
    public void onCreate() {
        new Thread(new Runnable() {

            @Override
            public void run() {
                while (!flag) {
                    try {
                        Thread.sleep(1000);
                    } catch (Exception e) {
                    }
                    x++;
                    Intent intent = new Intent();
                    intent.putExtra("x", x);
                    intent.setAction("com.hzu.countshow");
                    sendBroadcast(intent);
                }

            }
        }).start();
    }

    @Override
    public void onDestroy() {
        x = 0;
        flag = true;
        super.onDestroy();
```

```
38          }
39
40          @Override
41          public IBinder onBind(Intent intent) {
42
43                return null;
44          }
45
46    }
```

1）第 3～5 行引入了代码中需要使用的 Android 的相关类。

2）第 12～31 行重写了 Service 类的 onCreate()方法，在此方法中创建线程，第 19 行实现线程每隔 1 秒休眠 1 次，第 23～24 行表示封装 x 值到 Intent 中，第 25 行表示添加 Action，第 26 行表示发送广播。

3）第 34～38 行重写了 onDestroy()方法，关闭线程中的 while 循环。

4）第 41～44 行重写了 onBind()方法。

（3）在 src 的 com.hzu.countshow_test 包下创建 MainActivity.java 文件，编写内容如下：

```
package com.hzu.countshow_test;

import android.app.Activity;
import android.content.BroadcastReceiver;
import android.content.Context;
import android.content.Intent;
import android.content.IntentFilter;
import android.os.Bundle;
import android.view.View;
import android.widget.Button;
import android.widget.TextView;

public class MainActivity extends Activity {
      private Button bt_startCount;
      private TextView tv_show;
      private MyReceiver myReceiver;
      private Intent intent;

      @Override
      protected void onCreate(Bundle savedInstanceState) {
            super.onCreate(savedInstanceState);
            setContentView(R.layout.activity_main);
            bt_startCount = (Button) findViewById(R.id.button);
            tv_show = (TextView) findViewById(R.id.tv);

            myReceiver = new MyReceiver();
            IntentFilter filter = new IntentFilter();
```

```
28          filter.addAction("com.hzu.countshow");
29          // 动态注册 BroadcastReceiver
30          registerReceiver(myReceiver, filter);
31
32          bt_startCount.setOnClickListener(new View.OnClickListener() {
33
34              @Override
35              public void onClick(View v) {
36                  intent = new Intent(MainActivity.this, CountService.class);
37                  startService(intent);
38              }
39          });
40      }
41
42      @Override
43      protected void onDestroy() {
44          unregisterReceiver(myReceiver);
45          stopService(intent);
46          super.onDestroy();
47      }
48
49      class MyReceiver extends BroadcastReceiver {
50          @Override
51          public void onReceive(Context context, Intent intent) {
52              Bundle bundle = intent.getExtras();
53              int x = bundle.getInt("x");
54              tv_show.setText(x + "");
55          }
56      }
57  }
```

1）第 3～11 行引入了代码中需要使用的 Android 相关类。

2）第 20～40 行重写了 Activity 类的 onCreate()方法，第 26～28 行创建注册前需要的对象，第 30 行完成动态注册 BroadcastReceiver。第 32～39 行为“启动计数”按钮添加监听器，在 onClick()方法中启动服务。

3）第 43～47 行重写了 Activity 类的 onDestroy()方法，用此方法注销广播，停止服务。

4）第 49～56 行新建 MyReceiver 类，继承自 BroadcastReceiver，在 onReceive()方法中接收数据，更新数据后显示在界面上。

（4）在 CountShow_test 项目下的 AndroidManifest.xml 文件中添加 CountService 这个服务，由于比较简单，此处不再详细介绍。

**【运行结果】**在 Eclipse 中启动 Android 模拟器，接着运行 CountShow_test 项目，显示效果如图 9-12 所示，点击“启动计数”按钮，效果如图 9-13 所示。

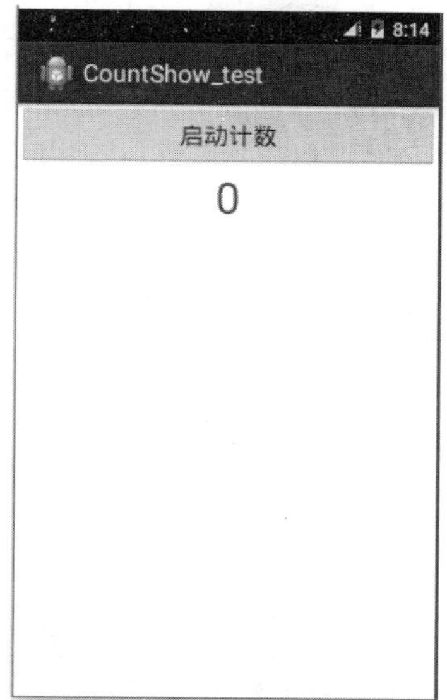

图 9-12　项目运行效果

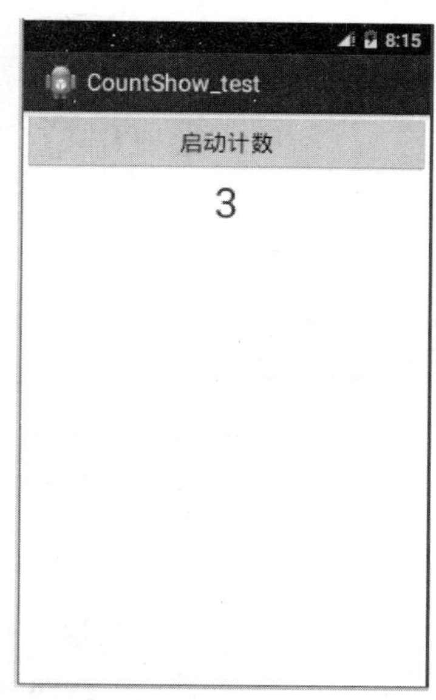


图 9-13　“启动计数”效果

## 9.4　本章小结

本章首先讲解了 Service 的用途，Service 是 Android 中重要的组件之一，在应用开发中使用频率仅次于 Activity；接着介绍了 Notification 的使用，根据不同的场景使用 Notification 或 Toast；最后介绍了 BroadcastReceiver。掌握这些后台编程组件，将为深层次的 Android 应用程序开发奠定基础。

# 第 10 章　Android 数据存储

（1）了解 SharedPreferences 与文件存储。

（2）掌握ContentProvider的基本使用。

（3）掌握 SQLite 的基本语法以及相关操作。

程序出现的目标就是处理数据。在操作系统与应用程序中有大量要处理的数据，因此存储数据是程序最重要的部分。在 Android 应用程序中也有要处理的数据，例如音乐、图片、视频以及短信、通讯录等。

在 Android 应用程序内部，所有数据都是私有的。为了更好地达到用户友好性的目标，Android 系统提供了应用程序之间数据通信的组件。Android 系统提供了以下几种数据存储与共享方式：SharedPreferences、ContentProvider、文件存储以及数据库等。

## 10.1　SharedPreferences使用

在 Android 提供的几种数据存储方式中，SharedPreferences 属于轻量级的键-值对存储方式，以 XML 文件方式保存数据，通常用来存储一些用户行为、开关状态等。SharedPreferences 一般存储类型是一些常见的基本数据类型。

使用 SharedPreferences 存储的 XML 数据可以通过以下方式查看：打开 DDMS 的 File Explorer 面板，在展开的文件列表中，查找/data/data/<package name>/shared_prefs 目录。SharedPreferences 对象本身只能获取数据而不支持存储与修改，存储与修改是通过 SharedPreferences.edit()方法获取内部接口 Editor 对象实现。由于 SharedPreferences 是一个接口，程序无法直接创建 SharedPreferences 实例，只能通过 Context 提供的 getSharedPreferences(String name, int mode)方法来获取 SharedPreferences 实例。该方法中的第一个参数 name 表示要操作的 xml 文件名，第二个参数 mode 表示操作模式，一共有以下三种操作模式：

- Context.MODE_PRIVATE：指定该 SharedPreferences 数据只能被本应用程序读写。
- Context.MODE_WORLD_READABLE：指定该 SharedPreferences 数据能被其他应用程序读，但不能写。
- Context.MODE_WORLD_WRITEABLE：指定该 SharedPreferences 数据能被其他应用程序读写。

SharedPreferences 接口与 SharedPreferences.Editor 内部接口包含一些常用的方法，这些常用的方法如表 10-1 与表 10-2 所示。

表 10-1 SharedPreferences 常用方法与说明

| 方法 | 说明 |
| --- | --- |
| contains(String key) | 判断是否包含该键值 |
| edit() | 获取 SharedPreferences.Editor |
| getAll() | 获取所有配置信息的 Map |
| getBoolean(String key, boolean defValue) | 获取一个 boolean 值 |
| getInt(String key, int defValue) | 获取一个 int 值 |
| getFloat(String key, float defValue) | 获取一个 float 值 |
| getLong(String key, long defValue) | 获取一个 long 值 |
| getString(String key, String defValue) | 获取一个 String 值 |

表 10-2 SharedPreferences.Editor 常用方法与说明

| 方法 | 说明 |
| --- | --- |
| clear() | 清空所有值 |
| commit() | 保存 |
| putBoolean(String key, boolean value) | 保存一个 boolean 值 |
| putInt(String key, int value) | 保存一个 int 值 |
| putFloat(String key, float value) | 保存一个 float 值 |
| putLong(String key, long value) | 保存一个 long 值 |
| putString(String key, String value) | 保存一个 String 值 |

SharedPreferences 对象写入数据的流程为：通过使用 edit()方法取得 SharedPreferences 的 Editor 对象，使用相应的 put()方法设置键值，然后使用 commit()方法，将数据写入 XML 中。

SharedPreferences 对象读取数据的流程为：通过获得 SharedPreferences 对象的键，然后根据不同的键使用相应的 get()方法就可以取得数据。

**【例 10.1】**在“贺州旅游”的用户登录界面上，增加一个“记住我”复选框。当勾选此复选框后，保存用户登录时输入的用户名与密码。

**【说明】**使用 SharedPreferences 保存用户输入的用户名与密码。

**【开发步骤】**

（1）创建一个名为 HZTour 的项目，Activity 组件名为 MainActivity。

（2）将 bg_login.png 等 4 张图片复制到本项目的 res/ drawable-hdpi 目录下。

（3）在 res/layout 下创建一个 activity_main.xml 文件，内容如下：

```
<RelativeLayout xmlns:android="http://schemas.android.com/apk/res/android"
    xmlns:tools="http://schemas.android.com/tools"
    android:id="@+id/rl_first"
    android:layout_width="match_parent"
    android:layout_height="match_parent"
    android:background="@drawable/bg_login">
```

```

<LinearLayout
        android:id="@+id/ll_first"
        android:layout_width="match_parent"
        android:layout_height="wrap_content"
        android:layout_marginLeft="20dp"
        android:layout_marginRight="20dp"
        android:layout_marginTop="60dp"
        android:background="#883366cc"
        android:orientation="vertical">

<LinearLayout
            android:layout_width="match_parent"
            android:layout_height="50dp"
            android:orientation="horizontal">

<ImageView
                android:layout_width="30dp"
                 android:layout_height="30dp"
                 android:src="@drawable/ic_user">
</ImageView>

<EditText
                android:id="@+id/et_user"
                android:layout_width="0dp"
                android:layout_height="wrap_content"
                android:layout_weight="1"
                android:hint="邮箱/手机号"
                android:textColorHint="#acacac" />
</LinearLayout>

<LinearLayout
            android:layout_width="match_parent"
            android:layout_height="50dp"
            android:orientation="horizontal">

<ImageView
                android:layout_width="30dp"
                android:layout_height="30dp"
                android:src="@drawable/ic_password">
</ImageView>

<EditText
                android:id="@+id/et_pwd"
                android:layout_width="0dp"
                android:layout_height="wrap_content"
```

```
                    android:layout_weight="1"
                    android:ems="10"
                    android:hint="密码"
                    android:password="true"
                    android:textColorHint="#acacac">

<requestFocus />
</EditText>
</LinearLayout>
</LinearLayout>

<RelativeLayout
        android:id="@+id/rl_second"
        android:layout_width="match_parent"
        android:layout_height="wrap_content"
        android:layout_below="@id/ll_first"
        android:layout_marginLeft="20dp"
        android:layout_marginRight="20dp">

<CheckBox
            android:id="@+id/cBox"
            android:layout_width="wrap_content"
            android:layout_height="wrap_content"
            android:layout_alignParentRight="true"
            android:checked="false"
            android:text="记住我"
            android:textColor="#5E5E5E" />
</RelativeLayout>

<Button
        android:id="@+id/button1"
        android:layout_width="match_parent"
        android:layout_height="wrap_content"
        android:layout_centerVertical="true"
        android:layout_marginLeft="20dp"
        android:layout_marginRight="20dp"
        android:layout_marginTop="30dp"
        android:background="#3399CC"
        android:text="登录"
        android:textColor="#ffffff" />

<TextView
        android:layout_width="wrap_content"
        android:layout_height="wrap_content"
        android:layout_alignParentBottom="true"
        android:layout_centerHorizontal="true"
```

```
99              android:text="贺州欢迎您"
100             android:textColor="#ffffff" />
101
102 </RelativeLayout>
```

1）第 1～102 行声明了一个相对布局，内部分为四个部分，第一部分用于布局用户名与密码，第二部分用于布局“记住我”，第三部分用于布局登录，第四部分用于布局“贺州欢迎您”。

2）第 8～62 行声明了一个线性布局，方向为垂直方向，内部再使用两个线性布局，分别用于摆放用户名图片和用户名输入框，密码图片和密码输入框。

3）第 64～80 行声明了一个相对布局，内部再添加一个 CheckBox。

4）第 82～92 行声明了一个 Button，通过 android:layout_centerVertical="true"使得 Button 位于整个屏幕中间的位置。

5）第 94～100 行声明了一个 TextView，通过 alignParentBottom 与 centerHorizontal 使此 TextView 位于屏幕底部中间的位置。

（4）在 src 的 com.hzu.hztour 包下创建 MainActivity.java 文件，编写内容如下：

```
package com.hzu.hztour;

import android.app.Activity;
import android.content.SharedPreferences;
import android.os.Bundle;
import android.view.View;
import android.widget.Button;
import android.widget.CheckBox;
import android.widget.EditText;
import android.widget.Toast;

public class MainActivity extends Activity {
        public static final String SP_NAME = "USER_FILE";
        public static final String USER = "User";
        public static final String PASSWORD = "Password";
        private EditText etUser;
        private EditText etPwd;
        private CheckBox cb;
        private Button bt;
        private String username;
        private String userpwd;

        @Override
        protected void onCreate(Bundle savedInstanceState) {
                super.onCreate(savedInstanceState);
                setContentView(R.layout.activity_main);
                etUser = (EditText) findViewById(R.id.et_user);
                etPwd = (EditText) findViewById(R.id.et_pwd);
                cb = (CheckBox) findViewById(R.id.cBox);
                bt = (Button) findViewById(R.id.button1);
```

```
31              checkData();
32              bt.setOnClickListener(new View.OnClickListener() {
33
34                  @Override
35                  public void onClick(View v) {
36
37                      if (cb.isChecked()) {
38                          username = etUser.getText().toString().trim();
39                          userpwd = etPwd.getText().toString().trim();
40                          rememberMe(username, userpwd);
41                          Toast.makeText(MainActivity.this, "用户名与密码已保存",
42                                  Toast.LENGTH_LONG).show();
43                      }
44                  }
45              });
46          }
47
48          public void checkData() {
49              SharedPreferences sp = getSharedPreferences(SP_NAME, MODE_PRIVATE);
50              username = sp.getString(USER, null);
51              userpwd = sp.getString(PASSWORD, null);
52              if (username != null && userpwd != null) {
53                  etUser.setText(username);
54                  etPwd.setText(userpwd);
55                  cb.setChecked(true);
56              }
57          }
58
59          public void rememberMe(String name, String pwd) {
60              SharedPreferences sp = getSharedPreferences(SP_NAME, MODE_PRIVATE);
61              SharedPreferences.Editor editor = sp.edit();
62              editor.putString(USER, name);
63              editor.putString(PASSWORD, pwd);
64              editor.commit();
65          }
66      }
```

1）第 13～15 行定义了三个字符串常量，分别表示配置文件名、用户名键名与密码键名。

2）第 24～46 行重写了 onCreate()方法，其中第 31 行调用 checkData()方法完成从 SharedPreferences 中读取用户名与密码。第 32～45 行使运行程序时，勾选“记住我”复选框，保存用户名与密码到 SharedPreferences 中。

3）第 48～57 行自定义 checkData()方法。其中第 49 行获得 SharedPreferences 对象 sp，第 50、51 行分别获得用户名与密码的值，第 53、54 行将用户名与密码的值传入到用户名输入框与密码输入框中，第 55 行将“记住我”复选框设置为选中状态。

4）第 59～65 行自定义 rememberMe(name,pwd)方法，将用户名和密码保存到 SharedPreferences 中。

【运行结果】在 Eclipse 中启动 Android 模拟器，接着运行 HZTour 项目，输入用户名与密码后显示效果如图 10-1 所示，退出程序后重新运行程序，显示效果如图 10-2 所示。

图 10-1 登录效果

图 10-2 重新运行效果

## 10.2 ContentProvider

ContentProvider为存储和读取数据提供了统一的接口，它的作用是实现应用程序之间的数据共享。Android系统中内置的许多数据都是使用 ContentProvider 形式，供开发者调用（如视频，音频，图片，通讯录等），通过这些定义好的 ContentProvider，数据的操作方式与获得的结果就像在操作数据库一样，可以进行增删查操作。注意在执行以上操作时要有相应的权限，在 AndroidManifest.xml 文件中添加权限即可。

1. ContentProvider

ContentProvider类位于 android.content 包下，应用程序通过实现ContentProvider抽象方法接口把自己的数据公开出去，其他程序就可以通过一组标准的接口访问本程序内部的数据。ContentProvider 类的主要方法如下：

- public boolean onCreate()：在 ContentProvider 创建后被调用。Android 开机后，ContentProvider 在其他应用第一次访问它时才会被创建。
- public Uri insert(Uri uri, ContentValues values)：用于供外部应用向 ContentProvider 添加数据。
- public int delete(Uri uri, String selection, String[] selectionArgs)：用于供外部应用从 ContentProvider 删除数据。
- public int update(Uri uri, ContentValues values, String selection, String[] selectionArgs)：用于供外部应用更新 ContentProvider 中的数据。

- public Cursor query(Uri uri, String[] projection, String selection, String[] selectionArgs, String sortOrder)：于外部应用从 ContentProvider 中获取数据。
- public String getType(Uri uri)：用于返回当前 URI 所代表数据的 MIME 类型。

2. ContentResolver

ContentResolver 是通过 URI 来查询 ContentProvider 中提供的数据。除了 URI 以外，还必须知道需要获取的数据段名称，以及此数据段的数据类型。在其他程序访问 ContentProvider 中的数据时，ContentResolver 提供的抽象方法与 ContentProvider 需要实现的方法对应，同样使用 insert()、delete()、query()、update 等方法来操作数据。

3. URI

URI（Universal Resource Identifier，通用资源标志符）可以帮助 ContentResolver 找到与之对应的 ContentProvider。

URI 由以下三部分组成：content://、数据的路径、标示 ID（可选）。

例如：

所有联系人的 URI：content://contacts/people。

所有图片的 URI：content://media/external。

4. 权限设置

使用系统资源时需在 AndroidManifest.xml 中添加访问权限，否则应用程序将不能访问由系统提供的 ContentProvider。

例如，对手机通讯录进行查询与修改操作的设置如下：

```
<uses-permission android:name="android.permission.READ_CONTACTS"/>
<uses-permission android:name="android.permission.WRITE_CONTACTS"/>
```

**【例 10.2】**编写一个查看手机联系人的程序。

**【说明】**通过使用 ContactsContract.CommonDataKinds.Phone.CONTENT_URI 常量来表示通讯录数据的 URI。

**【开发步骤】**

（1）创建一个名为 ContentProvider_test 的项目，Activity 组件名为 MainActivity。

（2）在 res/layout 下创建一个 activity_main.xml 文件，内容如下：

```
<RelativeLayout xmlns:android="http://schemas.android.com/apk/res/android"
        xmlns:tools="http://schemas.android.com/tools"
        android:layout_width="match_parent"
        android:layout_height="match_parent"
        android:paddingBottom="@dimen/activity_vertical_margin"
        android:paddingLeft="@dimen/activity_horizontal_margin"
        android:paddingRight="@dimen/activity_horizontal_margin"
        android:paddingTop="@dimen/activity_vertical_margin"
        android:background="#3CB371"
        tools:context="com.hzu.ntentprovider_test.MainActivity">

<Button
            android:id="@+id/button1"
            android:layout_width="wrap_content"
```

```
15              android:layout_height="wrap_content"
16              android:layout_alignParentLeft="true"
17              android:layout_alignParentTop="true"
18              android:layout_marginLeft="77dp"
19              android:layout_marginTop="14dp"
20              android:text="获取联系人" />
21
22      </RelativeLayout>
```

第 12～20 行添加一个 Button，其 id 为 button1。

（3）在 com.hzu.ntentprovider_test 包下，编写 MainActivity.java 文件，其内容如下：

```
1       package com.hzu.ntentprovider_test;
2
3       import android.app.Activity;
4       import android.content.ContentResolver;
5       import android.database.Cursor;
6       import android.os.Bundle;
7       import android.util.Log;
8       import android.view.View;
9       import android.widget.Button;
10      import android.provider.ContactsContract.CommonDataKinds.Phone;
11
12      public class MainActivity extends Activity {
13              private Button getBtn;
14              @Override
15              protected void onCreate(Bundle savedInstanceState) {
16                  super.onCreate(savedInstanceState);
17                  setContentView(R.layout.activity_main);
18                  getBtn=(Button) findViewById(R.id.button1);
19                  getBtn.setOnClickListener(new View.OnClickListener() {
20                      @Override
21                      public void onClick(View v) {
22                          ContentResolver resolver =getContentResolver();
23                          String info[]=new String[]{Phone.DISPLAY_NAME,Phone.NUMBER};
24                          Cursor c=resolver.query(Phone.CONTENT_URI, info, null, null, null);
25                        while(c.moveToNext()){
26                          String name=c.getString(c.getColumnIndex(info[0]));
27                          String number=c.getString(c.getColumnIndex(info[1]));
28                          Log.i("MainActivity", "name="+name+"|number="+number);}
29                      }
30                  });
31              }
32      }
```

1）第 22 行通过使用 getContentResolver()方法获取 ContentResolver 对象 resolver。

2）第 23 行声明了一个数组，用于定义姓名与电话。

3）第 24 行通过使用 resolver.query()方法来查询手机通讯录中的所有联系人信息，并返回给一个 Cursor 对象 c。其中 Phone.CONTENT_URI 是指向全体手机联系人的 URI。

4）第 25～29 行通过一个循环遍历出 c 中通讯录所包含的联系人姓名与电话号码。

（4）在 AndroidManifest.xml 文件中，添加相应权限。

```
1   <?xml version="1.0" encoding="utf-8"?>
2   <manifest xmlns:android="http://schemas.android.com/apk/res/android"
3       package="com.hzu.ntentprovider_test"
4       android:versionCode="1"
5       android:versionName="1.0">
6   
7   <uses-sdk
8           android:minSdkVersion="14"
9           android:targetSdkVersion="21" />
10  <uses-permission android:name="android.permission.READ_CONTACTS">
11  </uses-permission>
12  
13  <application
14          android:allowBackup="true"
15          android:icon="@drawable/ic_launcher"
16          android:label="@string/app_name"
17          android:theme="@style/AppTheme">
18  <activity
19              android:name=".MainActivity"
20              android:label="@string/app_name">
21  <intent-filter>
22  <action android:name="android.intent.action.MAIN" />
23  
24  <category android:name="android.intent.category.LAUNCHER" />
25  </intent-filter>
26  </activity>
27  </application>
28  
29  </manifest>
```

第 10～11 行添加对手机通讯录数据的可读权限。

**【运行结果】**如果模拟器中的手机通讯录没有任何联系人，请先在通讯录中添加联系人，步骤如下：

（1）按下模拟器上的 HOME 键，然后向左滑屏，找到如图 10-3 所示的 Contacts 应用程序。

（2）打开 Contacts 应用程序，选择右下角的添加联系人按钮，然后添加相应的联系人姓名与电话号码。

（3）重复上述第（2）步添加多个联系人，然后退出 Contacts 应用程序。

在 Eclipse 中启动 Android 模拟器，接着运行 ContentProvider_test 项目，显示效果如图 10-4 所示。

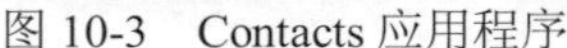
图 10-3 Contacts 应用程序

图 10-4 项目运行效果

点击"获取联系人"按钮后，在 MainActivity 过滤器中看到的内容如图 10-5 所示。

| Level | Time | PID | TID | Application | Tag | Text |
|---|---|---|---|---|---|---|
| I | 02-26 03:58:51.460 | 2107 | 2107 | com.hzu.ntentprovider_test | MainActivity | name=LiMing\|number=1 685-523-6985 |
| I | 02-26 03:58:51.461 | 2107 | 2107 | com.hzu.ntentprovider_test | MainActivity | name=Zhangsan\|number=1 689-636-5286 |

图 10-5 过滤器中的内容

## 10.3 文件存储

Android 系统是基于 Linux 的文件系统，操作 Android 下的文件可以像操作 Linux 中文件一样。文件同样具有访问权限，还可以用 Java 的 I/O 包中常用类来处理 Android 系统上的文件。在 Android 中对于复杂的数据或下载的数据一般使用文件形式来存储和读写。文件存储分为内部空间存储与外部空间存储两种方式。

内部存储是把数据存储在设备内部存储器上，存储在/data/data/<package name>/files 目录下。在默认情况下，这里存储的数据为应用程序的私有数据，文件管理器查看不到，其他应用程序不能访问。卸载应用程序后，内部存储器的/data/data/<package name>目录及其下子目录和文件一同被删除。一般应用于数据较少、数据私有等情况。

向内部存储空间中创建一个私有文件并向其中写入数据，其步骤如下：

（1）调用 openFileOutput(String fileName, int mode)方法，若 fileName 对应的文件存在，就打开该文件；若不存在，以 mode 权限创建该文件并打开，该方法返回一个指向 fileName

对应文件的 FileOutputStream，使用这个 FileOutputStream 可向文件中写入数据。

（2）调用 FileOutputStream 对象的 write()方法向文件中写入数据。

（3）调用 FileOutputStream 对象的 close()方法关闭文件写入流。

从内部存储空间读取私有文件数据的步骤如下：

（1）调用 openFileInput (String fileName)方法打开内部存储空间中 fileName 对应的文件，若该文件存在，则该方法返回一个指向 fileName 文件的 FileInputStream 对象。

（2）调用 FileInputStream 对象的 read()方法读取 fileName 文件中的内容。

（3）调用 FileInputStream 对象的 close()方法关闭文件读取流。

**【例 10.3】**在内部存储空间中写入与读取一组数据。

**【说明】**使用 openFileOutput(String fileName, int mode)方法与 openFileInput (String fileName)进行写与读操作。

**【开发步骤】**

（1）创建一个名为 InternalStorage_test 的工程，Activity 组件名为 MainActivity。

（2）在 res/layout 下创建一个 activity_main.xml 文件，内容如下：

```
<LinearLayout xmlns:android="http://schemas.android.com/apk/res/android"
      xmlns:tools="http://schemas.android.com/tools"
      android:layout_width="match_parent"
      android:layout_height="match_parent"
      android:layout_marginLeft="50dp"
      android:gravity="center_horizontal"
      android:orientation="vertical">

<LinearLayout
          android:layout_width="match_parent"
          android:layout_height="wrap_content"
          android:orientation="horizontal">

<EditText
              android:id="@+id/etwrite"
              android:layout_width="wrap_content"
              android:layout_height="wrap_content"
              android:text="请输入内容"
              android:maxLength="20"
              android:textSize="14sp" />

<Button
              android:id="@+id/btwrite"
              android:layout_width="wrap_content"
              android:layout_height="wrap_content"
              android:text="文件写入" />
</LinearLayout>
<LinearLayout
          android:layout_width="match_parent"
          android:layout_height="wrap_content"
```

```
            android:orientation="horizontal">

<Button
                android:id="@+id/btread"
                android:layout_width="wrap_content"
                android:layout_height="wrap_content"
                android:text="文件读取" />

<EditText
                android:id="@+id/tvshow"
                android:layout_width="wrap_content"
                android:layout_height="wrap_content"
                android:text="空"
                android:textSize="14sp" />
</LinearLayout>
</LinearLayout>
```

（3）在 src 的 com.hzu.internalstorage_test 包下创建 MainActivity.java 文件，内容如下：

```
package com.hzu.internalstorage_test;

import java.io.FileInputStream;
import java.io.FileNotFoundException;
import java.io.FileOutputStream;
import java.io.IOException;
import android.app.Activity;
import android.os.Bundle;
import android.view.View;
import android.view.View.OnClickListener;
import android.widget.Button;
import android.widget.EditText;
import android.widget.TextView;

public class MainActivity extends Activity implements OnClickListener {
        private TextView tv;
        private EditText et;
        private Button wButton, rButton;

        @Override
        protected void onCreate(Bundle savedInstanceState) {
                super.onCreate(savedInstanceState);
                setContentView(R.layout.activity_main);
                rButton = (Button) findViewById(R.id.btread);
                wButton = (Button) findViewById(R.id.btwrite);
                tv = (TextView) findViewById(R.id.tvshow);
                et = (EditText) findViewById(R.id.etwrite);
                rButton.setOnClickListener(this);
                wButton.setOnClickListener(this);
        }
```

```

        @Override
        public void onClick(View v) {
            switch (v.getId()) {
            case R.id.btwrite:
                String s = et.getText().toString();
                try {
                    FileOutputStream fo = openFileOutput("abc.txt",
                            MODE_WORLD_WRITEABLE);
                    fo.write(s.getBytes());
                    fo.close();
                } catch (FileNotFoundException e) {
                    e.printStackTrace();
                } catch (IOException e) {
                    e.printStackTrace();
                }
                break;
            case R.id.btread:
                try {
                    byte[] buffer = null;
                    FileInputStream fi = openFileInput("abc.txt");
                    buffer = new byte[fi.available()];
                    fi.read(buffer);
                    fi.close();
                    String data = new String(buffer);
                    tv.setText(data);

                } catch (FileNotFoundException e) {

                    e.printStackTrace();
                } catch (IOException e) {
                    // TODO Auto-generated catch block
                    e.printStackTrace();
                }
                break;
            }
        }
    }
```

**【运行结果】**在 Eclipse 中启动 Android 模拟器，接着运行 InternalStorage_test 项目，效果如图 10-5 所示，输入相应内容，点击“文件写入”和“文件读取”按钮，显示效果如图 10-6 所示。

外部存储在文件浏览器里是可以查看的，数据为共有的，所有人都可见和可用。一般应用于数据量较大的情况。

外部存储可能发生各种异常情况，通过 Environment.getExternalStorageState()方法查看外部存储的状态。通过 Environment.getExternalStorageDirectory 可以获得整个外部存储空间的路径。当应用程序被卸载后，外部存储中的数据不会随之被删除。

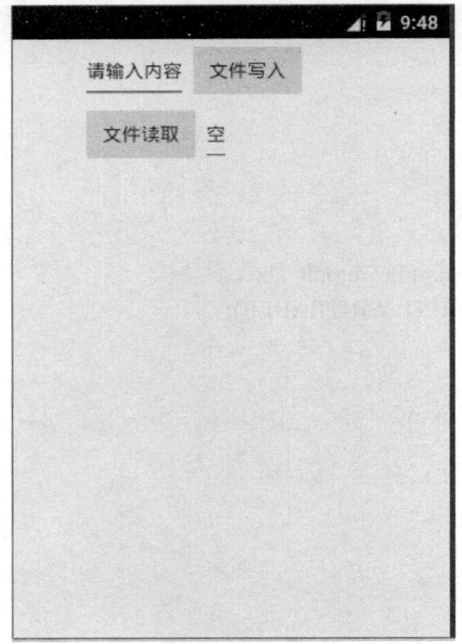


图 10-5　运行效果

图 10-6　读取数据效果

**【例 10.4】**在外部存储空间中写入与读取一组数据。

**【说明】**使用 Environment.getExternalStorageDirectory 获得外部路径，然后使用 openFileOutput(String fileName, int mode)与 openFileInput (String fileName)方法进行写与读操作。

（1）创建一个名为 ExtentalStorage_test 的项目，Activity 组件名为 MainActivity。

（2）在 res/layout 下创建一个 activity_main.xml 文件，内容如下：

```
<LinearLayout xmlns:android="http://schemas.android.com/apk/res/android"
      xmlns:tools="http://schemas.android.com/tools"
      android:layout_width="match_parent"
      android:layout_height="match_parent"
      android:layout_marginLeft="50dp"
      android:layout_marginTop="25dp"
      android:gravity="center_horizontal"
      android:orientation="vertical">

<LinearLayout
          android:layout_width="match_parent"
          android:layout_height="wrap_content"
          android:orientation="horizontal">

<EditText
              android:id="@+id/et_content"
              android:layout_width="wrap_content"
              android:layout_height="wrap_content"
              android:maxLength="20"
              android:text="请输入内容："
```

```
            android:textSize="14sp" />

    <Button
            android:id="@+id/btwrite"
            android:layout_width="wrap_content"
            android:layout_height="wrap_content"
            android:text="写操作" />
    </LinearLayout>

    <LinearLayout
        android:layout_width="match_parent"
        android:layout_height="wrap_content"
        android:orientation="horizontal">

    <Button
            android:id="@+id/btread"
            android:layout_width="wrap_content"
            android:layout_height="wrap_content"
            android:text="读操作" />

    <EditText
            android:id="@+id/tvshow"
            android:layout_width="wrap_content"
            android:layout_height="wrap_content"
            android:text="空"
            android:textSize="14sp" />
    </LinearLayout>

    </LinearLayout>
```

（3）在 src 的 com.hzu.extentalstorage_test 包下创建 MainActivity.java 文件，内容如下：

```
package com.hzu.extentalstorage_test;

import java.io.File;
import java.io.FileInputStream;
import java.io.FileNotFoundException;
import java.io.FileOutputStream;
import java.io.IOException;
import android.app.Activity;
import android.os.Bundle;
import android.os.Environment;
import android.view.View;
import android.view.View.OnClickListener;
import android.widget.Button;
import android.widget.EditText;
```

```
import android.widget.TextView;

public class MainActivity extends Activity implements OnClickListener {
    private EditText et;
    private TextView tv;
    private Button bt1, bt2;

    @Override
    protected void onCreate(Bundle savedInstanceState) {
        super.onCreate(savedInstanceState);
        setContentView(R.layout.activity_main);
        et = (EditText) findViewById(R.id.et_content);
        tv = (TextView) findViewById(R.id.tvshow);
        bt1 = (Button) findViewById(R.id.btwrite);

        bt2 = (Button) findViewById(R.id.btread);
        bt1.setOnClickListener(this);
        bt2.setOnClickListener(this);
    }

    @Override
    public void onClick(View v) {

        switch (v.getId()) {
        case R.id.btwrite:
            String state = Environment.getExternalStorageState();
            if (!state.equals(Environment.MEDIA_MOUNTED)) {
                return;
            }
            File f1 = Environment.getExternalStorageDirectory();
            String path1 = f1.getAbsolutePath() + "/abc.txt";
            File f2 = new File(path1);
            try {
                if (f2.exists()) {
                    f2.createNewFile();
                }
                FileOutputStream fo = new FileOutputStream(f2);
                String s = et.getText().toString();
                fo.write(s.getBytes());
                fo.close();
            } catch (IOException e) {
                e.printStackTrace();
            }
            break;
        case R.id.btread:
```

```
                    String path2 = Environment.getExternalStorageDirectory()
                            .getAbsolutePath() + "/abc.txt";
                    FileInputStream fi;
                    try {
                        fi = new FileInputStream(new File(path2));
                        byte[] buffer = null;
                        buffer = new byte[fi.available()];
                        fi.read(buffer);
                        fi.close();
                        String data = new String(buffer);
                        tv.setText(data);
                    } catch (FileNotFoundException e) {
                        e.printStackTrace();
                    } catch (IOException e) {
                        e.printStackTrace();
                    }
                    break;
                }
        }
}
```

（4）在项目的 AndroidManifest.xm 文件中添加如下内容：

```
<uses-permission android:name="android.permission.READ_EXTERNAL_STORAGE" />
<uses-permission android:name="android.permission.WRITE_EXTERNAL_STORAGE" />
```

**【运行结果】**在 Eclipse 中启动 Android 模拟器，接着运行 ExtentalStorage_test 项目，效果如图 10-7 所示，输入相应内容，点击“写操作”和“读操作”按钮，显示效果如图 10-8 所示，并且在 DDMS 下的 storage/sdcard 下生成 abc.txt 文件，如图 10-9 所示。

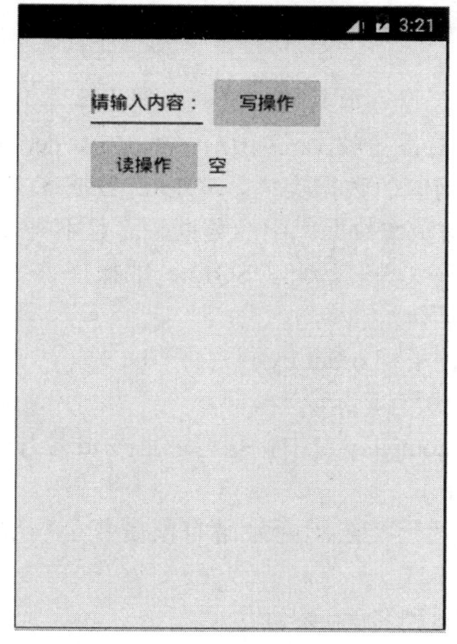


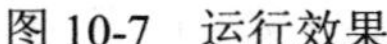
图 10-7 运行效果

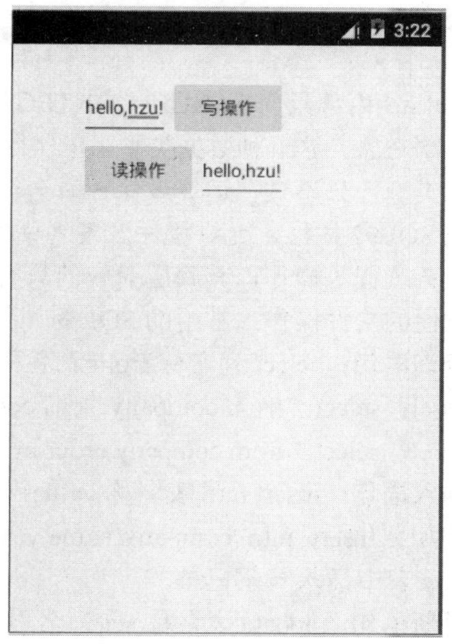


图 10-8 读取数据效果

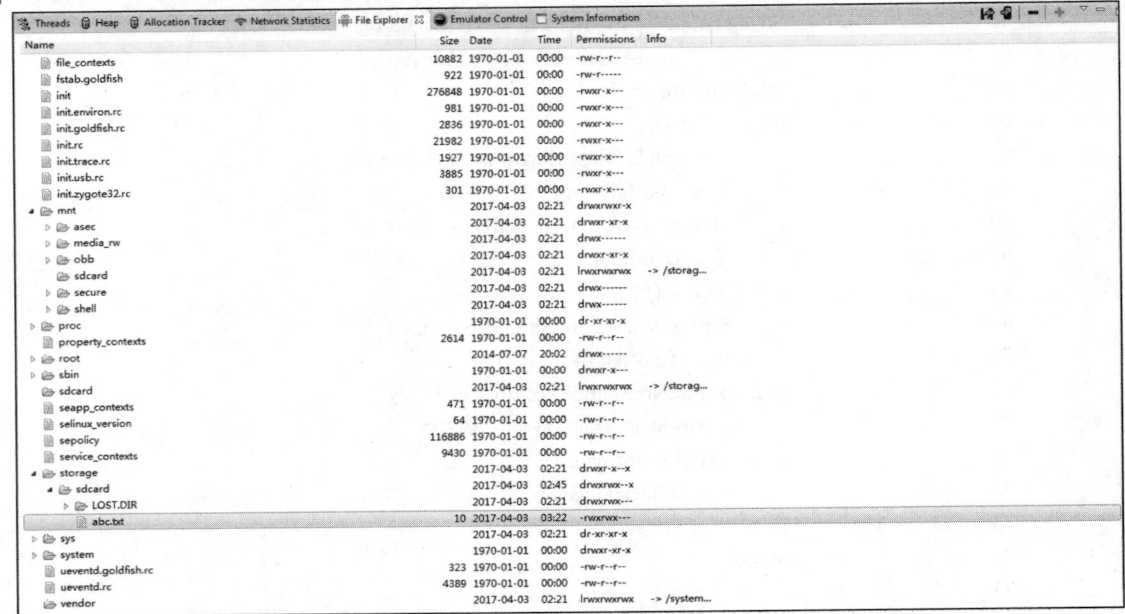


10-9　DDMS下的abc文件

## 10.4　SQLite数据库

SQLite数据库是一个用C语言编写的开源嵌入式数据库引擎，虽然没有包含大型客户/服务器数据库（如Oracle、SQL Server）的所有特性，但包含了操作本地数据的各种功能。它具有系统开销小、检索速度快的特点，适用于手机、平板电脑、机顶盒等设备。

### 10.4.1　SQLite数据库简单介绍

SQLite内部只支持NULL、INTEGER、REAL（浮点数）、TEXT（字符串文本）、BLOB（二进制对象）这五种数据类型，但实际上SQLite可以接收varchar(n)、char(n)、decimal(p,s)等数据类型，只不过在运算或保存时会转换成上面对应的数据类型。SQLite数据库支持绝大部分的SQL92语法，允许程序开发者使用SQL语句操作数据库中的数据，而且不需要安装、启动服务进程等操作，其底层对应的只有一个数据库文件。因此，SQLite的操作方式是一种更为方便的文件操作。常用的SQL标准语句示例如下：

**查询语句：** select 列名称 from 表名称 where 条件子句 order by 排序子句。

例如：select * from company 查询company表中所有记录。

select * from company order by id asc 查询company表中所有记录且按id号升序排列。

**插入语句：** insert into 表名称 values (值1，值2,...)。

例如：insert into company(name,year) values('贺州桂东国际旅游有限责任公司',2002)向company表中插入一条记录。

**更新语句：** update 表名称 set 列名称=新值 where 列名称=某值。

例如：update company set name='贺州华安国际旅行社'where id=20 将id为20的公司名称

改为贺州华安国际旅行社。

**删除语句：** delete from 表名称 where 列名称=值。

例如：delete from company where id=15 删除 company 表中 id 为 15 的记录。

### 10.4.2　SQLite 数据库相关类与接口

1．SQLiteDatabase 类

SQLiteDatabase 类位于 android.database.sqlite 包下，一个 SQLiteDatabase 对象代表一个数据库。该类封装了一些操作数据库的 API，使用该类可以完成对数据的添加（Create）、查询（Retrieve）、更新（Update）、删除（Delete）操作。

2．SQLiteOpenHelper 类

SQLiteOpenHelper 类位于 android.database.sqlite 包下，SQLiteOpenHelper 是 SQLiteDatabse 的一个帮助类，用来管理数据的创建和版本更新。一般用法是定义一个类继承于 SQLite OpenHelper，并通过两个回调方法 OnCreate(SQLiteDatabase db)和 onUpgrade(SQLiteDatabse, int oldVersion, int newVersion)来创建和更新数据库。

3．Cursor 接口

Cursor 位于 android.database 包下，它主要用于存放查询记录的接口，Cursor 是结果集游标，用于对结果集进行随机访问。

4．ContentValues 类

ContentValues 类位于 android.content 包下，它主要用来存放一些键-值对，并提供了数据库的列名、数据映射信息。它存储的键-值对当中的键是一个 String 类型，而值都是基本类型。一个 ContentValues 对象代表数据库中的一行记录。

### 10.4.3　管理 SQLite 数据库相关方法

在 Android 中，通过上述类的相关方法创建和管理 SQLite 数据库。下面分别介绍常用的方法。

SQLiteDatabase 类的常用方法与说明如表 10-3 所示。

表 10-3　SQLiteDatabase 类的常用方法与说明

| 方法 | 参数 | 说明 |
|---|---|---|
| openDatabase(String path, SQLiteDatabase.CursorFactory factory, int flags) | path：指定路径的数据库文件<br>factory：构造查询时返回的 Cursor 对象<br>flags：打开模式，主要模式的种类包括：<br>OPEN_READONLY（只读模式）<br>OPEN_READWRITE（读写模式）<br>CREATE_IF_NECESSARY（当数据库文件不存在时，创建数据库） | 打开或创建数据库 |
| openOrCreateDatabase(File file, SQLiteDatabase.CursorFactory factory) | 同上 | 相当于上述方法中 flags 的参数取值为 CREATE_IF_NECESSARY 的情况 |

续表

| 方法 | 参数 | 说明 |
| --- | --- | --- |
| create(SQLiteDatabase.CursorFactory factory) | 同上 | 创建数据库 |
| insert(String table, String nullColumnHack, ContentValues values) | table：数据表名称<br>nullColumnHack：空列的默认值<br>values：封装了列名与列值的 Map | 插入一条记录 |
| update(String table, ContentValues values, String whereClause, String[] whereArgs) | table：数据表名称<br>Values：更新列 ContentValues 类型键-值对<br>whereClause：更新条件<br>whereArgs：更新条件值数组 | 修改一条记录 |
| delete(String table, String whereClause, String[] whereArgs) | table：数据表名称<br>whereClause：删除条件<br>whereArgs：删除条件值数组 | 删除一条记录 |
| query(String table, String[] columns, String selection, String[] selectionArgs, String groupBy, String having,String orderBy) | table：数据表名称<br>columns：列名数组<br>selection：查询条件，可以使用通配符“？”<br>selectionArgs：参数数组，用于替换查询条件中的“？”<br>groupBy：分组的列名<br>having：分组的条件<br>orderBy：排序的列名 | 查询一条记录 |
| execSQL(String sql) | sql：SQL 语句 | 执行一条 SQL 语句 |
| close() | | 关闭数据库 |

Cursor 的常用方法与说明如表 10-4 所示。

表 10-4 Cursor 的常用方法与说明

| 方法 | 说明 |
| --- | --- |
| getCount() | 获取总记录条数 |
| isFirst() | 判断是否是第一条记录 |
| isLast() | 判断是否是最后一条记录 |
| moveToFirst() | 移动到第一条记录 |
| moveToLast() | 移动到最后一条记录 |
| move(int offset) | 移动到指定的记录 |
| moveToNext() | 移动到下一条记录 |
| moveToPrevious() | 移动到上一条记录 |
| getColumnIndex(String columnName) | 获得指定列索引的 int 类型 |

续表

| 方法 | 说明 |
| --- | --- |
| getInt(int columnIndex) | 根据列索引获得 int 类型值 |
| getFloat(int columnIndex) | 根据列索引获得 float 类型值 |
| getString(int columnIndex) | 根据列索引获得 String 类型值 |

SQLiteOpenHelper 的常用方法与说明如表 10-5 所示。

表 10-5　SQLiteOpenHelper 的常用方法与说明

| 方法 | 说明 |
| --- | --- |
| onCreate(SQLiteDatabase db) | 创建数据库时调用 |
| onUpgrade(SQLiteDatabase db, int oldVersion, int newVersion) | 升级数据库时调用 |
| getReadableDatabase() | 创建或打开只读数据库 |
| getWritableDatabase() | 创建或打开读写数据库 |

**【例 10.5】** 编写一个简易贺州旅游门票管理系统。

**【说明】** 本例使用 SQLiteDatabase、SQLiteOpenHelper、ContentValues、Cursor 类来实现将门票相关信息保存到本地的 SQLite 数据库功能，其中 SQLiteOpenHelper 是一个抽象类，使用子类来完成数据库表结构的定义。

**【开发步骤】**

（1）数据库设计。简易贺州旅游门票管理系统需要保存的信息较多，在此只设计与门票相关的数据表。在此案例中，数据库名为 HzuTour，数据表的名字为 TourTb。数据表 TourTb 的结构如表 10-6 所示。

表 10-6　TourTb 表结构

| 列名 | 类型 | 说明 |
| --- | --- | --- |
| _id | integer | 表 id，主键 |
| spot | varchar | 旅游景点名 |
| price | varchar | 旅游景点门票价 |
| locale | varchar | 旅游景点所在的地区 |

（2）创建一个名为 Sqlite_test 的项目，Activity 组件名为 MainActivity。

（3）在 res/layout 下创建一个 activity_main.xml 文件，内容如下：

```
<LinearLayout xmlns:android="http://schemas.android.com/apk/res/android"
        xmlns:tools="http://schemas.android.com/tools"
        android:layout_width="match_parent"
        android:layout_height="match_parent"
        android:orientation="vertical">

```

```
<TableLayout
        android:layout_width="match_parent"
        android:layout_height="wrap_content">

<TableRow>
<TextView
            android:layout_width="wrap_content"
            android:layout_height="wrap_content"
            android:layout_weight="1"
            android:gravity="center"
            android:text="景点:"
            android:textSize="20sp" />
<EditText
            android:id="@+id/ed_viewspot"
            android:layout_width="wrap_content"
            android:layout_height="wrap_content"
            android:layout_weight="5"
            android:singleLine="true" />
</TableRow>

<TableRow>
<TextView
            android:layout_width="wrap_content"
            android:layout_height="wrap_content"
            android:layout_weight="1"
            android:gravity="center"
            android:text="票价:"
            android:textSize="20sp" />
<EditText
            android:id="@+id/ed_price"
            android:layout_width="wrap_content"
            android:layout_height="wrap_content"
            android:layout_weight="5"
            android:singleLine="true" />
</TableRow>

<TableRow>
<TextView
            android:layout_width="wrap_content"
            android:layout_height="wrap_content"
            android:layout_weight="1"
            android:gravity="center"
            android:text="地点:"
            android:textSize="20sp" />
<EditText
            android:id="@+id/ed_locale"
```

```
                    android:layout_width="wrap_content"
                    android:layout_height="wrap_content"
                    android:layout_weight="5"
                    android:singleLine="true" />
</TableRow>
</TableLayout>

<LinearLayout
            android:layout_width="match_parent"
            android:layout_height="wrap_content"
            android:gravity="center_horizontal"
            android:orientation="horizontal">
<Button
                android:id="@+id/bt_add"
                android:layout_width="wrap_content"
                android:layout_height="wrap_content"
                android:text="添加" />
<Button
                android:id="@+id/bt_query"
                android:layout_width="wrap_content"
                android:layout_height="wrap_content"
                android:text="查询" />
</LinearLayout>

<LinearLayout
            android:id="@+id/l_title"
            android:layout_width="match_parent"
            android:layout_height="wrap_content"
            android:orientation="horizontal">
<TextView
                android:layout_width="0dp"
                android:layout_height="wrap_content"
                android:layout_weight="1"
                android:gravity="center"
                android:text="编号" />
<TextView
                android:layout_width="0dp"
                android:layout_height="wrap_content"
                android:layout_weight="2"
                android:gravity="center"
                android:text="景点" />
<TextView
                android:layout_width="0dp"
                android:layout_height="wrap_content"
                android:layout_weight="2"
                android:gravity="center"
```

```
99                  android:text="票价" />
100 <TextView
101                 android:layout_width="0dp"
102                 android:layout_height="wrap_content"
103                 android:layout_weight="2"
104                 android:gravity="center"
105                 android:text="地点" />
106 </LinearLayout>
107
108 <ListView
109         android:id="@+id/datashow"
110         android:layout_width="match_parent"
111         android:layout_height="wrap_content">
112 </ListView>
113
114 </LinearLayout>
```

1）第 1～114 行声明了一个线性布局，方向为垂直方向，内部再声明一个表格布局、一个子线性布局与一个列表视图。

2）第 11～25 行在 TableRow 中声明了一个文本框与文本输入框，为了界面上表现更为合理，使用了 android:layout_weight 与 android:gravity 等属性。

3）第 108～112 行声明了一个列表视图，其 id 为 datashow。

（4）在 res/layout 下创建一个 datashow.xml 文件，内容如下：

```
<?xml version="1.0" encoding="utf-8"?>
<LinearLayout xmlns:android="http://schemas.android.com/apk/res/android"
        android:layout_width="match_parent"
        android:layout_height="match_parent"
        android:orientation="horizontal">

<TextView
            android:id="@+id/tr_id"
            android:layout_width="0dp"
          android:layout_height="wrap_content"
          android:layout_weight="1"
          android:gravity="center"
          android:text="编号" />

<TextView
          android:id="@+id/tr_spot"
          android:layout_width="0dp"
          android:layout_height="wrap_content"
          android:layout_weight="2"
          android:gravity="center"
          android:text="景点" />

<TextView
```

```
24              android:id="@+id/tr_price"
25              android:layout_width="0dp"
26              android:layout_height="wrap_content"
27              android:layout_weight="2"
28              android:gravity="center"
29              android:text="票价" />
30
31      <TextView
32              android:id="@+id/tr_locale"
33              android:layout_width="0dp"
34              android:layout_height="wrap_content"
35              android:layout_weight="2"
36              android:gravity="center"
37              android:text="地点" />
38
39      </LinearLayout>
```

第 1～39 行声明了一个线性布局，方向为水平方向，内部声明了四个文本框，用于后期绑定数据库中的内容。

（5）在 src 的 com.hzu.sqlite_test 包下创建 MyOpenHelper.java 文件，内容如下：

```
package com.hzu.sqlite_test;

import android.content.Context;

import android.database.sqlite.SQLiteDatabase;
import android.database.sqlite.SQLiteDatabase.CursorFactory;
import android.database.sqlite.SQLiteOpenHelper;


public class MyOpenHelper extends SQLiteOpenHelper {
    public static final String DB_NAME = "HzuTour";
    public static final String TABLE_NAME = "TourTb";
    public static final String ID = "_id";
    public static final String SPOT = "spot";
    public static final String PRICE = "price";
    public static final String lOCALE = "locale";

    public MyOpenHelper(Context context, String name, CursorFactory factory,
        int version) {
        super(context, name, factory, version);
    }

    @Override
    public void onCreate(SQLiteDatabase db) {
        db.execSQL("create table if not exists " + TABLE_NAME + " ( " + ID +
" integer primary key," + SPOT + " varchar," + PRICE + " varchar," +
            lOCALE + " varchar)");
```

```
28          }
29
30          @Override
31          public void onUpgrade(SQLiteDatabase db, int oldVersion, int newVersion) {
32          }
33      }
```

1）第 8～33 行声明了一个 MyOpenHelper 类，该类继承于 SQLiteOpenHelper 类，并重写它的 onCreate()与 onUpgrade()方法，当数据库创建时会调用 onCreate()方法。

2）第 11～16 行定义了一些字符串常量，分别用来表示数据库名、数据表名以及表中的列名。

（6）在 src 的 com.hzu.sqlite_test 包下创建 MainActivity.java 文件，内容如下：

```
package com.hzu.sqlite_test;

import android.app.Activity;
import android.content.ContentValues;
import android.database.Cursor;
import android.database.sqlite.SQLiteDatabase;
import android.os.Bundle;
import android.view.View;
import android.view.View.OnClickListener;
import android.widget.Button;
import android.widget.EditText;
import android.widget.LinearLayout;
import android.widget.ListView;
import android.widget.SimpleCursorAdapter;
import android.widget.Toast;
import static com.hzu.sqlite_test.MyOpenHelper.DB_NAME;
import static com.hzu.sqlite_test.MyOpenHelper.ID;
import static com.hzu.sqlite_test.MyOpenHelper.LOCALE;
import static com.hzu.sqlite_test.MyOpenHelper.PRICE;
import static com.hzu.sqlite_test.MyOpenHelper.SPOT;
import static com.hzu.sqlite_test.MyOpenHelper.TABLE_NAME;

public class MainActivity extends Activity implements OnClickListener {
    MyOpenHelper myOpenHelper;
    EditText et_viewspot;
    EditText et_price;
    EditText et_locale;
    Button add;
    Button query;
    ListView show;
    LinearLayout title;

    @Override
    protected void onCreate(Bundle savedInstanceState) {
```

```
        super.onCreate(savedInstanceState);
        setContentView(R.layout.activity_main);
        et_viewspot = (EditText) findViewById(R.id.ed_viewspot);
        et_price = (EditText) findViewById(R.id.ed_price);
        et_locale = (EditText) findViewById(R.id.ed_locale);
        show = (ListView) findViewById(R.id.datashow);
        add = (Button) findViewById(R.id.bt_add);
        query = (Button) findViewById(R.id.bt_query);
        title = (LinearLayout) findViewById(R.id.l_title);
        title.setVisibility(View.INVISIBLE);
        add.setOnClickListener(this);
        query.setOnClickListener(this);
    }

    @Override
    public void onClick(View v) {
        myOpenHelper = new MyOpenHelper(this, DB_NAME, null, 1);
        SQLiteDatabase db = myOpenHelper.getWritableDatabase();
        String viewspotStr = et_viewspot.getText().toString();
        String priceStr = et_price.getText().toString();
        String localeStr = et_locale.getText().toString();
        switch (v.getId()) {
        case R.id.bt_add:
            addData(db, viewspotStr, priceStr, localeStr);
            Toast.makeText(MainActivity.this, 添加记录成功！", 1000).show();
            et_viewspot.setText("");
            et_price.setText("");
            et_locale.setText("");
            break;
        case R.id.bt_query:
            title.setVisibility(View.VISIBLE);
            Cursor cursor = queryData(db, viewspotStr, priceStr, localeStr);
            if (cursor.getCount() <= 0) {
                Toast.makeText(MainActivity.this, "数据库暂时没有数据！", 1000).show();
                return;
            }

            SimpleCursorAdapter resultAdapter = new SimpleCursorAdapter(
                    MainActivity.this,
                    R.layout.datashow, cursor,
                    new String[] { ID, SPOT, PRICE, LOCALE },
                    new int[] {
                    R.id.tr_id, R.id.tr_spot, R.id.tr_price, R.id.tr_locale
                    });
            show.setAdapter(resultAdapter);
            break;
```

```
81              }
82          }
83
84          public void addData(SQLiteDatabase db, String viewspotStr, String priceStr,
85                  String localeStr) {
86              ContentValues values = new ContentValues();
87              values.put(SPOT, viewspotStr);
88              values.put(PRICE, priceStr);
89              values.put(LOCALE, localeStr);
90              db.insert(TABLE_NAME, ID, values);
91              db.close();
92          }
93
94          public Cursor queryData(SQLiteDatabase db, String viewspotStr,
95                  String priceStr, String localeStr) {
96              Cursor cursor = db.rawQuery("select * from " + TABLE_NAME + " where " +
97                          SPOT + " like ? and " + PRICE + " like ? and " + LOCALE +
98                          " like ?", new String[] {"%" + viewspotStr + "%", "%" + priceStr + "%",
99                          "%" + localeStr + "%"
100                     });
101             return cursor;
102         }
103
104         @Override
105         protected void onDestroy() {
106             if (myOpenHelper != null) {
107                 myOpenHelper.close();
108             }
109         }
110 }
```

1）第 23～110 行声明了一个 MainActivity 类，使其继承 Activity 类并实现 OnClickListener 接口。

2）第 34～47 行重写了 onCreate()方法，其中通过“title.setVisibility(View.INVISIBLE);”语句使其在运行项目时看不到表头，只有通过用户点击“查询”按钮才显示出来。

3）第 57～63 行管理用户点击“添加”按钮的行为，第 58 行调用 addData()方法将用户填写的内容添加到数据库。第 59 行表示添加成功后给用户的友好提示。第 60～62 行清空用户填写的各个字段值。

4）第 64～80 行管理用户点击“查询”按钮的行为，第 65 行显示表头，第 66 行获取用户要查询的结果集，第 67～70 行判断数据库内容是否为空。第 72～79 行使用 SimpleCursorAdapter 适配器将数据库查询的内容绑定到用户界面。

5）第 84～92 行声明了 addData()方法，将用户填写的内容先存放到 ContentValues 对象中，然后通过 SQLiteDatabase 类的 insert()方法存放到数据库。

6）第 96～100 行通过 SQLiteDatabase 类的 rawQuery()方法查询用户需要的内容。

7）第 105～109 行重写了 onDestroy()方法，通过 close()方法关闭数据库连接。

【运行结果】在 Eclipse 中启动 Android 模拟器，接着运行 Sqlite_test 项目，在图 10-10 所示界面添加相关记录，点击“查询”按钮后，显示效果如图 10-11 所示。

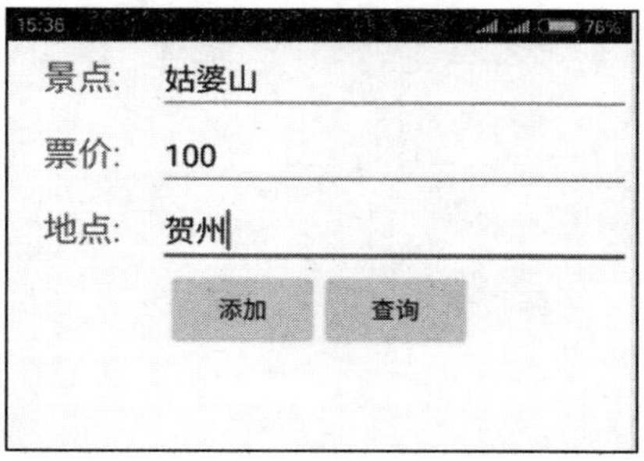

图 10-10　添加记录

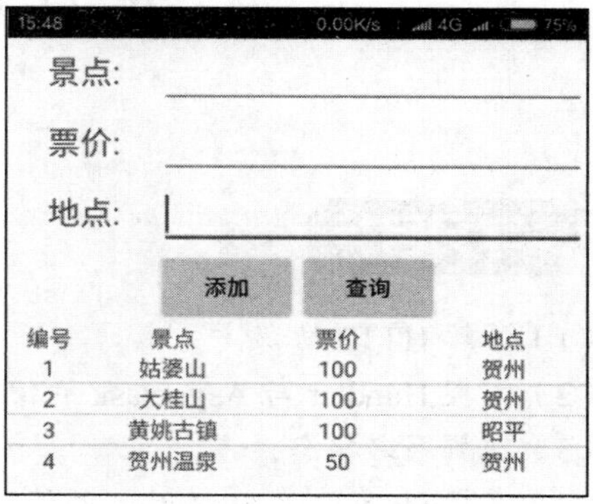

图 10-11　查询记录

## 10.5　本章小结

本章主要介绍了 SharedPreferences 存取、SQLite 数据库操作的方法、应用程序间的数据共享ContentProvider组件和使用文件存储的基本步骤，为使用互联网数据库的操作提供了前期基础。

# 第 11 章　网络编程

（1）了解 HTTP 协议。

（2）掌握 Handler 与 Asynctask 的使用场景。

（3）理解网络状态。

（4）掌握 HttpURLConnection 的使用。

（5）掌握 JSON 的基本语法格式。

随着移动互联网时代的到来，智能手机普及率的提高，人们可以不受时间、空间的限制进行自由的通信，经常使用的网络产品有手机微信、手机银行、手机地图、手机音乐、手机收发邮件、手机打车服务等，这些网络产品都离不开网络编程。

## 11.1　HTTP 协议

WWW 是以 Internet 作为传输媒介的一个应用系统，WWW 网上基本的传输单位是 Web 网页。WWW 的工作是基于客户机/服务器计算模型，由 Web 浏览器和 Web 服务器构成，两者之间采用超文本传输协议 HTTP 进行通信。HTTP 协议是基于 TCP/IP 协议之上的协议，是 Web 浏览器和 Web 服务器之间的应用层协议，是通用的、无状态的面向对象协议。例如在日常生活中，当人们想看新闻时在手机浏览器输入www.163.com，即可进入网易新闻，此访问过程是通过 HTTP 协议完成的，手机端访问服务器端的图解过程如图 11-1 所示。

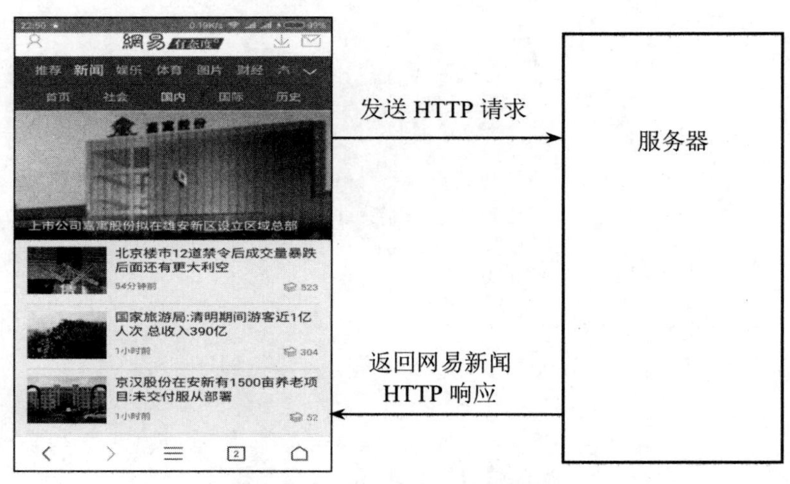


图 11-1　HTTP 请求与响应

从图 11-1 可以看到，使用手机访问网易时，手机先发出一个 HTTP 请求，服务器接收到请求后做出响应，返回需要的数据，此请求与响应的过程就是 HTTP 的通信过程。

## 11.2　Handler 消息机制原理

基于 Android 系统整体优化性能的考虑，Android 的 UI 操作是线程不安全的，这样导致有多个线程并发操作 UI 组件时，可能出现线程安全问题。当应用程序启动时，会开启一个主线程（也就是 UI 线程），由它来管理 UI，监听用户点击行为、响应用户并分发事件等。一般在主线程中不要执行比较耗时的操作，如联网下载数据等，否则会出现 ANR 错误。因此应该将耗时操作放在子线程中，由于 Android UI 线程是不安全的，所以只能在主线程中更新 UI。为了很好地解决这个问题，在 Android 4.0 后规定子线程不能更新 UI 界面，UI 线程中不访问网络，通过使用 Handler 通信机制来实现线程间通信。

**注意：**

- 线程安全是指多线程访问时，采用加锁机制，当一个线程访问该类的某个数据时，对数据进行保护，其他线程不能进行访问，直到该线程读取完，这样做的目的是不会出现数据不一致或者数据污染。
- 线程不安全是指多线程访问时，不提供数据访问保护，有可能出现多个线程先后更改数据造成所得到的数据可能是脏数据。

Handler 类主要做两件事：一是在新启动的子线程中发送消息；二是在主线程中获取、处理消息。

主线程处理新的子线程发送过来的消息，主要是通过回调方法来实现——程序开发者重写 Handler 类处理消息方法，当新的子线程发送消息过来后，消息会与关联的 MessageQueue 绑定，而 Handler 将从 MessageQueue 中获取消息。

Handler 机制主要包括 4 个对象：Message、MessageQueue、Handler、Looper，下面分别对这 4 个对象进行介绍。

Message：包含描述任意数据对象的消息，用于发送给 Handler，它主要是在不同线程间进行数据交换。Message 的 obj 字段可以携带一个 Object 对象，what 字段是用户自定义的消息代码，这样接收者可以了解这个消息的信息。每个 Handler 各自包含自己的消息代码，所以不用担心自定义的消息跟其他 Handler 有冲突。

MessageQueue：是指消息队列，它主要用来接收 Handler 发送过来的消息，这些消息存放在 MessageQueue 中等待处理。每个线程只有一个 MessageQueue 对象。

Handler：是指发送消息与处理消息的对象。Handler 处理消息的步骤如下：

（1）在 Activity 中声明 Handler 对象，然后重写 handleMessage()方法。

（2）在新启动的线程中调用 sendMessage()或 sendEmptyMessage()方法向 Handler 发送消息。

（3）在 Handler 对象中使用 handleMessage()方法接收消息，然后根据消息做后续操作处理。

Handler 的常用方法及说明如表 11-1 所示。

表 11-1 Handler 常用方法及说明

| 方法 | 说明 |
|---|---|
| void handleMessage(Message msg) | 消息发送后在这个方法中接收处理 |
| boolean sendMessage(Message msg) | 发送消息到 Handler |
| boolean sendEmptyMessage(int what) | 发送只有一个 what 值的消息 |
| boolean sendMessageDelayed(Message msg, long delayMillis) | 延时发送消息 |
| void removeMessages(int what) | 删除消息/取消定时消息 |

Looper：是每个线程中的 MessageQueue 管理者。程序在初始化 Looper 时会创建一个与之关联的 MessageQueue，然后进入一个无限循环中。当 MessageQueue 有消息进来时，将它取出并传递给 Handler 的 handleMessage()方法。在主线程创建 Handler 对象时，系统已经创建了 Looper 对象，不需要程序开发者手动创建。Handler 消息处理流程如图 11-2 所示。

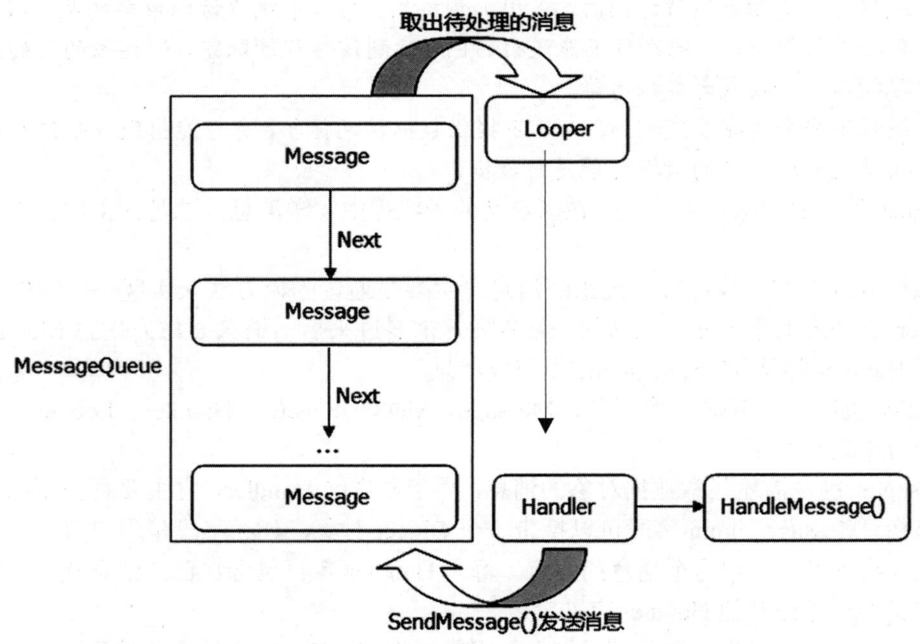


图 11-2 Handler 消息处理流程

从图 11-2 可以看到整个 Handler 消息处理流程。先在 UI 线程中创建一个 Handler 对象，然后在子线程中使用 Handler 的 sendMessage()方法，接着将此消息存放到 MessageQueue 中，之后使用 Looper 对象取出 MessageQueue 中的消息，最后发送给 Handler 对象的 handleMessage()方法做后续操作处理。

**【例 11.1】**通过 Handler 实现每隔 1s 更换一次 TextView 背景色，并在 TextView 中设置文字显示的内容为当前颜色。

**【说明】**本例主要使用两个类，一个是继承 Activity 的类，在其中声明 Handler 对象，并重写 handleMessage()方法，在此方法中根据消息的值来更改 TextView 背景色以及文字的内容；

另一个是继承 Thread 的类，并重写 run()方法，在此类下使用 sendEmptyMessage()方法向 Handler 发送消息。

**【开发步骤】**

（1）创建一个名为 Handler_test 的项目，Activity 组件名为 MainActivity。

（2）在 res/layout 的 activity_main.xml 文件的内容如下：

```
1  <RelativeLayout xmlns:android="http://schemas.android.com/apk/res/android"
2          xmlns:tools="http://schemas.android.com/tools"
3          android:layout_width="match_parent"
4          android:layout_height="match_parent"
5          android:paddingBottom="@dimen/activity_vertical_margin"
6          android:paddingLeft="@dimen/activity_horizontal_margin"
7          android:paddingRight="@dimen/activity_horizontal_margin"
8          android:paddingTop="@dimen/activity_vertical_margin"
9          tools:context="com.hzu.handler_test.MainActivity">
10
11 <TextView
12             android:id="@+id/tv"
13             android:layout_width="match_parent"
14             android:layout_height="300dp"
15             android:gravity="center"
16             android:text="动态改动背景色" />
17
18 <Button
19             android:id="@+id/bt"
20             android:layout_width="wrap_content"
21             android:layout_height="wrap_content"
22             android:layout_below="@+id/tv"
23             android:layout_centerHorizontal="true"
24             android:layout_marginTop="21dp"
25             android:text="启动" />
26
27 </RelativeLayout>
```

第 1～27 行声明了一个相对布局，其内部包含有一个 TextView 和一个 Button。

（3）在 src 的 com.hzu.handler_test 包下的 MainActivity.java 文件内容如下：

```
package com.hzu.handler_test;

import android.app.Activity;
import android.os.Bundle;
import android.os.Handler;
import android.os.Message;
import android.view.View;
import android.widget.Button;
import android.widget.TextView;

public class MainActivity extends Activity {
    TextView tv_bg;
```

```
        Button bt_start;
        int currentId = 0;
        Handler myHandler;

        @Override
        protected void onCreate(Bundle savedInstanceState) {
            super.onCreate(savedInstanceState);
            setContentView(R.layout.activity_main);
            tv_bg = (TextView) findViewById(R.id.tv);
            bt_start = (Button) findViewById(R.id.bt);

            final int[] bgid = new int[] {
                    0xFFFF1493, 0xFF800080, 0xFF0000FF, 0xFF008000
                };
            final String[] txt = new String[] {"粉红", "紫色", "蓝色", "绿色" };
            myHandler = new Handler() {
                        @Override
                        public void handleMessage(Message msg) {
                            if (msg.what == 0x123) {
                            // 动态修改图片的信息
                                tv_bg.setBackgroundColor(bgid[currentId % bgid.length]);
                            }
                            tv_bg.setText(txt[currentId % txt.length]);
                            currentId++;
                        }
                    };
            bt_start.setOnClickListener(new View.OnClickListener() {
                    @Override
                    public void onClick(View v) {
                        new myThread().start();
                    }
                });
        }

        private class myThread extends Thread {
            @Override
            public void run() {
                while (true) {
                    myHandler.sendEmptyMessage(0x123);
                    try {
                        Thread.sleep(1000);
                    } catch (Exception e) {
                    }
                }
            }
        }
    }
```

1）第 28～38 行创建了一个 Handler 对象，并重写了 handleMessage()方法，在其内部根据传过来的消息，更改 TextView 背景色以及其文字的内容。

2）第 39～44 行通过一个 Button 启动线程。

3）第 49～57 行在 run()方法中执行一个 while 循环，使用 sendEmptyMessage()方法每隔 1s 向 MainActivity 中的 myHandler 对象发送一次消息。

**【运行结果】**在 Eclipse 中启动 Android 模拟器，接着运行 Handler_test 项目，显示效果如图 11-3 所示，点击“启动”按钮后的效果如图 11-4 所示。

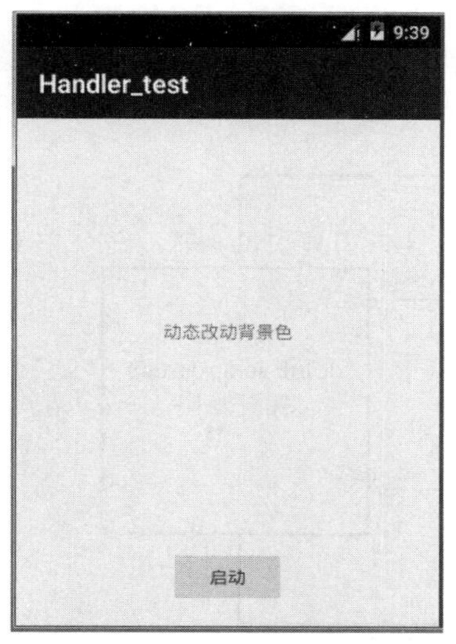


图 11-3　运行 Handler_test 后的效果

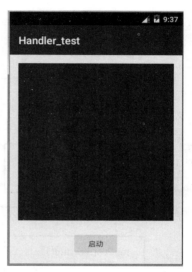


图 11-4　点击“启动”按钮后的效果

## 11.3　Asynctask

上一小节的 Handler 与 Thread 的组合很好地解决了因耗时操作刷新界面困难的问题，但两者组合实现过程相对繁琐。Asynctask 抽象出这个过程成为另一种解决方案，它基于异步通信机制，轻松实现子线程切换到主线程。

Asynctask 是抽象类，创建一个类继承此类时可以指定三个泛型参数，这三个参数的作用分别如下：

- Params：在执行 Asynctask 时需要传入的参数，用于在后台任务中使用，比如 HTTP 请求的 URL。
- Progress：在后台任务执行时，如果需要在界面上显示进度，此参数将作为进度的百分比。
- Result：后台执行任务最终返回的结果，为 String 类型。

Asynctask 的执行过程分以下四个步骤，每一步对应一个回调方法。

（1）onPreExecute()，此方法在执行后台操作前被主线程调用。可以在此方法中做一些前期准备工作，如显示一个进度条。

（2）doInBackground(Params...)，此方法在 onPreExecute()方法执行后马上执行，在后台

线程中运行，这里主要用来处理一些耗时操作。注意，此方法是抽象方法，子类必须实现它。在这个方法中可以使用 publicProgress(Progress…)来更新任务的进度。

（3）onProgressUpdate(Progress…)，在 publishProgress(Progress…)方法被调用后，UI 线程将调用此方法来更新界面展示任务的进度，方法中携带的参数是后台任务中传递过来的数据。

（4）onPostExecute(Result)，在 doInBackground(Params…)方法执行完成后返回的参数传递到此方法中，进行后续处理，如提醒当前任务已经完成。

Asynctask 的执行过程如图 11-5 所示。

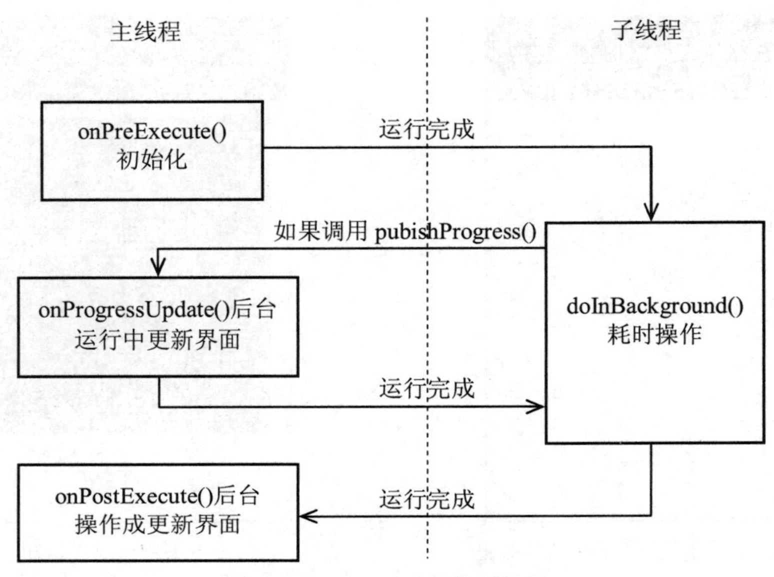


图 11-5 Asynctask 的执行过程

Asynctask 的使用还要用到一些其他方法，如表 11-2 所示。

**表 11-2 Asynctask 的常用方法**

| 方法 | 说明 |
|---|---|
| AsyncTask execute(Params... params) | 开启任务 |
| boolean cancel(boolean mayInterruptIfRunning) | 取消任务的执行 |
| AsyncTask.Status getStatus() | 获取任务的当前状态 |
| boolean isCancelled() | 如果任务正常后取消任务返回 true，否则为 false |

**【例 11.2】**通过 Asynctask 求用户输入的一个数与 1 之间所包含的所有合数，并把这些合数显示在用户界面上。

**【说明】**本例主要使用 Asynctask 类的四个方法，onPreExecute()方法控制按钮的可用状态，onPostExecute(Result) 方法输出最终结果，doInBackground(Params…) 与 onProgressUpdate (Progress…) 实时处理得到的合数并显示在用户界面上。

**【开发步骤】**

（1）创建一个名为 Asynctask_test 的项目，Activity 组件名为 MainActivity。

（2）在 res/layout 下的 activity_main.xml 文件的内容如下：

```
1	<LinearLayout xmlns:android="http://schemas.android.com/apk/res/android"
2	      xmlns:tools="http://schemas.android.com/tools"
3	      android:layout_width="match_parent"
4	      android:layout_height="match_parent"
5	      android:layout_marginTop="20dp"
6	      android:orientation="vertical">
7
8	<EditText
9	          android:id="@+id/et"
10	          android:layout_width="wrap_content"
11	          android:layout_height="wrap_content"
12	          android:layout_gravity="center"
13	          android:hint="请输入一个整数" />
14
15	<Button
16	          android:id="@+id/bt"
17	          android:layout_width="match_parent"
18	          android:layout_height="wrap_content"
19	          android:text="开始计算" />
20
21	<TextView
22	          android:id="@+id/tv1"
23	          android:layout_width="wrap_content"
24	          android:layout_height="wrap_content" />
25
26	<TextView
27	          android:id="@+id/tv2"
28	          android:layout_width="wrap_content"
29	          android:layout_height="wrap_content" />
30
31	</LinearLayout>
```

第 1～31 行声明了一个线性布局，其内部包含一个 EditText、一个 Button 和两个 TextView。

（3）在 src 的 com.hzu.asynctask_test 包下的 MainActivity.java 文件内容如下：

```java
package com.hzu.asynctask_test;

import android.app.Activity;
import android.os.AsyncTask;
import android.os.Bundle;
import android.view.View;
import android.widget.Button;
import android.widget.EditText;
import android.widget.TextView;
import android.widget.Toast;

public class MainActivity extends Activity {
```

```
    EditText etext;
    TextView tView1;
    TextView tView2;
    Button button;
    String s;

    @Override
    protected void onCreate(Bundle savedInstanceState) {
        super.onCreate(savedInstanceState);
        setContentView(R.layout.activity_main);
        etext = (EditText) findViewById(R.id.et);
        tView1 = (TextView) findViewById(R.id.tv1);
        tView2 = (TextView) findViewById(R.id.tv2);
        button = (Button) findViewById(R.id.bt);
        button.setOnClickListener(new View.OnClickListener() {
                @Override
                public void onClick(View v) {
                    s = etext.getText().toString().trim();
                    if ((s == null) || s.equals("")) {
                        Toast.makeText(MainActivity.this,"请输入一个整数",
                            Toast.LENGTH_LONG).show();
                        return;
                    }
                    new MyAsynctask().execute(Integer.parseInt(s));
                }
            });
    }

    class MyAsynctask extends AsyncTask<Integer, Integer, Integer> {
        @Override
        protected void onPreExecute() {
            button.setEnabled(false);
            super.onPreExecute();
        }

        @Override
        protected Integer doInBackground(Integer... params) {
            int num = params[0];
            int count = 0;
            for (int i = 1; i < num; i++) {
                for (int j = 2; j < i; j++) {
                    if ((i % j) == 0) {
                        publishProgress(i);
                        count++;
                        break;
                    }
                }
```

```
60                  }
61                  return count;
62              }
63
64              @Override
65              protected void onProgressUpdate(Integer... values) {
66                  tView1.setText(tView1.getText() + "-" + values[0]);
67                  super.onProgressUpdate(values);
68              }
69
70              @Override
71              protected void onPostExecute(Integer result) {
72                  tView2.setText("一共有" + result + "合数");
73                  button.setEnabled(true);
74                  super.onPostExecute(result);
75              }
76          }
77      }
```

1）第 27～38 行为 button 设置监听器，在其内部获取用户输入的数据，然后启动 MyAsynctask，并将用户数据传入到 MyAsynctask 内部。

2）第 41～76 行声明了 MyAsynctask 内部类，让其继承 Asynctask 类，并重写了 onPreExecute() 方法、doInBackground(Integer...params)方法、onProgressUpdate(Integer...values)方法和 onPostExecute (Integer result)方法。其中第 55 行的 publishProgress(i)方法将参数传给 onProgressUpdate (Integer... values)。

**【运行结果】**在 Eclipse 中启动 Android 模拟器，接着运行 Asynctask_test 项目，显示效果如图 11-6 所示，输入 100，点击“开始计算”按钮后的效果如图 11-7 所示。

图 11-6 运行后的显示效果

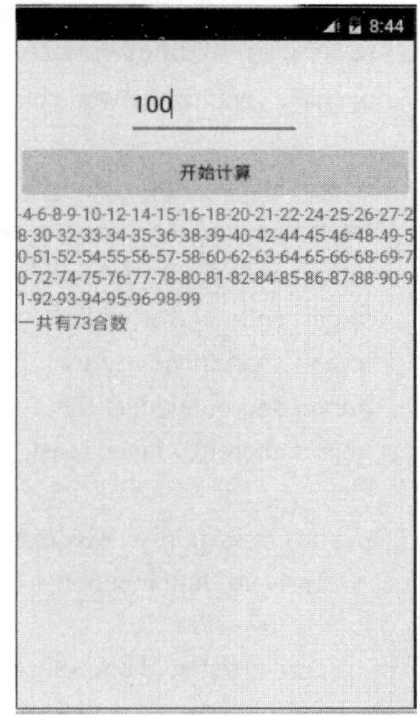


图 11-7 点击“开始计算”按钮后的效果

## 11.4 网络状态

现在大量的 Android 应用程序需要在网络上获取数据，但是在获取网络数据之前，先要对数据状态进行判断，这样可以提升用户的体验。用户在访问一些网络音乐与视频等大文件数据时，使用手机 WiFi 或手机移动数据进行访问时，后者所产生的费用是比较高的。所以在进行网络访问之前，先判断网络连接是否可用，再次是判断当前是使用 WiFi 还是手机移动数据进行访问。

网络使用的场景：

- 用户界面上需要表述当前网络的状态。
- Android 应用程序后台要根据不同的网络状态来处理数据。
- Android 应用程序后台要根据网络情况来开启相关服务。

获取网络服务的步骤：

（1）获得系统服务。ConnectivityManager 负责管理所有的连接服务。使用 getSystemService() 获得系统服务，系统服务包括 3G/4G、WiFi、蓝牙服务。

（2）获得网络服务。通过 ConnectivityManager 的 getActiveNetworkInfo()方法获得 NetworkInfo 类，此类即可用来检测网络的状态与类型。

**【例 11.3】**判断当前手机的网络连接状态与类型。

**【说明】**本例通过使用 ConnectivityManager.getActiveNetworkInfo()方法获得 NetworkInfo 类，然后检测网络连接状态与类型。

**【开发步骤】**

（1）创建一个 NetworkConnectivity_test 的项目，Activity 组件名为 MainActivity。

（2）在 res/layout 的 activity_main.xml 文件中的线性布局下声明一个 Button，此处比较简单，不再介绍。

（3）在 src 的 com.hzu.networkconnectivity_test 包下的 MainActivity.java 文件的内容如下：

```
package com.hzu.networkconnectivity_test;

import android.app.Activity;
import android.net.ConnectivityManager;
import android.net.NetworkInfo;
import android.os.Bundle;
import android.view.View;
import android.widget.Button;
import android.widget.Toast;

public class MainActivity extends Activity {
    private Button button;
    @Override
    protected void onCreate(Bundle savedInstanceState) {
        super.onCreate(savedInstanceState);
        setContentView(R.layout.activity_main);
```

```
            button = (Button) findViewById(R.id.bt_checknetwork);
            button.setOnClickListener(new View.OnClickListener() {
                @Override
                public void onClick(View v) {
ConnectivityManager cm = (ConnectivityManager) getSystemService(CONNECTIVITY_SERVICE);
                    NetworkInfo ni = cm.getActiveNetworkInfo();
                    if (ni == null) {
                        Toast.makeText(MainActivity.this, "无网络连接",
                            Toast.LENGTH_LONG).show();
                        return;
                    }
                    if (ni.getState() != NetworkInfo.State.CONNECTED) {
                        Toast.makeText(MainActivity.this, "网络不可用",
                            Toast.LENGTH_LONG).show();
                        return;
                    }
                    switch (ni.getType()) {
                    case ConnectivityManager.TYPE_WIFI:
                        Toast.makeText(MainActivity.this, "WiFi",
                            Toast.LENGTH_LONG).show();
                        break;
                    case ConnectivityManager.TYPE_MOBILE:
                        Toast.makeText(MainActivity.this, "移动数据",
                            Toast.LENGTH_LONG).show();
                        break;
                    default:
                        Toast.makeText(MainActivity.this, "其他类型",
                            Toast.LENGTH_LONG).show();
                        break;
                    }
                }
            });
        }
}
```

（4）在项目下的 AndroidManifest.xml 文件中添加如下内容：

```
<uses-permission android:name="android.permission.ACCESS_NETWORK_STATE"/>
```

**【运行结果】**在 Eclipse 中启动 Android 模拟器，接着运行 NetworkConnectivity_test 项目，移动网络打开时的显示效果如图 11-8 所示，WiFi 打开时的显示效果如图 11-9 所示。

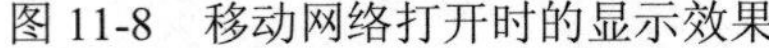
图 11-8 移动网络打开时的显示效果

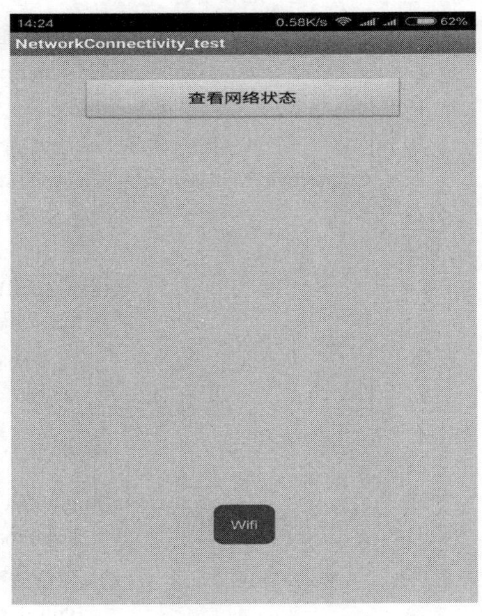


图 11-9 WiFi 打开时的显示效果

## 11.5 HttpURLConnection 访问网络

在 Android 开发应用中，大量应用程序需要与服务器进行数据交互，此时就可以使用 HttpURLConnection 对象。HttpURLConnection 是一个标准的 Java 类。访问网络资源的操作与文件读写操作一样，通过 InputStream 对象来接收网络数据，通过 OutputStream 对象发送网络数据。

HttpURLConnection 的通信流程如下：

（1）创建 URL。

（2）创建 HttpURLConnection。

（3）设置 HttpURLConnection 请求参数。

（4）HttpURLConnection 建立连接。

（5）HttpURLConnection 发送请求。

（6）HttpURLConneciton 获取响应。

**【例 11.4】**从网络中获取一个图片并显示在 Android 应用程序界面上。

**【说明】**访问网络资源使用 HttpURLConnection 对象来进行通信。

**【开发步骤】**

（1）创建一个名为 HttpURLConnection_test 的项目，Activity 组件名为 MainActivity。

（2）在 res/layout 的 activity_main.xml 文件中先声明一个线性布局，然后在线性布局中声明一个 Button 和一个 ImageView，此处比较简单，不再介绍。

（3）在 src 的 com.hzu.httpurlconnection_test 包下的 MainActivity.java 文件的内容如下：

```
package com.hzu.httpurlconnection_test;

```

```
import android.app.Activity;
import android.graphics.Bitmap;
import android.graphics.BitmapFactory;
import android.os.Bundle;
import android.os.Handler;
import android.os.Message;
import android.view.View;
import android.widget.Button;
import android.widget.ImageView;
import android.widget.Toast;

import java.io.InputStream;
import java.net.HttpURLConnection;
import java.net.URL;

public class MainActivity extends Activity {
    private Button button;
    private ImageView image;
    private String path = "https://timgsa.baidu.com/timg?image&quality" +
"=80&size=b9999_10000&sec=1491929723880&di=1fd69bb00f60634" +
"7b64f323e9c43b24d&imgtype=0&src=http%3A%2F%2Fimg.51ztzj.c" +
"om%2Fupload%2Fimage%2F20150120%2Fsj201501191024_279x419.jpg";
    private HttpURLConnection conn;
    private Handler handler;

    @Override
    protected void onCreate(Bundle savedInstanceState) {
      super.onCreate(savedInstanceState);
      setContentView(R.layout.activity_main);
      button = (Button) findViewById(R.id.bt);
      image = (ImageView) findViewById(R.id.iv);
      handler = new Handler() {
                  public void handleMessage(Message msg) {
                      if (msg.what == 0x123) {
                          Bitmap b = (Bitmap) msg.obj;
                          image.setImageBitmap(b);
                      } else if (msg.what == 0x124) {
                          Toast.makeText(MainActivity.this, "显示图片失败",
                              Toast.LENGTH_LONG).show();
                      }
                  };
              };

      button.setOnClickListener(new View.OnClickListener() {
              @Override
              public void onClick(View v) {
```

```
49                  new Thread() {
50                      public void run() {
51                          try {
52                              URL url = new URL(path);
53                              conn = (HttpURLConnection) url.openConnection();
54                              conn.setRequestMethod("GET");
55                              conn.setConnectTimeout(8000);
56                              conn.setReadTimeout(8000);
57                              int code = conn.getResponseCode();
58                              if (code == 200) {
59                                  InputStream in = conn.getInputStream();
60                                  Bitmap bitmap1 = BitmapFactory.decodeStream(in);
61                                  Message m = new Message();
62                                  m.what = 0x123;
63                                  m.obj = bitmap1;
64                                  handler.sendMessage(m);
65                              } else {
66                                  Message m = new Message();
67                                  m.what = 0x124;
68                                  handler.sendMessage(m);
69                              }
70                          } catch (Exception e) {
71                              Message m = new Message();
72                              m.what = 0x124;
73                              handler.sendMessage(m);
74                          }
75                      };
76                  }.start();
77              }
78          });
79      }
80  }
```

1）第 21～24 行设置网络图片地址。

2）第 34～43 行创建 Handler 对象，用来接收 handler.sendMessage()方法发送的消息。

3）第 46～78 行为 button 设置监听器，第 52 行为 URL 传入要访问的资源路径。第 53 行创建 HttpURLConnection 对象。第 54 行设置请求方式。第 55 行设置连接服务器超时时间。第 56 行设置从服务器读取数据的超时时间。第 57 行获取网络状态码。第 59 行获取服务器返回的输入流。第 60 行通过 BitmapFactory 的 decodeStream()方法解析网络图片资源。第 61～63 行创建 Message 对象，并封装相应的信息。第 64 行发送消息。第 66～68 行与第 71～73 行表示获取网络图片失败时发送的消息。

（4）在项目下的 AndroidManifest.xml 文件中添加如下内容：

```
<uses-permission  android:name="android.permission.INTERNET"/>
```

**【运行结果】**在 Eclipse 中启动 Android 模拟器，接着运行 HttpURLConnection_test 项目，运行时显示效果如图 11-10 所示，点击“提交”按钮后的显示效果如图 11-11 所示。

图 11-10　运行项目时的显示效果

图 11-11　点击“提交”按钮后的显示效果

## 11.6　数据提交方式

HTTP/1.1 协议规定的 HTTP 请求方式有 OPTIONS、GET、HEAD、POST、PUT、DELETE、TRACE、CONNECT 这八种，其中 GET 和 POST 两种请求方式是最常用的，下面对这两种请求方式进行介绍。

1. GET 与 POST 两种请求方式的区别

- GET 方式是以实体方式得到由请求 URL 所指向的资源信息，它向服务器提交的参数紧跟在 URL 后面。使用 GET 方式访问网络时 URL 的长度是有限制的，一般情况下请求 URL 的长度不超过 1KB。
- POST 方式是向服务器发送请求，然后接收附在请求后的数据。它向服务器提交数据是以流的方式直接写给服务器的，这种方式对 URL 的长度没有限制。

2. POST 方式提交

案例 11.4 已经演示使用 GET 方式提交数据的案例。下面使用 HttpURLConnection 的 POST 方式提交数据，具体关键代码如下：

```
try {
        String PATH = "http://10.0.2.2/login.php";
        String info="usrname="+URLEncoder.encode("admin")
                +"$password="+URLEncoder.encode("123456");
        URL url=new URL(PATH);
        HttpURLConnection urlConnection = (HttpURLConnection)
                url.openConnection();
        urlConnection.setConnectTimeout(3000);
        urlConnection.setRequestMethod("POST");
        urlConnection.setDoOutput(true);
```

```
11              urlConnection.setRequestProperty("Content-Type",
12                          "application/x-www-form-urlencoded");
13              urlConnection.setRequestProperty("Content-Length",
14                          String.valueOf(info.length()));
15              OutputStream os=urlConnection.getOutputStream();
16              os.write(info.getBytes());
17              int code = urlConnection.getResponseCode();
18              if (code == 200) {
19              //取得服务器信息
20              }
21      } catch (Exception e) {
22              // TODO: handle exception
23      }
```

1）第 2 行设置网络访问地址。

2）第 3 行准备数据并给参数进行编码。

3）第 9 行设置请求方式。

4）第 10 行设置允许向外写数据。

5）第 11～12 行设置请求头——数据以 form 表单形式提交。

6）第 13～14 行设置提交数据的长度。

7）第 15 行向服务器写数据的输出流。

8）第 16 行将数据写向服务器。

## 11.7 JSON

1．JSON 定义

JSON(JavaScript Object Notation)是 JavaScript 对象标记，是一种基于文本的、独立于语言的轻量级数据交换格式，易于阅读和编写，易于机器解析和生成。

JSON 数据的书写格式是：名称/值对。名称/值对包括字段名称（在双引号中），后面写一个冒号，然后是值，如"name":"zhangsan"。

JSON 对象在花括号中书写，对象可以包含多个名称/值对。

如：{ "name":" zhangsan", "age":"21" }。

JSON 数组在方括号中书写，数组可包含多个对象。

如：

```
{ "employees":
[
{ "name":"zhangsan" , " age":"21" },
{ "name":"lisi" , "age" :"30" },
{"name":"wangwu" , "age" :"25" }
]
}
```

2．JSON 解析

JSONObject 代表了一个待解析的名称/值对集合，使用“JSONObject jsonObject = new

JSONObject(data)”语句获得 JSONObject 对象。如果值为基本数据类型，使用相应的 getX()方法取值；如果值为对象，使用 getJSONObject()方法取值；如果值为数组，使用 getJSONArray()方法取值。

**【例 11.5】** 编程解析下面 JSON 数据：

```
{
"name": "zhangsan",
"age": 25,
"address": {
"country": "china",
"province": "guangxi"
    },
"married": false,
"friend": [
        {
"name": "liming",
"age": "24"
        },
        {
"name": "hanmeimei",
"age": "26"
        }
    ]
}
```

**【说明】** 使用 JSONObject 封装上述 JSON 数据，然后使用相应的 getX()方法取得各个具体的值。

**【开发步骤】**

（1）创建一个名为 JSONObject_test 的项目，Activity 组件名为 MainActivity。

（2）在 res/layout 的 activity_main.xml 文件中，先添加一个线性布局，之后在该线性布局中声明一个 Button 和一个 ImageView，此处比较简单，不再介绍。

（3）在 src 的 com.hzu.jsonobject_test 包下的 MainActivity.java 文件中的内容如下：

```
package com.hzu.jsonobject_test;

import android.app.Activity;
import android.os.Bundle;
import android.view.View;
import android.widget.Button;
import android.widget.TextView;
import org.json.JSONArray;
import org.json.JSONException;
import org.json.JSONObject;

public class MainActivity extends Activity {
    String s = "{\"name\": \"zhangsan\",\"age" +
"\": \"25\",\"address\": {\"country\": \"china" +
```

```
"\",\"province\": \"guangxi\"},\"married" +
"\": false,\"friend\": [{ \"name\": \"liming" +
"\", \"age\": \"24\"},{\"name\": \"hanmeimei" +
"\",\"age\": \"26\"}]}";
    Button button;
    TextView textView;
    StringBuilder sb = new StringBuilder();

    @Override
    protected void onCreate(Bundle savedInstanceState) {
        super.onCreate(savedInstanceState);
        setContentView(R.layout.activity_main);
        button = (Button) findViewById(R.id.bt);
        textView = (TextView) findViewById(R.id.tv);
        button.setOnClickListener(new View.OnClickListener() {
                @Override
                public void onClick(View v) {
                    try {
                        JSONObject object = new JSONObject(s);
                        String name = object.getString("name");
                        sb.append("name=" + name + "\n");

                        int age = object.getInt("age");
                        sb.append("age=" + age + "\n");
                        JSONObject object1 = object.getJSONObject("address");
                        String country = object1.getString("country");
                        sb.append("country=" + country + "");
                        String province = object1.getString("province");
                        sb.append("province=" + province + "\n");
                        boolean married = object.getBoolean("married");
                        sb.append("married=" + married + "\n");
                        JSONArray array2 = object.getJSONArray("friend");
                        for (int i = 0; i < array2.length(); i++) {
                            JSONObject friend = array2.getJSONObject(i);
                            String name1 = friend.getString("name");
                            int age1 = friend.getInt("age");
                            sb.append(" friend name=" + name1 + " age=" + age1 +
                                "\n");
                        }

                        textView.setText(textView.getText().toString() + "\n" +
                            sb.toString());
                    } catch (JSONException e) {
                        e.printStackTrace();
                    }
                }
```

```
61                    });
62            }
63    }
```

1）第 13～18 行所有的“\”表示转意字符。

2）第 33 行将字符串 s 封装成 JSONObject 对象。

3）第 47～53 行使用 JSONObject 来解析 JSON 中的 friend，然后通过循环遍历其中包含的内容。

**【运行结果】** 在 Eclipse 中启动 Android 模拟器，接着运行 JSONObject_test 项目，运行时显示效果如图 11-12 所示，点击“开始解析”按钮的显示效果如图 11-13 所示。

图 11-12　运行时显示效果

图 11-13　点击“开始解析”的显示效果

## 11.8　本章小结

本章主要讲解 Android 应用程序开发过程中的网络连接协议、网络连接状态、网络连接数据交互、网络数据处理等知识，同时介绍了两种网络请求方式——GET 方式与 POST 方式的区别，以及 JSON 的定义，并通过案例讲解了 JSON 的解析过程。

# 第 12 章　应用项目开发实例

（1）了解本章项目开发的基本环境。

（2）掌握本章项目开发的开发组件。

（3）掌握贺州旅游新闻系统的开发步骤。

贺州是桂湘粤三省结合部，风景秀丽，汇聚了多个少数民族，拥有黄姚古镇、姑婆山国家森林公园、十八水景区、紫云仙境、贺州玉石林等旅游景区，并且享有“长寿之市”的美誉，让我们来开发 Android 应用程序——贺州旅游新闻，更好地宣传贺州。

## 12.1　开发环境

开发工具：WAMP5 1.7.4、Chrome 58.0.3029.96、HBuilder 8.0.2，以及整本书用的 Eclipse JUNO。

### 12.1.1　Chrome 浏览器

Chrome 是由Google公司开发的网页浏览器，该浏览器是基于其他开源软件所撰写的，包括WebKit，目标是提升稳定性、提高速度和增加安全性，并创造出简单且有效率的使用者界面。Chrome 对 HTML5、jquery、MUI、PHP 都能很好地支持。在网上可以自行下载 Chrome 浏览器的安装包，也可在作者百度云盘上（地址为http://pan.baidu.com/s/1bo7e5J5）下载。

### 12.1.2　HBuilder

HBuilder 是DCloud（数字天堂）推出的一款支持HTML5的Web开发IDE。HBuilder 由 Java 编写，它基于 Eclipse，所以顺其自然地兼容了Eclipse的插件，HBuilder 对 HTML5、jquery、MUI、PHP 的代码编写能很好地支持，可及时显示程序开发者编写的基本视图界面。在网上可以自行下载 HBuilder 的安装包，也可在作者百度云盘上（地址为 http://pan.baidu.com/s/1nuJBtcH）下载。

### 12.1.3　WAMP

当前的项目开发环境，使用的是 WAMP5 1.7.4 版本。WAMP5 集成了 Apache+PHP+Mysql 环境，拥有简单的图形、菜单安装和配置环境等特点。在网上可以自行下载 WAMP5 的安装包，也可在作者百度云盘上（地址为http://pan.baidu.com/s/1c2Eeu0S）下载，下载后的安装过程如下所示：

（1）双击 wamp5_1.7.4.exe 安装文件。

（2）出现安装界面后，点击 Next 按钮，如图 12-1 所示。

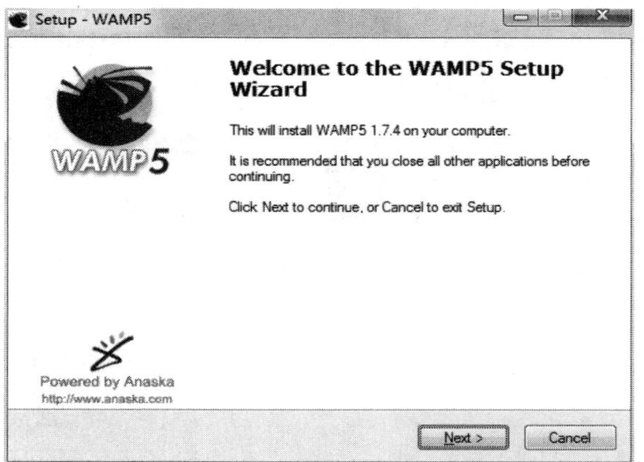


图 12-1　WAMP5 安装界面 1

（3）选中 I accept the agreement 选项，点击 Next 按钮，图 12-2 所示是指定 WAMP5 程序安装的位置。

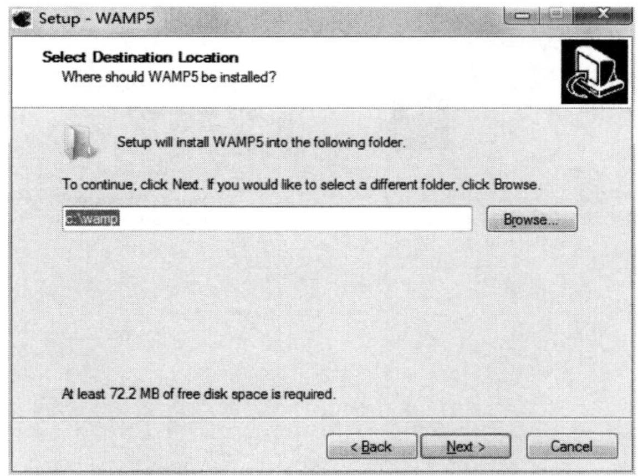


图 12-2　WAMP5 安装界面 2

（4）依次点击 Next 按钮，出现如图 12-3 和图 12-4 所示的界面。

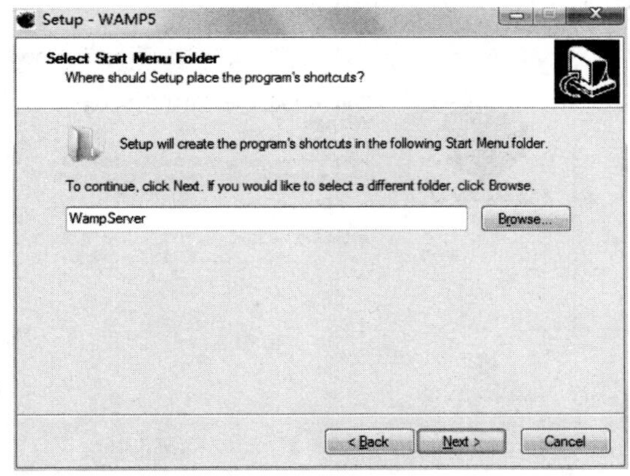


图 12-3　WAMP5 安装界面 3

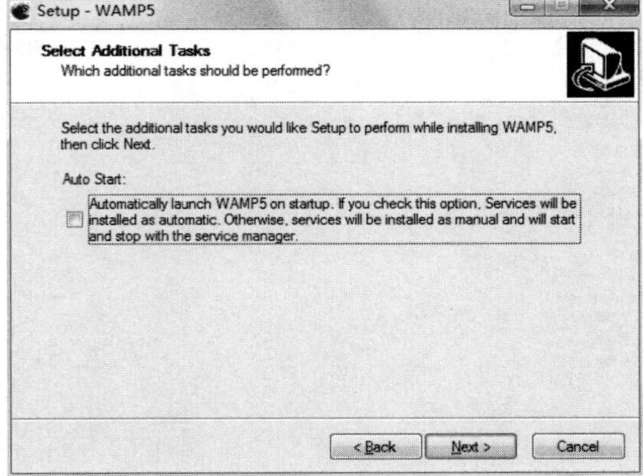


图 12-4　WAMP5 安装界面 4

（5）点击 Next 按钮，出现如图 12-5 所示的界面。点击 Install 按钮，出现如图 12-6 所示的表示程序开发者开发的代码存放位置的界面。

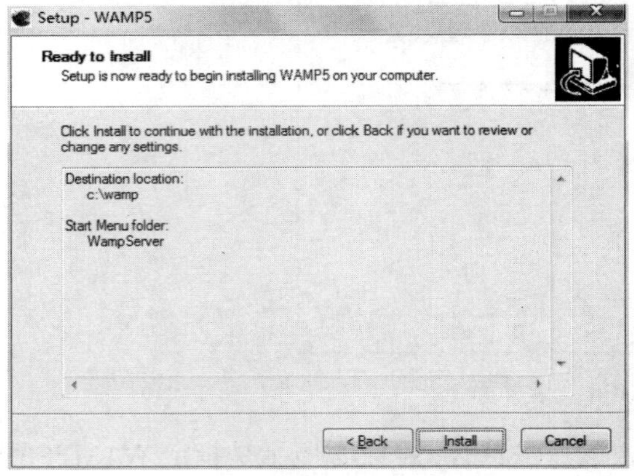


图 12-5　WAMP5 安装界面 5

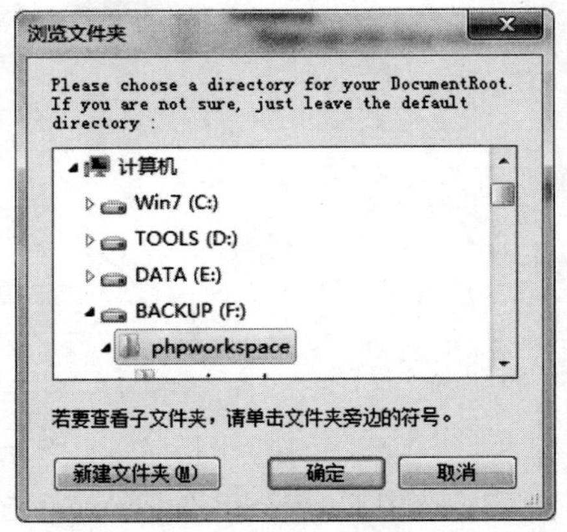


图 12-6　WAMP5 安装界面 6

（6）依次点击“确定”和 Next 按钮，出现图 12-7 和图 12-8 所示的界面，选择默认设置。

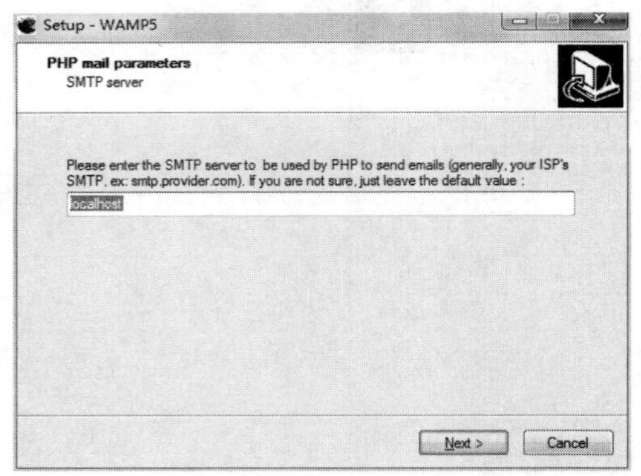


图 12-7　WAMP5 安装界面 7

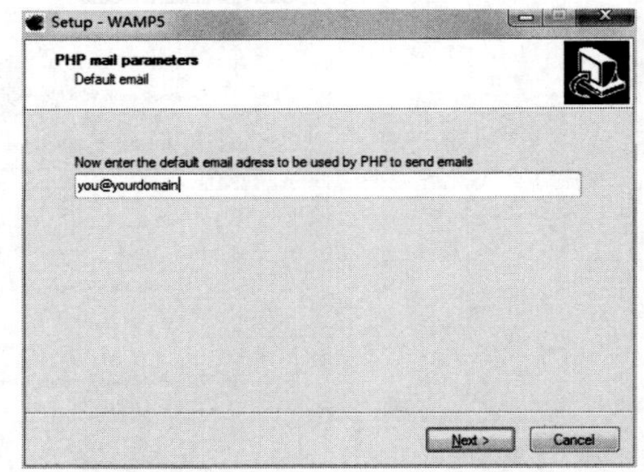


图 12-8　WAMP5 安装界面 8

（7）点击 Next 按钮，出现图 12-9 所示的选择测试网页所使用的浏览器界面，点击“打开”按钮，在图 12-10 所示的界面点击 Finish 按钮，即可启动 WAMP5。

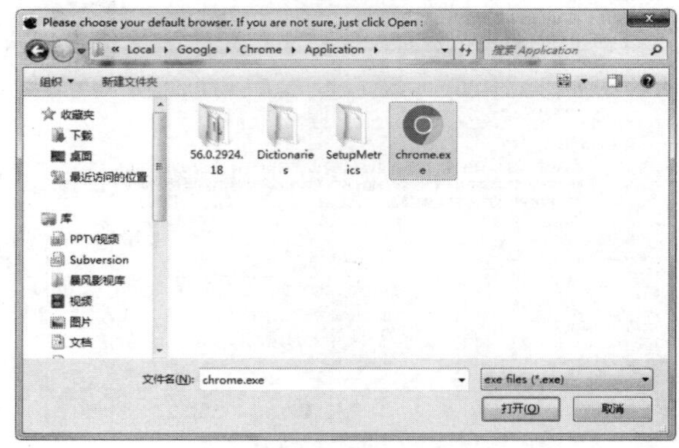


图 12-9　WAMP5 安装界面 9

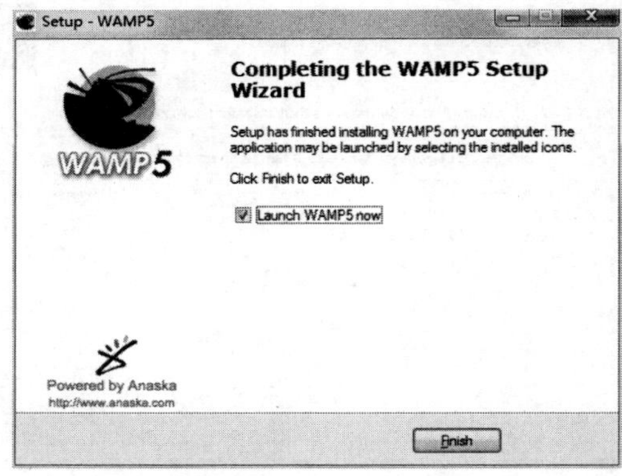


图 12-10　WAMP5 安装界面 10

## 12.2　开发组件

jQuery-2.1.1 和 MUI 是两大网页前端显示控件，通过使用这些控件使得网页可以自适应手机页面与计算机网页，具有极强的可移植性。

### 12.2.1　jQuery

jQuery 是一个快速、简洁的 JavaScript 框架，jQuery 具有独特的链式语法和短小清晰的多功能接口，具有高效灵活的 CSS 选择器，并且可对 CSS 选择器进行扩展，拥有便捷的插件扩展机制和丰富的插件。特别是使用 jQuery 的 ajax 方法，可以在不重新载入整个页面的情况下更新网页的一部分，在与服务器进行数据交互时带来极大的方便。在网上可以自行下载 jQuery 2.1.1，也可在作者百度云盘上（地址为 http://pan.baidu.com/s/1mhHB9aC）下载。

### 12.2.2　MUI

MUI 是一个轻量级的 HTML、CSS 和 JS 框架，遵循 Google 的 Material Design 设计思路。MUI 组件已被封装成HBuilder代码块，只需要简单几个字符，就可以快速生成各个组件对应的 HTML 代码。MUI 要引入的代码会在创建 HBuilder 项目时自动引入。

## 12.3　贺州旅游新闻系统

**【例 12.1】**开发的系统具有如下功能：在网页后台通过管理员账户可以添加新闻、查看新闻、接收手机客户端的用户注册数据等；在手机客户端可以注册用户、查看新闻等。

**【说明】**后台使用 PHP 结合 jQuery、MUI、JSON 来实现各种功能，前台使用 Android 的相关控件以及 JSON 来实现各种功能。

**【开发步骤】**

1．后台功能开发。

（1）在 MySQL 中创建数据库 hztourdb。

（2）在数据库中生成如表 12-1 至表 12-3 所示的 3 张表。

表 12-1　useradmin 表

| 字段名 | 数据类型 | 是否主键 | 描述 |
|---|---|---|---|
| id | integer | 是 | 编号 |
| username | varchar(20) | 否 | 账号 |
| password | varchar(20) | 否 | 密码 |

表 12-2 user表

| 字段名 | 数据类型 | 是否主键 | 描述 |
|---|---|---|---|
| id | integer | 是 | 编号 |
| username | varchar(20) | 否 | 账号 |
| password | varchar(20) | 否 | 密码 |

表 12-3 news表

| 字段名 | 数据类型 | 是否主键 | 描述 |
|---|---|---|---|
| id | integer | 是 | 编号 |
| title | varchar(20) | 否 | 新闻标题 |
| content | varchar(800) | 否 | 新闻内容 |
| imgpath | varchar(50) | 否 | 新闻图片路径 |
| newsdate | varchar(20) | 否 | 新闻日期 |

（3）在HBuilder中点击“文件”→“新建”→“移动 App”，创建名为 servicecode 的项目，选择 mui 项目，如图 12-11 所示。

（4）在生成的项目 js 目录下添加 jquery-2.1.1.min.js 文件，并创建如图 12-12 所示的文件和目录。

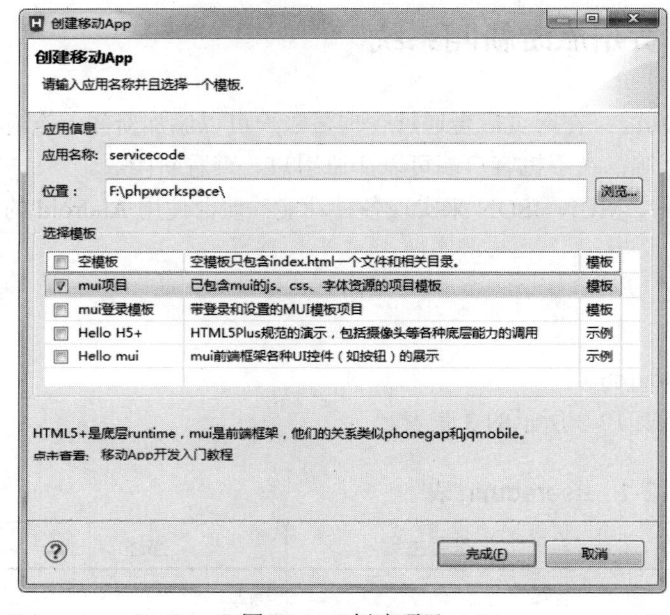


图 12-11 创建项目

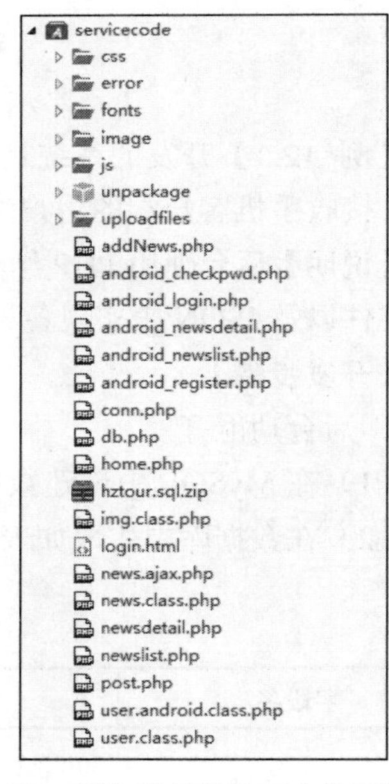


图 12-12 后台项目结构（文件和目录）

（5）导入相关的图片到项目的 image 文件夹下，编写好 db.php 文件内容和 conn.php 文件内容，实现数据库的连接、增、删、查、改操作。编写好的后台代码放到 WAMP5 的 www 目录下。

（6）文件 login.html 为登录后台代码，其界面如图 12-13 所示。

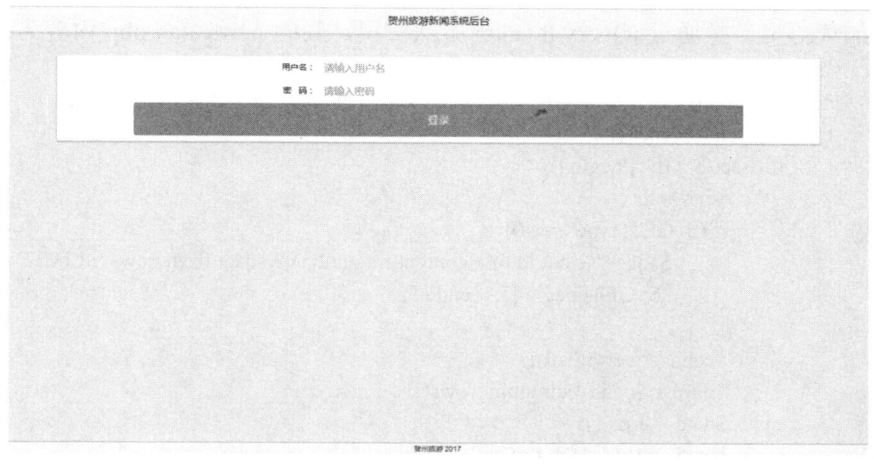


图 12-13　后台登录界面

（7）文件 home.php 为主页界面，其界面如图 12-14 所示。

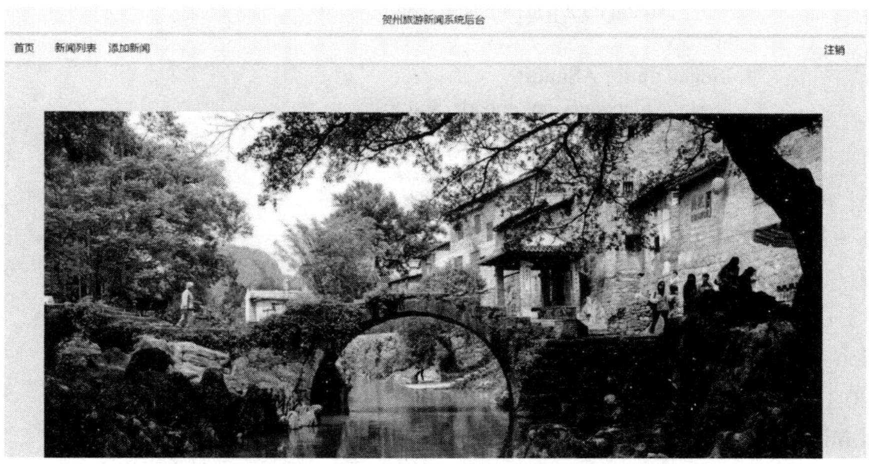


图 12-14　主页界面

（8）文件 newslist.php 为新闻列表界面，其界面如图 12-15 所示。

图 12-15　新闻列表界面

“点击加载更多”是通过 jQuery 的 ajax 方法实现，其中 news.ajax.php 内容为：

```
1	<?php
2	include "conn.php";
3	if (isset($_GET['type'])) {
4		if (isset($_GET['begin'])) {
5			$end = 8;
6			if ($_GET['type'] == 0) {
7				$sql = "select id,title,content,imgpath,newsdate from news    LIMIT
8				{$_GET['begin']},{$end} ";
9			}
10			$conn->query($sql);
11			$num = $conn->db_num_rows();
12			$data = array();
13			for ($i = 0;$i < $num;$i++) {
14				$row = $conn->fetch_array();
15				$json_row = array("id" => $row['id'], "title" => $row['title'],
16	"content" => urlencode($row['content']),"imgpath" => $row['imgpath'], "newsdate" => $row['newsdate']);
17				$data['data'][$i] = $json_row;
18			}
19			$data['num'] = $num;
20			echo urldecode(json_encode($data));
21		}
22	}
23	?>
```

1）第 1～23 行实现了网页上新闻列表后台数据加载。

2）第 7～8 行每次从数据库取出 8 条新闻记录。

3）第 13～18 行把从数据库取得新闻的 id、title、content、imgpath、newsdate 字段上的内容存储到$json_row 变量中。

4）第 20 行将取得的数据以 JSON 形式返回给显示界面。

newslist.php 文件内容为：

```
<?php
session_start();
include "conn.php";
	if(empty($_SESSION['User']))
	{
		Header("Location: login.html");
}
?>
<!DOCTYPE html>
<html>
<head>
<meta charset="utf-8">
<meta name="viewport" content="width=device-width,initial-scale=1,minimum-scale=1,
maximum-scale=1,user-scalable=no" />
<title>贺州旅游新闻</title>
<script src="js/mui.min.js"></script>
<script src="js/jquery-2.1.1.min.js"></script>
```

```
<link href="css/mui.min.css" rel="stylesheet"/>
<script type="text/javascript" charset="utf-8">
        mui.init();
</script>
<style>
        .clrear{ clear:both; height:0.5em;}
        ul,li{list-style:none; margin:0; padding:0;}
        nav{margin: 3em auto 1em auto; width: 94%;}
        .newsmore{ width:80%; margin:0 auto; height:1.5em; text-align:center; border:1px solid #999;
cursor:pointer; border-radius:10px;}
</style>
<script>
        $(document).ready(function(e) {
                var num=0;
                $(".newsmore").click(function(e) {
                        $.ajax({
                                type:"GET",
                                url:"news.ajax.php",
                                async:true,
                                dataType:"json",
                                data:{type:0,begin:num},
                                success: function(data){
                                        var str=$("#item1mobile ul").html();
                                        var i=0;
                                        for(i=0;i<data.num;i++,++num){
                                        str+="<li class='mui-table-view-cell'
                                        id='"+(data.data[i].id)+"'><a
                                        href='newsdetail.php?id="+(data.data[i].id)
                                        +"';'>"+data.data[i].title+"<br />时间：
                                        ["+ data.data[i].newsdate +"]</a></li>";
                                        }
                                        $("#item1mobile ul").html(str);
                                }
                        });
                });

                $.ajax({
                        type:"GET",
                        url:"news.ajax.php",
                        async:true,
                        dataType:"json",
                        data:{type:0,begin:0},
                        success: function(data){
                                var str="";
                                var i=0;
                                for(i=0;i<data.num;i++,++num){
                                        str+="<li class='mui-table-view-cell' id='"+(data.data[i].id)+"'><a
                                        href='newsdetail.php?id="+(data.data[i].id)
```

```
66                                  +";'>"+data.data[i].title+"<br />时间：
67                                  ["+ data.data[i].newsdate +"]</a></li>";
68                              }
69                              $("#item1mobile ul").html(str);
70                          }
71                      });
72              });
73      </script>
74      </head>
75      <body>
76              <header class="mui-bar mui-bar-nav">
77              <aclass="mui-icon mui-icon-left-nav mui-pull-left"
78      href="javascript:history.go(-1)"></a>
79                      <div class="mui-title">贺州旅游新闻</div>
80              </header>
81
82              <nav>
83
84              </nav>
85              <div class="mui-content-padded">
86                      <div id="item1mobile" class="mui-control-content mui-active">
87                              <ul class="mui-table-view">
88                              </ul>
89      <div class="clrear"></div>
90      <div class="newsmore">点击加载更多</div>
91                      </div>
92              </div>
93      </body>
94      </html>
```

1）第 1～94 行实现了网页上新闻列表前台界面显示。

2）第 3～8 行实现了登录用户的 Session 检测。

3）第 32～52 行实现了“点击加载更多”的新闻列表的 8 条数据加载。

4）第 54～71 行实现新闻列表界面默认 8 条数据加载。

（9）文件 newsdetail.php 为具体新闻界面，其界面如图 12-16 所示：

图 12-16　具体新闻界面

（10）文件 addNews.php 为添加新闻界面，其界面如图 12-17 所示：

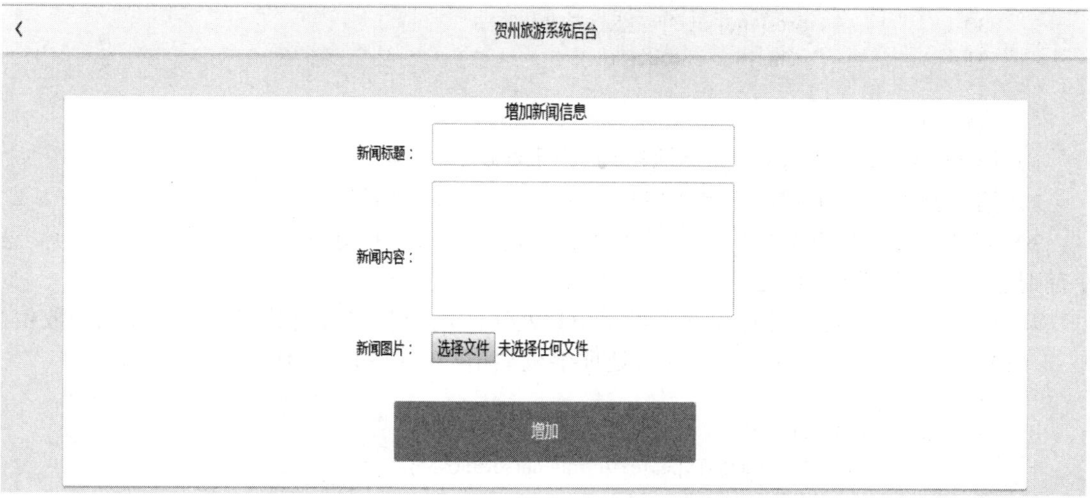

图 12-17　添加新闻界面

（11）文件 android_login.php 为处理手机端用户登录的后台代码。

```
<?php
session_start();
        header("Content-Type: text/html;charset=utf-8");
        include "conn.php";
       include "user.android.class.php";
       $res=array();
       if(isset($_POST['user_name'])&&isset($_POST['user_password'])){
        $user_name=$_POST['user_name'];
        $user_password=$_POST['user_password'];
        $user1=new userAndroid($conn);
            $user1->GetAndroidData($user_name);
                    if($user_password==$user1->password)
                    {
                            $_SESSION['user_name']=$user->username;
                            $_SESSION['user_password']=$user->password;

                             $res['success']=1;
                             $res['message']="登录成功！";
                             echo json_encode($res);

                    }else{
                             $res['success']=0;
                             $res['message']="登录失败！";
                             echo json_encode($res);
                    }
       }

```

```
28        else{
29          $res['success']=0;
30                $res['message']="登录失败！";
31                echo json_encode($res);
32        }
33    ?>
```

1）第1～33行实现访问后台数据的PHP代码。

2）第7行判断手机端用户是否发送了用户名与密码数据。

3）第12～25行查询数据库，检索数据库是否有相应的用户，有则返回登录成功信息，没有则返回登录失败信息。

4）第28～32行表示没有收到手机端用户发送的用户名与密码数据，返回登录失败信息。

（12）文件android_register.php为处理手机端用户注册的后台代码。

```
1    <?php
2    session_start();
3         header("Content-Type: text/html;charset=utf-8");
4         include "conn.php";
5        include "user.android.class.php";
6        $res=array();
7        if(isset($_POST['user_name'])&&isset($_POST['user_password'])){
8         $user_name=$_POST['user_name'];
9         $user_password=$_POST['user_password'];
10        $user=new userAndroid($conn);
11        $flag=$user->AndroidAdd($user_name,$user_password);
12                if($flag==true)
13                {
14                        $res['success']=1;
15                        $res['message']="注册成功！";
16                        echo json_encode($res);
17
18                }else{
19                        $res['success']=0;
20                        $res['message']="此用户已经注册！";
21                        echo json_encode($res);
22                }
23       }
24       else{
25         $res['success']=0;
26              $res['message']="注册失败！";
27              echo json_encode($res);
28       }
29   ?>
```

1）第1～29行实现了手机端用户注册访问后台数据的PHP代码。

2）第7行判断手机端用户是否发送了注册用的用户名与密码数据。

3）第12～22行查询数据库，检索数据库是否有相同的用户名，有则返回“此用户已经注册”，没有则返回“注册成功”信息。

4）第 24～28 行表示没有收到手机端用户发送的用于注册用的用户名与密码数据，返回“注册失败”信息。

（13）文件 android_newslist.php 为手机端新闻列表的后台代码。

```
1    <?php
2    include "conn.php";
3            $sql="select id,title,content,imgpath,newsdate from news ";
4            $conn->query($sql);
5            $num=$conn->db_num_rows();
6            $data=array();
7            for($i=0;$i<$num;$i++)
8            {
9                $row=$conn->fetch_array();
10               $json_row=array(
11                       "id"=>$row['id'],
12                       "title"=>$row['title'],
13                       "content"=>urlencode($row['content']),
14                       "imgpath"=>$row['imgpath'],
15                       "newsdate"=>$row['newsdate']
16               );
17               $data['data'][$i]=$json_row;
18           }
19           $data['num']=$num;
20           echo urldecode(json_encode($data));
21   ?>
```

1）第 1～21 行实现了手机端的新闻列表在后台数据加载。

2）第 3 行从数据库取出所有新闻记录。

3）第 10～16 行把从数据库取得新闻的 id、title、content、imgpath、newsdate 字段上的内容存储到$json_row 变量中。

4）第 20 行将取得的数据以 JSON 形式返回给显示界面。

（14）文件 android_checkpwd.php 为手机端用户修改个人密码的后台代码。

```
<?php
session_start();
     header("Content-Type: text/html;charset=utf-8");
     include "conn.php";
    include "user.android.class.php";
    $res=array();
    if(isset($_POST['user_name'])&&isset($_POST['newpassword'])){
     $username=$_POST['user_name'];
     $password=$_POST['newpassword'];
     $user2=new userAndroid($conn);
         $flag= $user2->AndroidChangePwd($username,$password);
             if($flag==true)
             {
                 $res['success']=1;
                 $res['message']="修改密码成功！";
```

```
16                      echo json_encode($res);
17
18                  }else{
19                      $res['success']=0;
20                      $res['message']="修改密码失败！ ";
21                      echo json_encode($res);
22                  }
23          }
24          else{
25            $res['success']=0;
26                $res['message']="修改密码失败！ ";
27                echo json_encode($res);
28          }
29      ?>
```

1）第 1～29 行实现手机端用户更改登录密码来访问后台数据的 PHP 代码。

2）第 7 行判断手机端用户是否发送了更改密码用的用户名与密码数据。

3）第 11 行在数据库中执行修改密码操作，并返回操作码。

4）第 12～22 如果返回成功操作码，则返回“修改密码成功！”信息，否则返回“修改密码失败！”信息。

5）第 24～28 行表示没有收到手机端用户发送的用于修改密码用的用户名和密码数据，返回“修改密码失败！”信息。

（15）文件 user.android.class.php 为处理手机端用户各种请求的后台功能函数代码。

```
<?php
class userAndroid{
      public $username;
      public $password;
      private $conn;
      public function __construct($c)
      {
            $this->conn=$c;
      }

            public function GetAndroidData($user_name)
      {
            $sql="select * from user    where username='{$user_name}' ";
            $this->conn->query($sql);
            if($this->conn->db_num_rows()>0)
            {
                  $row=$this->conn->fetch_array();
                  $this->username=$row['username'];
                  $this->password=$row['password'];
                  return true;
            }else{
                  return false;
            }
```

```
24          }
25
26
27          public function AndroidAdd($username,$password)
28          {
29
30                  $sql="select * from user    where username='{$username}' ";
31                  $this->conn->query($sql);
32                  if($this->conn->db_num_rows()>0)
33                  {
34                          return false;
35                  }
36  $sql="insertinto user
37  (username,password)values('{$username}','{$password}')";
38                  $this->conn->query($sql);
39                  if($this->conn->db_affected_rows()>0)
40                  {
41                          return true;
42                  }else{
43                          return false;
44                  }
45          }
46
47
48          public function AndroidChangePwd($username,$password)
49          {
50
51                  $sql="select * from user    where username='{$username}'";
52                  $this->conn->query($sql);
53                  if($this->conn->db_num_rows()==0)
54                  {
55                          return false;
56                  }
57          $sql=" update user set password ='{$password}' where
58  username='{$username}'";
59                  $this->conn->query($sql);
60                  if($this->conn->db_affected_rows()>0)
61                  {
62                          return true;
63                  }else{
64                          return false;
65                  }
66          }
67  }
68  ?>
```

1）第 1～68 行表示处理手机端用户各种请求的后台功能函数代码。

2）第 11～24 行处理手机端用户登录的后台 PHP 代码。

3）第 27～45 行处理手机端用户注册的后台 PHP 代码。

4）第 48～66 行处理手机端用户修改密码的后台 PHP 代码。

2．手机端功能开发

（1）在 Eclipse 中创建项目名为 HZTour 的项目，导入相关图片到 res 下的 drawable-hdpi 目录下，然后分别创建 com.hzu.adapter、com.hzu.fragment、com.hzu.hztour、com.hzu.model、com.hzu.util 五个包，其中 src 目录下的文件结构如图 12-18 所示。

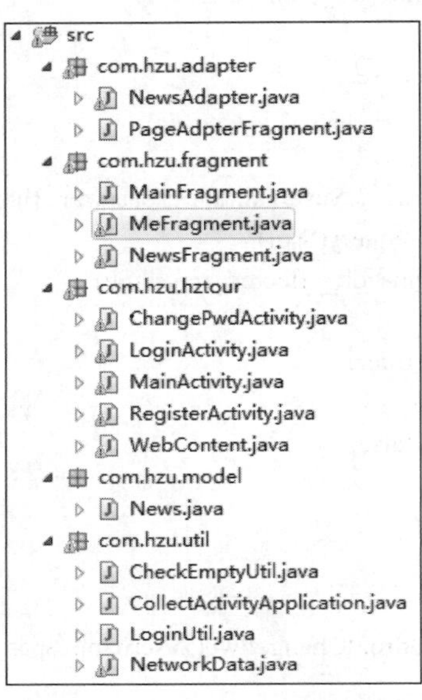


图 12-18　src 目录下的文件结构

（2）创建用户登录界面，如图 12-19 所示；创建用户注册界面，如图 12-20 所示。

图 12-19 用户登录界面

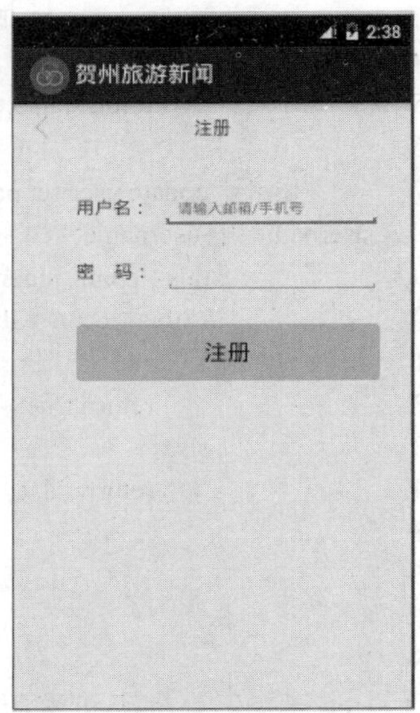


图 12-20　用户注册界面

用户登录的 activity 代码为：

```
package com.hzu.hztour;

import java.util.ArrayList;
import java.util.List;

import org.apache.http.NameValuePair;
import org.apache.http.message.BasicNameValuePair;
import org.json.JSONObject;

import com.hzu.util.CheckEmptyUtil;
import com.hzu.util.CollectActivityApplication;
import com.hzu.util.NetworkData;

import android.app.Activity;
import android.app.ProgressDialog;
import android.content.Intent;
import android.os.AsyncTask;
import android.os.Bundle;
import android.view.View;
import android.widget.Button;
import android.widget.EditText;
import android.widget.Toast;

public class LoginActivity extends Activity {
	private EditText user;
	private EditText pwd;
	private Button btRegister, btLogin;
	private String username;
	private String userpwd;
	private String jsonData;
	private ProgressDialog pDialog;
	public static String LoginUrl = NetworkData.HomeUrl +
"android_login.php";
	public String mes;
	public int suc = 2;
	NetworkData networkData = new NetworkData();
	CollectActivityApplication   CollectActivity=new
CollectActivityApplication();
	@Override
	protected void onCreate(Bundle savedInstanceState) {
		super.onCreate(savedInstanceState);
		CollectActivity.getInstance().addActivity(this);
		setContentView(R.layout.activity_login);
		user = (EditText) findViewById(R.id.et_user);
		pwd = (EditText) findViewById(R.id.et_pwd);
```

```
        btLogin = (Button) findViewById(R.id.bt_login);
        btLogin.setOnClickListener(new View.OnClickListener() {

            @Override
            public void onClick(View v) {
                username = user.getText().toString().trim();
                userpwd = pwd.getText().toString().trim();
                if (!CheckEmptyUtil.checkEmpty(username)) {
                    Toast.makeText(LoginActivity.this, "用户名不能为空！",
                            Toast.LENGTH_SHORT).show();
                    return;
                }
                if (!CheckEmptyUtil.checkEmpty(userpwd)) {
                    Toast.makeText(LoginActivity.this, "密码不能为空！",
                            Toast.LENGTH_SHORT).show();
                    return;
                }
                new LoginUser().execute();
            }
        });

        btRegister=(Button) findViewById(R.id.bt_register);
        btRegister.setOnClickListener(new View.OnClickListener() {

            @Override
            public void onClick(View v) {
                Intent intent = new Intent( LoginActivity.this,
                        RegisterActivity.class);
                startActivity(intent);
            }
        });
    }

    // 用户登录的后台处理
    class LoginUser extends AsyncTask<String, String, String> {
        @Override
        protected void onPreExecute() {
            pDialog = new ProgressDialog(LoginActivity.this);
            pDialog.setMessage("正在登录...请稍等！");
            pDialog.setIndeterminate(false);
            pDialog.setCancelable(true);
            pDialog.show();
        }

        @Override
        protected String doInBackground(String... params) {
```

```
92                          List<NameValuePair> paramsdata = new
93      ArrayList<NameValuePair>();
94                          paramsdata.add(new BasicNameValuePair("user_name", username));
95                  paramsdata.add(new BasicNameValuePair("user_password", userpwd));
96
97                          try {
98                                  jsonData = networkData.makeHttpRequest(LoginUrl, "post",
99                                                  paramsdata);
100                         } catch (Exception e) {
101                                 e.printStackTrace();
102                         }
103                         try {
104                                 JSONObject jSONObject = new JSONObject(jsonData);
105                                 mes = jSONObject.getString("message");
106                                 suc = jSONObject.getInt("success");
107                         } catch (Exception e) {
108                                 e.printStackTrace();
109                         }
110                         return null;
111                 }
112
113                 @Override
114                 protected void onPostExecute(String result) {
115                         pDialog.dismiss();
116                 Toast.makeText(LoginActivity.this, "返回的信息为:" + suc + ";" + mes,
117                                         Toast.LENGTH_SHORT).show();
118                         if (suc == 1) {
119                                 Intent intent = new Intent(LoginActivity.this,
120                                                 MainActivity.class);
121                                 intent.putExtra("username", username);
122                                 intent.putExtra("userpwd", userpwd);
123                                 startActivity(intent);
124                         }
125                 }
126         }
127 }
```

1）第 1～127 行表示处理手机端用户登录的 Android 代码。

2）第 37 行、38 行和 42 行将当前的 Activity 添加到 CollectActivity 对象中，用于退出 App 时销毁当前的 Activity。

3）第 46～62 行表示用户点击“登录”按钮后，判断用户名与密码是否为空。

4）第 63 行表示执行异步任务操作。

5）第 67～76 表示用户点击“注册”按钮后，跳转到注册页面。

6）第 80～126 行用于检查用户输入的用户名和密码是否存在数据库中。第 82～88 行表示执行 onPreExecute()方法，加载进度条。第 85 行表示没有明确值的加载方式，第 86 行表示可以通过点击 Back 键取消加载。第 91～111 行表示 doInBackground()方法，获取网络上的登录信息的状态。第 114～125 行表示执行 onPostExecute()方法，其中通过判断 suc 的值来决定是否跳转到主界面。

用户注册的 activity 代码为：

```
package com.hzu.hztour;

import java.util.ArrayList;
import java.util.List;

import org.apache.http.NameValuePair;
import org.apache.http.message.BasicNameValuePair;
import org.json.JSONObject;

import android.app.Activity;
import android.app.ProgressDialog;
import android.content.Intent;
import android.os.AsyncTask;
import android.os.Bundle;
import android.view.View;
import android.widget.Button;
import android.widget.EditText;
import android.widget.ImageView;
import android.widget.Toast;

import com.hzu.util.CheckEmptyUtil;
import com.hzu.util.CollectActivityApplication;
import com.hzu.util.NetworkData;

public class RegisterActivity extends Activity {
        private EditText etuser;
        private EditText etpwd;
        private String rgusername;
        private String rguserpwd;
        private Button bt_regiter;
        private ProgressDialog pDialog;
        private ImageView ivrt;
        public static String RegisterUrl =    NetworkData.HomeUrl +
"android_register.php";
        NetworkData networkData = new NetworkData();
        private String jsonData;
        public String mes;
        public int suc = 2;
        CollectActivityApplication CollectActivity=new
CollectActivityApplication();
        @Override
```

```
    protected void onCreate(Bundle savedInstanceState) {
        super.onCreate(savedInstanceState);
        CollectActivity.getInstance().addActivity(this);
        setContentView(R.layout.activity_register);
        ivrt=(ImageView) findViewById(R.id.imrt_register);
        ivrt.setOnClickListener(new View.OnClickListener() {

            @Override
            public void onClick(View v) {
                Intent intent =new Intent(RegisterActivity.this,
LoginActivity.class);
                startActivity(intent);
            }
        });
        etuser = (EditText) findViewById(R.id.et_rguser);
        etpwd = (EditText) findViewById(R.id.et_rgpwd);
        bt_regiter = (Button) findViewById(R.id.bt_rguser);
        bt_regiter.setOnClickListener(new View.OnClickListener() {
            @Override
            public void onClick(View v) {
                rgusername = etuser.getText().toString().trim();
                rguserpwd = etpwd.getText().toString().trim();
                if (!CheckEmptyUtil.checkEmpty(rgusername)) {
                    Toast.makeText(RegisterActivity.this, "用户名不能为空！",
                            Toast.LENGTH_SHORT).show();
                    return;
                }
                if (!CheckEmptyUtil.checkEmpty(rguserpwd)) {
                    Toast.makeText(RegisterActivity.this, "密码不能为空！",
                            Toast.LENGTH_SHORT).show();
                    return;
                }
                new RegisterUser().execute();
            }
        });
    }

    // 用户注册的后台处理
        class RegisterUser extends AsyncTask<String, String, String> {
            @Override
            protected void onPreExecute() {
                pDialog = new ProgressDialog(RegisterActivity.this);
                pDialog.setMessage("正在注册...请稍等！");
                pDialog.setIndeterminate(false);
                pDialog.setCancelable(true);
```

```
87                     pDialog.show();
88                 }
89 
90                 @Override
91                 protected String doInBackground(String... params) {
92                     List<NameValuePair> paramsdata = new
93     ArrayList<NameValuePair>();
94         paramsdata.add(new BasicNameValuePair("user_name", rgusername));
95         paramsdata.add(new BasicNameValuePair("user_password", rguserpwd));
96 
97                     try {
98                     jsonData = networkData.makeHttpRequest(RegisterUrl, "post",
99                                 paramsdata);
100                    } catch (Exception e) {
101                        e.printStackTrace();
102                    }
103                    try {
104                        JSONObject jSONObject = new JSONObject(jsonData);
105                        mes = jSONObject.getString("message");
106                        suc = jSONObject.getInt("success");
107                    } catch (Exception e) {
108                        e.printStackTrace();
109                    }
110                    return null;
111                }
112 
113                @Override
114                protected void onPostExecute(String result) {
115                    pDialog.dismiss();
116        Toast.makeText(RegisterActivity.this, "返回的信息为:" + suc + ";" + mes,
117                                Toast.LENGTH_LONG).show();
118                    if (suc == 1) {
119                        Intent intent = new Intent(RegisterActivity.this,
120                                LoginActivity.class);
121                        startActivity(intent);
122                    }
123                }
124            }
125    }
```

1）第 1～125 行表示处理手机端注册的 Android 代码。

2）第 39 行、40 行和 44 行表示将当前的 Activity 添加到 CollectActivity 对象中，用于退出 App 时销毁当前的 Activity。

3）第 36～55 行用于左上角的返回图标的监听，用于返回到登录界面。

4）第 64～73 行表示用户点击“注册”按钮后，判断用户名和密码是否为空。

5）第 74 行表示执行异步任务操作。

6）第 80～124 行用于将用户输入的用户名和密码存到数据库中。第 82～88 行表示执行 onPreExecute()方法，加载进度条。第 85 行表示没有明确值的加载方式，第 86 行表示可以通过点击 Back 键取消加载。第 91～111 行表示 doInBackground()方法，获取网络上的注册信息。第 114～125 行表示执行 onPostExecute()方法，其中通过判断 suc 的值来决定是否跳转到登录界面。

用户登录与注册请求网络的 NetworkData.java 文件的内容为：

```
package com.hzu.util;

import java.io.BufferedReader;
import java.io.InputStream;
import java.io.InputStreamReader;

import java.util.List;

import org.apache.http.HttpEntity;
import org.apache.http.HttpResponse;
import org.apache.http.NameValuePair;
import org.apache.http.client.CookieStore;
import org.apache.http.client.entity.UrlEncodedFormEntity;
import org.apache.http.client.methods.HttpPost;
import org.apache.http.cookie.Cookie;
import org.apache.http.impl.client.DefaultHttpClient;
import org.apache.http.protocol.HTTP;

public class NetworkData {
        public static InputStream inputStream;
        public static String SESSIONID = null;
        public static String jsondata = "";
        public static String HomeUrl = "http://10.0.2.2/servicecode/";

        public String makeHttpRequest(String url, String method,
                        List<NameValuePair> params) {
                try {
                        HttpPost httpPost = new HttpPost(url);
                httpPost.setEntity(new UrlEncodedFormEntity(params, HTTP.UTF_8));
                        if (SESSIONID == null) {
                                httpPost.setHeader("Cookie", "SESSIONID=" + SESSIONID);
                        }
                        DefaultHttpClient defaultHttpClient = new DefaultHttpClient();
                        HttpResponse httpResponse =
defaultHttpClient.execute(httpPost);
                        HttpEntity httpEntity = httpResponse.getEntity();
```

```
37                  inputStream = httpEntity.getContent();
38                  CookieStore cookieStore = defaultHttpClient.getCookieStore();
39                  List<Cookie> cookies = cookieStore.getCookies();
40                  for (int i = 0; i < cookies.size(); i++) {
41                      if ("SESSIONID".equals(cookies.get(i).getName())) {
42                          SESSIONID = cookies.get(i).getName();
43                          break;
44                      }
45                  }
46              } catch (Exception e) {
47                  e.printStackTrace();
48              }
49
50              try {
51                  BufferedReader bufferedReader = new BufferedReader(
52                          new InputStreamReader(inputStream, "UTF-8"));
53                  StringBuilder stringBuilder = new StringBuilder();
54                  String line = null;
55                  while ((line = bufferedReader.readLine()) != null) {
56                      stringBuilder.append(line + "\n");
57                  }
58                  inputStream.close();
59                  jsondata = stringBuilder.toString();
60              } catch (Exception e) {
61                  e.printStackTrace();
62              }
63              return jsondata;
64          }
65      }
```

1）第 1～65 用于相关的网络请求数据过程处理。

2）第 23 行表示访问网络的原始主页地址。

3）第 28 行表示使用 HttpPOST 方式连接指定的 URL。

4）第 29 行设置请求参数。

5）第 30～32 行将 SESSIONID 发送给服务器。

6）第 33 行设置默认的 HTTP 客户端。

7）第 34～35 行执行 HTTP 连接。

8）第 36～37 行获得请求响应实体，并获得 InputStream 对象。

9）第 38～45 行获取 Cookie[‘SESSIONID’]的值，并保证每次都是同一个值。

10）第 50～62 行将从网络获取的数据转换成字符串。

（3）用户登录成功，进入的主界面如图 12-21 所示。点击新闻菜单显示的界面如图 12-22 所示。

图 12-21　App 主界面

图 12-22　新闻界面

主界面对应的文件为 MainActivity.java，其内容如下：

```
package com.hzu.hztour;

import java.util.ArrayList;
import java.util.List;

import com.hzu.adapter.PageAdpterFragment;
import com.hzu.fragment.MainFragment;
import com.hzu.fragment.MeFragment;
import com.hzu.fragment.NewsFragment;
import com.hzu.util.CollectActivityApplication;

import android.graphics.Color;
import android.os.Bundle;
import android.support.v4.app.Fragment;
import android.support.v4.app.FragmentActivity;
import android.support.v4.view.ViewPager;
import android.view.View;
import android.view.Window;
import android.view.View.OnClickListener;
import android.widget.ImageView;
import android.widget.LinearLayout;
import android.widget.TextView;
```

```

public class MainActivity extends FragmentActivity implements
OnClickListener {
    private LinearLayout lMain, lNews, lMe;
    private ImageView imMain, imNews, imMe;
    private TextView tvMian, tvnews, tvMe;
    private ViewPager mViewpage;
    private List<Fragment> fragments;
    private PageAdpterFragment pageAdpterFragment;
    private String username;
    private String userpwd;
    CollectActivityApplication CollectActivity = new
                                CollectActivityApplication();
    @Override
    protected void onCreate(Bundle savedInstanceState) {
        super.onCreate(savedInstanceState);
        requestWindowFeature(Window.FEATURE_NO_TITLE);
        CollectActivity.getInstance().addActivity(this);
        setContentView(R.layout.main_framework);
        lMain = (LinearLayout) findViewById(R.id.first);
        lNews = (LinearLayout) findViewById(R.id.second);
        lMe = (LinearLayout) findViewById(R.id.third);

        imMain = (ImageView) findViewById(R.id.imMain);
        imNews = (ImageView) findViewById(R.id.imNews);
        imMe = (ImageView) findViewById(R.id.imMe);

        tvMian = (TextView) findViewById(R.id.tvMian);
        tvnews = (TextView) findViewById(R.id.tvnews);
        tvMe = (TextView) findViewById(R.id.tvMe);

        username = getIntent().getStringExtra("username");
        userpwd = getIntent().getStringExtra("userpwd");

        mViewpage = (ViewPager) findViewById(R.id.view_pager);

        fragments = new ArrayList<Fragment>();
        Fragment fMain = new MainFragment();
        Fragment fNews = new NewsFragment();
        Fragment fMe = new MeFragment();
        Bundle bundle = new Bundle();
        bundle.putString("username", username);
        bundle.putString("userpwd", userpwd);
        fMe.setArguments(bundle);
        fragments.add(fMain);
        fragments.add(fNews);
```

```
            fragments.add(fMe);
            pageAdpterFragment = new PageAdpterFragment(
                    getSupportFragmentManager(), fragments);
            mViewpage.setAdapter(pageAdpterFragment);
            mViewpage.setOnPageChangeListener(new
            ViewPager.OnPageChangeListener() {
                @Override
                public void onPageSelected(int arg0) {
                    initContent();
                    int item = mViewpage.getCurrentItem();
                    switch (item) {
                    case 0:
                        imMain.setImageResource(R.drawable.home);
                        tvMian.setTextColor(Color.parseColor("#FF32CD32"));
                        break;
                    case 1:
                        imNews.setImageResource(R.drawable.news);
                        tvnews.setTextColor(Color.parseColor("#FF32CD32"));
                        break;
                    case 2:
                        imMe.setImageResource(R.drawable.me);
                        tvMe.setTextColor(Color.parseColor("#FF32CD32"));
                        break;
                    }
                }

                @Override
                public void onPageScrolled(int arg0, float arg1, int arg2) {
                    // TODO Auto-generated method stub
                }

                @Override
                public void onPageScrollStateChanged(int arg0) {
                    // TODO Auto-generated method stub
                }
            });

            lMain.setOnClickListener(this);
            lNews.setOnClickListener(this);
            lMe.setOnClickListener(this);
        }

        public void initContent() {
            imMain.setImageResource(R.drawable.home_unselect);
            imNews.setImageResource(R.drawable.news_unselected);
            imMe.setImageResource(R.drawable.me_unselect);
```

```
115                 tvMian.setTextColor(Color.parseColor("#FFD3D3D3"));
116                 tvnews.setTextColor(Color.parseColor("#FFD3D3D3"));
117                 tvMe.setTextColor(Color.parseColor("#FFD3D3D3"));
118         }
119
120         @Override
121         public void onClick(View v) {
122                 switch (v.getId()) {
123                 case R.id.first:
124                         setSelected(0);
125                         break;
126                 case R.id.second:
127                         setSelected(1);
128                         break;
129                 case R.id.third:
130                         setSelected(2);
131                         break;
132                 }
133
134         }
135
136         public void setSelected(int x) {
137                 initContent();
138                 switch (x) {
139                 case 0:
140                         imMain.setImageResource(R.drawable.home);
141                         tvMian.setTextColor(Color.parseColor("#ff32CD32"));
142                         break;
143                 case 1:
144                         imNews.setImageResource(R.drawable.news);
145                         tvnews.setTextColor(Color.parseColor("#32CD32"));
146                         break;
147                 case 2:
148                         imMe.setImageResource(R.drawable.me);
149                         tvMe.setTextColor(Color.parseColor("#32CD32"));
150                         break;
151
152                 }
153                 mViewpage.setCurrentItem(x);
154     }
155 }
```

1）第 1～155 行表示用户登录成功后，进入到主界面的代码。

2）第 39 行设置当前 Activity 不再显示标题。

3）第 34 行、35 行和 40 行表示将当前的 Activity 添加到 CollectActivity 对象中，用于退出 App 时销毁当前的 Activity。

4）第 42～52 行用于获取底部菜单各个控件。

5）第 54～55 行用于接收从登录成功的 Activity 中取得的用户名和密码。

6）第 59～104 行，将底部三个菜单与三个 Fragment 进行绑定，然后把三个 Fragment 传到 PageAdpterFragment 中，用于实现用户滑动时底部菜单也相应改变图标和颜色。第 63～66 行把用户名和密码传到“我的” Fragment 上。

7）第 106～108 行为底部三个菜单设置监听器。

8）第 111～118 实现底部三个菜单的图标与颜色的初始化。

9）第 121～154 行实现用户点击不同菜单显示不同的 Fragment。

（4）新闻界面中显示的新闻列表，如图 12-23 所示。点击某条新闻后，显示的新闻详情页面如图 12-24 所示。

图 12-23　新闻列表

图 12-24　新闻详情

新闻列表对应的文件为 NewsFragment.java，代码如下：

```
package com.hzu.fragment;

import java.util.ArrayList;
import java.util.List;

import org.apache.http.NameValuePair;
import org.json.JSONArray;
import org.json.JSONObject;

import com.hzu.adapter.NewsAdapter;

```

```
import com.hzu.hztour.R;
import com.hzu.hztour.WebContent;
import com.hzu.model.News;
import com.hzu.util.NetworkData;

import android.app.ProgressDialog;
import android.content.Intent;
import android.os.AsyncTask;
import android.os.Bundle;
import android.support.v4.app.Fragment;
import android.view.LayoutInflater;
import android.view.View;
import android.view.ViewGroup;
import android.widget.AdapterView;
import android.widget.ListView;

public class NewsFragment extends Fragment {
	public ListView listView;
	public NewsAdapter newsadapter;
	public String jsonData;
	public ProgressDialog pDialog;
	public static String NewsUrl = NetworkData.HomeUrl +
"android_newslist.php";
	NetworkData networkData = new NetworkData();
	public ArrayList<News> newsList = new ArrayList<News>();

	@Override
	public View onCreateView(LayoutInflater inflater, ViewGroup container,
			Bundle savedInstanceState) {
		View view = inflater.inflate(R.layout.newslist_fragment, container,
				false);
		listView = (ListView) view.findViewById(R.id.lvnews);
		listView.setOnItemClickListener(new
AdapterView.OnItemClickListener() {
			@Override
			public void onItemClick(AdapterView<?> parent, View view,
					int position, long id) {
				News news = newsList.get(position);
				Intent intent = new Intent(getActivity(), WebContent.class);
				intent.putExtra("id", news.getId() + "");
				startActivity(intent);
			}
		});
		return view;
	}

```

```
    @Override
    public void onResume() {
        super.onResume();
        new NewsLoad().execute();
    }

    // 新闻列表的后台处理
    class NewsLoad extends AsyncTask<String, String, ArrayList<News>> {

        @Override
        protected void onPreExecute() {
            pDialog = new ProgressDialog(getActivity());
            pDialog.setMessage("正在加载，请稍等！");
            pDialog.setIndeterminate(false);
            pDialog.setCancelable(true);
            pDialog.show();
        }

        @Override
        protected ArrayList<News> doInBackground(String... params) {
            List<NameValuePair> paramsdata = new
                                    ArrayList<NameValuePair>();
            try {
                jsonData = networkData.makeHttpRequest(NewsUrl, "post",
                        paramsdata);
            } catch (Exception e) {
                e.printStackTrace();
            }
            try {
                JSONObject object = new JSONObject(jsonData);
                JSONArray jsonArray = object.getJSONArray("data");
                newsList = new ArrayList<News>();
                for (int i = 0; i < jsonArray.length(); i++) {
                    JSONObject object1 = jsonArray.getJSONObject(i);
                    int id = object1.getInt("id");
                    String title = object1.getString("title");
                    String content = object1.getString("content");
                    String newsdate = object1.getString("newsdate");
                    newsList.add(new News(id, title, content, newsdate));
                }
            } catch (Exception e) {
                e.printStackTrace();
            }
            return newsList;
        }

```

```
104             @Override
105             protected void onPostExecute(final ArrayList<News> newsList1) {
106                     pDialog.dismiss();
107                     newsadapter = new NewsAdapter(getActivity(), newsList1);
108                     listView.setAdapter(newsadapter);
109                     newsadapter.notifyDataSetChanged();
110             }
111         }
112 }
```

1）第 1～112 行实现了手机端的新闻列表功能。

2）第 38～56 行在 onCreateView()方法中实现动态加载新闻列表，第 46～54 行实现了用户点击某一条新闻时，跳转到新闻详情页面。

3）第 58～62 行实现当前 Fragment 可见时，执行网络加载新闻数据。

4）第 65～111 行实现网络请求加载新闻数据的请求过程。第 67～74 行表示执行 onPreExecute()方法，加载进度条。第 71 行表示没有明确值的加载方式，第 72 行表示可以通过点击 Back 键取消加载。第 76～102 行表示 doInBackground()方法，获取网络上的新闻数据。第 105～111 行表示执行 onPostExecute()方法，将网络上的新闻数据绑定到 listView 对象上，第 109 行更新适配器内容。

新闻详情对应的文件为 WebContent.java，代码如下：

```
package com.hzu.hztour;

import com.hzu.util.CollectActivityApplication;
import com.hzu.util.NetworkData;

import android.app.Activity;
import android.os.Bundle;
import android.view.Window;
import android.webkit.WebSettings;
import android.webkit.WebView;

public class WebContent extends Activity {
    public WebView webView;
    public String path = NetworkData.HomeUrl + "android_newsdetail.php?id=";
    CollectActivityApplication CollectActivity = new
                            CollectActivityApplication();
    @Override
    protected void onCreate(Bundle savedInstanceState) {
        CollectActivity.getInstance().addActivity(this);
        super.onCreate(savedInstanceState);
        requestWindowFeature(Window.FEATURE_NO_TITLE);
        setContentView(R.layout.webcontent);
        webView = (WebView) findViewById(R.id.webshow);
        String id = getIntent().getStringExtra("id");
        webView.loadUrl(path + id);
```

```
26                    webView.getSettings().setJavaScriptEnabled(true);
27                    webView.getSettings().setCacheMode(
28                                    WebSettings.LOAD_CACHE_ELSE_NETWORK);
29            }
30    }
```

1）第 1～30 行实现新闻详情页面。

2）第 15 行、16 行和 19 行用于将当前的 Activity 添加到 CollectActivity 对象中，用于退出 App 时销毁当前的 Activity。

3）第 25 行实现加载网络请求地址。

4）第 26 行设置 WebView 属性，能够执行 JavaScript 脚本。

5）第 27 行与 28 行实现使用缓存。

（5）用户点击“我的”，显示界面如图 12-25 所示。点击“修改密码”界面如图 12-26 所示。

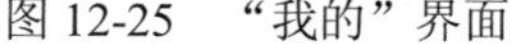
图 12-25　“我的”界面

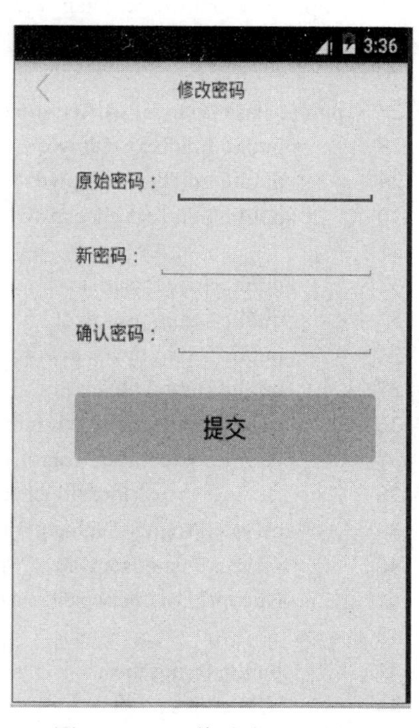


图 12-26　“修改密码”界面

修改密码对应的 ChangePwdActivity.java 文件，其内容为：

```
package com.hzu.hztour;

import java.util.ArrayList;
import java.util.List;

import org.apache.http.NameValuePair;
import org.apache.http.message.BasicNameValuePair;
import org.json.JSONObject;

import com.hzu.util.CheckEmptyUtil;
```

```
import com.hzu.util.CollectActivityApplication;
import com.hzu.util.NetworkData;

import android.app.Activity;
import android.app.ProgressDialog;
import android.content.Intent;
import android.os.AsyncTask;
import android.os.Bundle;
import android.util.Log;
import android.view.View;
import android.view.Window;
import android.widget.Button;
import android.widget.EditText;
import android.widget.ImageView;
import android.widget.Toast;

public class ChangePwdActivity extends Activity {
	public EditText oldpwd;
	public EditText newpwd;
	public EditText checkpwd;

	public String username;
	public String userpwd;
	public String newpassword;
	public Button btcgpwd;
	public ImageView returnimg;
	public static String LoginUrl = NetworkData.HomeUrl
			+ "android_checkpwd.php";
	private ProgressDialog pDialog;
	private String jsonData;
	NetworkData networkData = new NetworkData();

	public String mes;
	public int suc = 2;

	CollectActivityApplication CollectActivity = new
CollectActivityApplication();
	@Override
	protected void onCreate(Bundle savedInstanceState) {
		super.onCreate(savedInstanceState);
		requestWindowFeature(Window.FEATURE_NO_TITLE);
		setContentView(R.layout.activity_cgpwd);
		CollectActivity.getInstance().addActivity(this);
		username = getIntent().getStringExtra("username");
		userpwd = getIntent().getStringExtra("userpwd");

```

```
            oldpwd = (EditText) findViewById(R.id.old_pwd);
            newpwd = (EditText) findViewById(R.id.new_pwd);
            checkpwd = (EditText) findViewById(R.id.check_pwd);

            returnimg = (ImageView) findViewById(R.id.rtimg);
            returnimg.setOnClickListener(new View.OnClickListener() {
                @Override
                public void onClick(View v) {
                    Intent intent = new Intent(ChangePwdActivity.this,
                            MainActivity.class);
                    startActivity(intent);
                }
            });

            btcgpwd = (Button) findViewById(R.id.bt_cgpwdsubmit);
            btcgpwd.setOnClickListener(new View.OnClickListener() {

                @Override
                public void onClick(View v) {
                    if (!CheckEmptyUtil.checkEmpty(oldpwd.getText().toString()
                            .trim())) {
                    Toast.makeText(ChangePwdActivity.this, "原始密码不能为空！",
                                Toast.LENGTH_SHORT).show();
                        return;
                    }
                    if (!CheckEmptyUtil.checkEmpty(newpwd.getText().toString()
                            .trim())) {
                Toast.makeText(ChangePwdActivity.this, "新密码不能为空！",
                                Toast.LENGTH_SHORT).show();
                        return;
                    }
if (!CheckEmptyUtil.checkEmpty(checkpwd.getText().toString().trim())) {
                Toast.makeText(ChangePwdActivity.this, "确认密码不能为空！",
                                Toast.LENGTH_SHORT).show();
                        return;
                    }
if (!userpwd.equals(oldpwd.getText().toString().trim())) {
                    Toast.makeText(ChangePwdActivity.this, "原始密码不正确！",
                                Toast.LENGTH_SHORT).show();
                        return;
                    }
if (!(newpwd.getText().toString().trim()).equals(checkpwd
                            .getText().toString().trim())) {
                Toast.makeText(ChangePwdActivity.this, "新密码与确认密码不一致！",
                                Toast.LENGTH_SHORT).show();
                        return;
```

```
                }
                newpassword = checkpwd.getText().toString().trim();
                new ChageUserPWD().execute();
            }
        });
    }

    // 修改密码的后台处理
    class ChageUserPWD extends AsyncTask<String, String, String> {

        @Override
        protected void onPreExecute() {
            pDialog = new ProgressDialog(ChangePwdActivity.this);
            pDialog.setMessage("正在更改密码，请稍等！ ");
            pDialog.setIndeterminate(false);
            pDialog.setCancelable(true);
            pDialog.show();
        }

        @Override
        protected String doInBackground(String... params) {
            List<NameValuePair> paramsdata = new
                    ArrayList<NameValuePair>();
    paramsdata.add(new BasicNameValuePair("user_name", username));
    paramsdata.add(new BasicNameValuePair("newpassword", newpassword));
            try {
                jsonData = networkData.makeHttpRequest(LoginUrl, "post",
                        paramsdata);
            } catch (Exception e) {
                e.printStackTrace();
            }
            try {
                JSONObject jSONObject = new JSONObject(jsonData);
                mes = jSONObject.getString("message");
                suc = jSONObject.getInt("success");
            } catch (Exception e) {
                e.printStackTrace();
            }
            return null;
        }

        @Override
        protected void onPostExecute(String result) {
            pDialog.dismiss();
            Toast.makeText(ChangePwdActivity.this,
                    "返回的信息为： " + suc + ";" + mes + "修改密码成功！ ",
```

```
149  Toast.LENGTH_LONG)
150                              .show();
151                  if (suc == 1) {
152                      Intent intent = new Intent(ChangePwdActivity.this,
153                              MainActivity.class);
154                      startActivity(intent);
155                  }
156              }
157          }
158  }
```

1）第 61～69 行表示用户点击图 12-26 中左上角返回按钮后，跳转到主界面。

2）第 71～107 行实现新旧密码的判空与一致性操作，第 105 行执行异步任务操作。

3）第 111～157 行实现修改密码的网络请求，124～127 封装用户名和新密码到 List 对象上，129～130 发送请求数据到 NetworkData 的 makeHttpRequest()方法上。第 135～137 将网络获取的 JSON 数据进行解析。第 144～156 行执行 onPostExecute()方法，将修改密码请求的网络数据返回到手机端显示。第 151～155 表示如果密码修改成功，则跳转到主界面。

整个 HZTour 项目还有一些非关键性代码没有在本书中进行详细介绍，请登录中国水利水电出版社网站（http://www.waterpub.com.cn/softdown/）和万水书苑（http://www.wsbookshow.com）下载源代码进行查看。

## 12.4　本章小结

本章主要以贺州旅游为特色，开发贺州旅游新闻 App。首先对开发本项目的环境搭建做了基本介绍，然后对开发后台所使用的组件进行了介绍，最后综合本书前面各个章节所学的知识，并运用了一些新的技术进行该 App 的开发。